The sense of smell is probably the most important sense to a large number of the animal species on earth. The aim of the study of olfactory cell biology is to understand the cellular basis on which olfactory-driven behavior is based.

In this book the author presents a critical analysis of what is known about the olfactory sensory cells in both the nasal cavity and the vomeronasal organ: their structure, their connections in the main and accessory olfactory bulb, receptor biology, transduction mechanisms, cell replacement, developmental biology and plasticity. Also discussed are the trigeminal and terminal nerves in the nasal cavity. The book concludes with a section highlighting some of the gaps in our knowledge of olfaction and stresses the need for continued research.

Although the emphasis is on mammalian olfaction, basic issues that have been addressed by research on other vertebrates and invertebrates are also discussed. Throughout the book the premise is that we can more thoroughly comprehend normal behavior and dysfunction only if we have a detailed understanding of the cells and tissues involved in the sense of smell.

The book will be of interest to scientists and graduate students working in the fields of sensory neuroscience and cell, developmental, and behavioral biology.

DEVELOPMENTAL AND CELL BIOLOGY SERIES

EDITORS

P.W. BARLOW   D. BRAY   P.B. GREEN   J.M.W. SLACK

CELL BIOLOGY OF OLFACTION

Developmental and cell biology series

SERIES EDITORS
Dr. P.W. Barlow, *Long Ashton Research Station, Bristol*
Dr. D. Bray, *King's College, London*
Dr. P.B. Green, *Dept. of Biology, Stanford University*
Dr. J.M.W. Slack, *ICRF Developmental Biology Unit, Oxford*

The aim of the series is to present relatively short critical accounts of areas of developmental and cell biology where sufficient information has accumulated to allow a considered distillation of the subject. The fine structure of the cells, embryology, morphology, physiology, genetics, biochemistry and biophysics are subjects within the scope of the series. The books are intended to interest and instruct advanced undergraduates and graduate students and to make an important contribution to teaching cell and developmental biology. At the same time, they should be of value to biologists who, while not working directly in the area of a particular volume's subject matter, wish to keep abreast of developments relative to their particular interests.

BOOKS IN THE SERIES

R. Maksymowych *Analysis of leaf development*
L. Roberts *Cytodifferentiation in plants: xylogenesis as a model system*
P. Sengel *Morphogenesis of skin*
A. McLaren *Mammalian chimaeras*
E. Roosen-Runge *The process of spermatogenesis in animals*
F. D'Amato *Nuclear cytology in relation to development*
P. Nieuwkoop & L. Sutasurya *Primordial germ cells in the chordates*
J. Vasiliev & I. Gelfand *Neoplatic and normal cells in culture*
R. Chaleff *Genetics of higher plants*
P. Nieuwkoop & L. Sutasurya *Primordial germ cells in the invertebrates*
K. Sauer *The biology of Physarum*
N. Le Douarin *The neural crest*
M.H. Kaufman *Early mammalian development: parthenogenic studies*
V.Y. Brodsky & I.V. Uryvaeva *Genome multiplication in growth and development*
P. Nieuwkoop, A.G. Johnen & B. Albers *The epigenetic nature of early chordate development*
V. Raghavan *Embryogenesis in angiosperms: a developmental and experimental study*
C.J. Epstein *The consequences of chromosome imbalance: principles, mechanisms, and models*
L. Saxen *Organogenesis of the kidney*
V. Raghaven *Developmental biology of fern gametophytes*
R. Maksymowych *Analysis of growth and development in Xanthium*
B. John *Meiosis*
J. Bard *Morphogenesis: the cellular and molecular processes of developmental anatomy*
R. Wall *This side up: spatial determination in the early development of animals*
T. Sachs *Pattern formation in plant tissues*
J.M.W. Slack *From egg to embryo: regional specification in early development*

CELL BIOLOGY
OF OLFACTION

ALBERT I. FARBMAN

Department of Neurobiology and Physiology
Northwestern University

CAMBRIDGE
UNIVERSITY PRESS

#25282725

Published by the Press Syndicate of the University of Cambridge
The Pitt Building, Trumpington Street, Cambridge CB2 1RP
40 West 20th Street, New York, NY 10011–4211, USA
10 Stamford Road, Oakleigh, Victoria 3166, Australia

© Cambridge University Press 1992

First published 1992

Printed in Canada

Library of Congress Cataloging-in-Publication Data
Farbman, Albert I.
Cell biology of olfaction / Albert I. Farbman.
p. cm. – (Developmental and cell biology series)
Includes bibliographical references and index.
ISBN 0-521-36438-8(hardback)
1. Smell. 2. Cytology. 3. Olfactory mucosa. I. Title.
II. Series.
QP458.F37   1992
599'.01826 – dc20   92–4160

A catalog record for this book is available from the British Library.

ISBN 0–521–36438–8 hardback

To Winifred,
for so many reasons

Contents

Preface   *page* xi

1 Introduction   1
  1.1 The chemical senses   2
  1.2 Responses to chemicals in simple organisms   5
  1.3 Origins of special chemoreceptor neurons in
       invertebrates   8
  1.4 The olfactory system in vertebrates   11
  1.5 The olfactory sensory apparatus   18
  1.6 Vulnerability of olfactory sensory cells   18
  1.7 Some unique features of the olfactory system   19
  1.8 Some aspects of olfactory behavior   20

2 Structure of olfactory mucous membrane   24
  2.1 Structure of the olfactory mucosa in the main
       nasal cavity   24
  2.2 Cytology of the sensory neuron   28
  2.3 Cytology of supporting cells   42
  2.4 Basal cells   49
  2.5 Bowman's glands   49
  2.6 Does response specificity exist in the olfactory
       system?   50
  2.7 Neurogenesis in olfactory epithelium   55
  2.8 Structure of the olfactory nerve   63
  2.9 Are there topographical projections from epithelium
       to bulb?   68
  2.10 The septal olfactory organ   69
  2.11 Trigeminal innervation of the nasal cavity   70
  2.12 The terminal nerve   72

3 Functional aspects of olfaction   75
  3.1 Characteristics of an olfactory stimulus molecule   75

3.2 The initial events in the response of the sensory cell
      to the odorant stimulus   76
3.3 The role of second messengers in olfactory
      transduction   91
3.4 Physiology of the olfactory epithelium   100
3.5 Physiological properties of the olfactory sensory
      neuron   103
3.6 Electrical properties of the supporting cell   112
3.7 Nonolfactory sensation in the nose: function of the
      trigeminal innervation   113
3.8 Olfactory dysfunction in humans   115

4 Vomeronasal organ   118
4.1 Distribution and location   118
4.2 Stimulus access   119
4.3 A dual olfactory system   122
4.4 Cellular structure   123
4.5 Function   127

5 Olfactory bulb   132
5.1 Size of the bulb and its significance   132
5.2 Neural input to the olfactory bulb   134
5.3 Structure of the main olfactory bulb   134
5.4 Foci of activity in the olfactory bulb in response to
      stimuli (2-deoxyglucose studies)   140
5.5 Principal neurons of the olfactory bulb: the tufted
      and mitral cells   142
5.6 Interneurons of the olfactory bulb   150
5.7 Synaptology of the olfactory bulb   153
5.8 Bulbar connections with the olfactory cortex   158
5.9 Accessory olfactory bulb   161
5.10 Olfactory bulb in fish   163

6 Development and plasticity   167
6.1 Early development of the sensory epithelium   167
6.2 Postnatal growth of epithelium   179
6.3 Development of the olfactory bulb   181
6.4 Cellular interactions between the sensory epithelium
      and bulb during development   186
6.5 Functional maturation   194
6.6 Plasticity in the olfactory system   196

7 Concluding remarks   207
7.1 Odor receptor molecules   207

7.2 Coding in the olfactory system   208
7.3 Transduction   209
7.4 Olfactory mucus and the role of supporting cells   209
7.5 Neurotransmitter   210
7.6 Sensory cell replacement   210
7.7 Growth of olfactory nerves   211
7.8 Behavior   212
7.9 Parallel sensory systems   214
7.10 Final thoughts   214

References   215
Index   279

Preface

The ultimate goal of the study of olfactory cell biology is to understand the cellular substrate on which olfactory-driven behavior is based. Olfaction is one of the most important, if not the most important, of the senses driving basic patterns of behavior in most of the earth's animal species. The phenomenology of olfaction has been examined by students of animal behavior extensively, but until recently little detail was known about the biology of the cells that respond to odor stimuli. Because of recent advances in our knowledge of the biology of olfactory sensory cells, it seemed a good idea to try to review and analyze what is known. The writing has taken more time than anticipated, and because so much new information has been brought to light since the book was begun, it has been revised so many times it barely resembles the original outline. Although more than eleven hundred references have been cited, it will be obvious to the knowledgeable reader that many have not been included. I take full responsibility for the selection, and apologize in advance to those colleagues and friends who might feel their work should have been more prominently featured.

In writing this monograph, I have tried to make it useful not only for cell or molecular biologists, but for behavioral biologists as well. We still do not know much about the networks of cells driving behavior in response to odor stimuli, nor to what extent the behavioral responses are dependent on the context in which they occur (i.e., the relative importance of other senses in driving certain behaviors). The important thing at this stage in the development of our knowledge is to try to draft the questions that must be asked and to design carefully the experiments that can furnish the answers. A sound knowledge of cell biology can provide a framework for this, and continued research will lead us to understand the cellular substrate for olfactory-driven behavior.

In this book, the term "olfactory sensory cell" is used in preference to "olfactory receptor cell," as has been the case in most publications, including my own. The reason is that the word "receptor" has come to mean a molecule in a cell that binds to some sort of chemical substance applied from outside the cell. In the context of the olfactory sensory cell the word refers to the molecule(s) within the cell that responds to an odor stimulus. Therefore, to avoid misunderstanding, I refer to the cell as a sensory neuron, or a sensory cell, and to the molecule as a receptor.

Many people have helped me prepare this book. In particular, I should like to thank those colleagues with whom I now work most closely, Drs. Bert Menco, Richard Bruch, Virginia Carr, Michael Shipley, Richard Akeson, Robert Gesteland, and Sarah Pixley, for our many interesting and enjoyable conversations about several of the topics covered. Drs. Menco and Shipley in particular brought to my attention several important papers I might otherwise have overlooked. I thank also those with whom I have collaborated in the past, Drs. Frank Margolis, James Morgan, Peter Brunjes, and Fred Gonzales, and the many students who have worked in my laboratory. I have been inspired by the wisdom and diligence of all those named and unnamed. I am grateful to many colleagues and friends who have generously provided very useful critical comments on the drafts of parts of the book discussing aspects of olfaction in which they are considerably more expert than I. They are, in alphabetical order, Drs. Richard Bruch, Peter Brunjes, Robert Gesteland, Thomas Getchell, Bert Menco, Michael Meredith, Larry Pinto, and Michael Shipley. I thank Ms. Brenda Rentfro for the excellent drawings, Ms. Judy Buchholz for her expert technical assistance, and Mr. Eugene Minner for his help with the electron microscopy and photography. The hundreds of scientists whose works are cited have made significant contributions to my knowledge about olfactory cell biology. My thanks to them is acknowledged by my promulgating their work in this book. Finally, I am grateful to the National Institutes of Health which have supported my research for nearly three decades.

October, 1991

1
Introduction

This book explores several aspects of the cell biology of the olfactory systems of vertebrates, particularly mammals. The emphasis is on the biology of the cells and tissues constituting the primary olfactory pathway. By this is meant the olfactory sensory neuron and its first synaptic contact in the central nervous system (i.e., in the olfactory bulb). In the following pages, the recent findings regarding the structure and function of the olfactory system are reviewed, with an eye toward how the findings may relate to the olfactory-driven behavior of the animal.

Before examining the cellular details, we begin with a few of the diverse external structures associated with olfaction in some invertebrates and vertebrates. Figures 1.1 and 1.2 show schematically some external structures in which olfactory cells are found in a few species of invertebrates and vertebrates, respectively. In each case, the organ associated with the olfactory sense is on the head or an extension from the head and is located in a position that permits ready access to odor molecules reaching the organism.

Although the housing of the olfactory sensory nerve cells varies depending on the species, there is remarkable conservation of the morphology of the primary olfactory pathway. As we shall see later, in both vertebrates and invertebrates the sensory neurons are bipolar cells, each with a relatively short dendrite and a longer axon (Figure 1.3). The cell body is located in the animal's periphery – for example, housed within the cuticle of a sensillum, as in arthropods, or within an epithelium lining the nasal cavity or a nasal (olfactory) pit, as in vertebrates. The dendritic terminals have several slender appendages, either microvilli or cilia, that amplify the cell surface and contain the receptor molecules and transduction apparatus. The axons enter the central nervous system and terminate at synapses on secondary neurons in an olfactory lobe or olfactory bulb located in the most anterior or rostral region of the central nervous system. From the mollusk to the mammal, the basic plan of this system is remarkably consistent. The modifications from one species to another are relatively minor variations on the common theme.

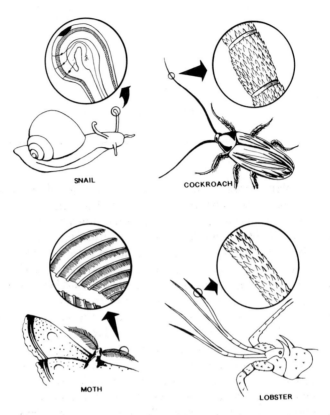

Figure 1.1. Schematic of the housing in which olfactory sensory neurons are found in four invertebrates. In snails, the olfactory organ is located on the surface tentacles extending from the head. (Adapted from Chase & Tolloczko 1986.) The bipolar olfactory neurons are drawn within the circle at higher magnification in relation to the surface of the tentacle. In cockroaches, the olfactory organ is located in the segmented antennae within sensilla, the fine bristle-like structures projecting from the antennae (shown at higher magnification within the circle). In moths, the two antennae resemble feathers. The sensilla containing olfactory sensory cells are the fine, hairlike projections from the antennal branches shown encircled in the higher magnification of the antenna. In lobsters, only the distal regions of lateral branches of antennae contain olfactory sensory cells. They are within the fine, bristle-like projections of the antenna.

1.1 The chemical senses

The sense of smell, as we currently understand it, is based on the ability of specialized cells to respond to environmental chemicals by initiating a nerve impulse. We can think of these specialized cells, the olfactory sensory neurons, as having two primary properties: (1) the ability to react to one or perhaps a few specific chemical molecules; this means

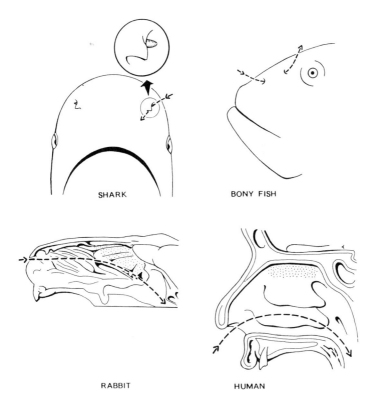

SHARK   BONY FISH

RABBIT   HUMAN

Figure 1.2. Diagrammatic representation of where olfactory organs are located in two aquatic and two terrestrial vertebrates. The arrows indicate the direction of flow of the medium, water or air, carrying olfactory stimuli. In elasmobranchs, the cartilaginous fishes (exemplified here by the lemon shark), the paired olfactory organs are placed on the ventral aspect of the snout. An anterior opening (naris) permits water to enter the organ, and a slit or posterior naris permits water to exit. The ridged olfactory organ is shown at higher magnification in the circle (see Figure 1.7). In many teleosts (bony fishes), the olfactory organs are in recesses on the dorsal aspect of the snout, and often there is an anterior naris and a posterior naris, as in the elasmobranchs, although there is variability in both groups of fishes. For example, in some the outlet may open into the pharynx. In lower mammals, such as the rabbit, the nasal passage is highly complex, containing nonolfactory folds, the maxillary turbinates, and, deeper in the nasal cavity, the ectoturbinates and endoturbinates, lined mostly with olfactory epithelium. In this diagram the olfactory region of the nasal cavity is marked with dots. In humans the nasal cavity is much less complex, with only a few turbinates and a relatively small area of olfactory tissue (marked with dots).

that the sensory cell can discriminate among molecules to which it is reactive and others to which it is not; moreover, it can respond to the intensity of the stimulus as well as its quality; (2) the ability to transform

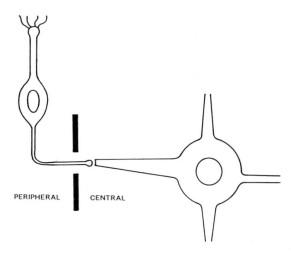

Figure 1.3. The basic plan of the primary olfactory pathway is the bipolar sensory cell, with a cell body and dendrite in the periphery. The dendritic terminal contains fluid- or mucus-bathed tiny appendages that have access to odorants from the outside world. The axon enters the central nervous system to terminate on a synapse with a secondary neuron. The secondary neuron, in turn, projects its axon to other regions of the central nervous system.

(transduce) the chemical signal into a meaningful response; in the olfactory sensory cell, the meaningful response is a nerve impulse that is transmitted to the brain.

Selective responsiveness to some chemicals and not to others is based on the presence of molecules, called receptors, to which the odorant chemicals are specifically bound (see Chapter 3). These receptors are on the sensory cell surface, where they are accessible to environmental chemicals. The receptors are linked to a transducing mechanism that ultimately elicits the effect. Thus, when an odorant chemical is bound to an appropriate receptor on an olfactory sensory neuron, the transducing system is activated and a nerve impulse is generated. Absence of a receptor for a particular chemical stimulus will preclude the cell from responding to that chemical.

The olfactory sensory neuron, however, is not unique in its ability to react to chemical signals. Other chemical sense organs are placed elsewhere in the body. Taste organs in most animals are located in or near the oral cavity (especially the tongue and soft palate), pharynx, and larynx. In some fish they are found on the external surface of the body.

Most people are familiar with the four basic taste qualities: salt, sour, sweet, and bitter. Sweet and bitter stimuli apparently bind to receptors that are linked to a transducing system, which in turn is linked to a means for changing the electric potential of the cell, much as in the olfactory cells. Responsiveness to salt and sour tastes is more directly linked to membrane channels and apparently does not work through specific receptors and a second messenger system, as do the others (reviewed by Kinnamon 1988). A major difference between taste sensory cells and olfactory sensory cells is that the former do not have an axon and therefore do not carry a nerve impulse directly to the central nervous system. They are modified epithelial cells innervated by sensory nerves; this means that the taste impulse, generated in the sensory cell, must cross a synapse to the sensory nerve, which then carries the information to the brain.

Other chemical sense organs are found within the carotid and aortic bodies. These are clusters of chemically responsive cells in the wall of the common carotid artery, near the bifurcation of its internal and external branches, and in the wall of the aorta, near or at the branch of the subclavian artery. In contrast to olfactory and taste organs, which are exteroreceptors (i.e., they respond to chemicals originating in the outside world), these sense organs respond to internal chemical changes in the body. They are responsive to the concentrations of carbon dioxide and oxygen in the blood, and possibly to the pH of the blood. As in the case of taste sensory cells, their responses are transmitted to nerve endings that carry the information to the central nervous system.

Many textbooks consider only the olfactory and taste organs and the carotid bodies as the chemical sensing organs of the body. That, however, is a rather limited view. The two important properties of chemosensory cells, namely, their ability to respond selectively to chemical signals and their ability to transduce a chemical signal into a significant reaction, are shared by *all* living cells, from humble unicellular organisms, such as yeasts or amoebae, to the complex nerve cells in the human brain. Indeed, the ability to detect chemical signals in the environment is as old as the first living cells.

This chapter presents a brief overview of responsiveness to chemicals in several organisms. More detailed accounts of the structures and functions of some of these chemical sense organs can be found in four special issues on olfaction and two on taste in the journal, *Microscopy Research and Technique* (1992 and 1993).

1.2 Responses to chemicals in simple organisms

Unicellular animals, which possess neither a nervous system nor a special sensory apparatus, can respond to chemical cues in their environment and locate food or avoid toxic substances (Kleene 1986; Adler 1987; Van Houten & Preston 1987; Manson 1990). In bacteria, for example, about 25 different receptors have been identified for attractant substances. "Attraction" is defined by the behavior of the unicellular animal (i.e., the cell moves in the direction of the substance). Amino acids, such as aspartate and serine, and sugars, such as ribose, galactose, and maltose, are potent attractants for the well-studied enteric bacterium *Escherichia coli* (reviewed by Manson 1990). The attractant binds to a receptor on the surface of the organism, and this binding results in the methylation of a transducing protein by the action of an enzyme, methyltransferase. The end result is movement of the bacterium toward the source of the attractant chemical.

The membrane of another unicellular organism, the ciliated *Paramecium,* has high-affinity receptors for certain chemicals, including folic acid and cyclic adenosine monophosphate (cyclic AMP or cAMP) (Smith et al. 1987). In *Paramecium,* an attractant such as folic acid induces an increase in the voltage difference across the membrane (between inside and outside the cell), resulting in hyperpolarization. Cells that are hyperpolarized in response to attractants swim faster and move toward the stimulus (Van Houten & Preston 1987). On the other hand, repellents decrease the voltage difference across the membrane, and the depolarized cells swim more slowly, make more turns, and move away from the repelling stimulus.

Even yeasts, such as *Saccharomyces cerevisiae,* the common brewer's yeast, have primitive chemosensory systems. This organism exists in three forms, two haploid cell types, **a** and **α** cells, and a third that results from the fusion of an **a** and an **α** to form a diploid cell type, **a/α.** The ability of the haploid cells to conjugate or "mate" depends on diffusible chemical attractants, called pheromones, secreted into the medium by each haploid cell type (reviewed by Sprague, Blair, & Thorner 1983). Thus, each haploid cell type secretes a peptide pheromone that affects the other. The pheromones are not, strictly speaking, attractants, because yeast cells, unlike *E. coli* or *Paramecium,* are nonmotile and therefore unable to move toward the source of the pheromone. Conjugation requires direct contact between the two haploid cells, but the efficiency of their fusion depends on random collision and their ability to adhere to one another after contact. One of the actions of the pheromone in the haploid yeast cells is to induce production of cell surface molecules that increase adherence of the conjugating cells to one another when they are brought into contact by random movement. Thus, in yeasts, the pheromone signal activates a transduction mechanism that results in production of new gene products that enhance adhesion.

In the aquatic fungus *Allomyces macrogynus,* pheromone interactions between the male and female gametes have been studied. The female gamete produces a compound named sirenin [molecular weight = 236 daltons (Da)] (Figure 1.4), which acts as a sperm attractant in very low

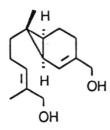

Figure 1.4. Structure of L-sirenin, a sesquiterpenediol (molecular mass 236 Da) produced by female gametes of *Allomyces macrogynus*. Only the *l*-enantiomer, depicted here, is active in attracting male gametes of the species.

concentrations: 10 picomolar (1 pM = 10^{-12} M). The presence of threshold amounts of sirenin results in influx of calcium ions into the sperm cytoplasm and a change in the swimming pattern, resulting in movement toward the source of the pheromone (reviewed by Pommerville, Strickland, & Harding 1990). The male gametes, in turn, produce an attractant, named parisin, that acts on female gametes. Although female gametes do not swim as rapidly or as efficiently as sperm, the presence of parisin in the water results in their movement toward the source of the pheromone. The molecular nature of parisin is not as well characterized as that of sirenin. It is known to retain its activity after freezing, after boiling for 10 min, and after autoclaving for 20 min (Pommerville et al. 1990).

The slime mold *Dictyostelium discoideum* exists as a single amoeboid cell during the first phase of its developmental cycle. These amoebae feed on bacteria and are attracted to them by a chemical signal, folic acid, released by the bacteria. In a later stage of development, or in response to a decrease in availability of food, the amoebae aggregate, in response to a chemical attractant, cyclic AMP (reviewed by Gerisch 1982; Van Haastert, Jannsens, & Erneux 1991). The membrane of the *Dictyostelium* amoeba has high-affinity receptors for folic acid and cyclic AMP. The transduction apparatus in *Dictyostelium* involves production of at least two types of second messengers: cyclic guanosine monophosphate (cGMP) and the products of the activity of the enzyme phospholipase C, namely, diacylglycerol and inositol trisphosphate.

These simple organisms are concerned mostly with behaviors that promote their survival. Their basic requirements are to feed, to avoid harmful situations, and to reproduce themselves. In motile organisms, transduction of chemical signals might evoke changes in ionic flux across the cell membrane, resulting in changes in direction and/or velocity of movement. In nonmotile organisms, other kinds of changes involving selective gene expression of molecules associated with differentiation, development, or reproduction might result from appropriate chemical signals. Chemical receptors and transduction mechanisms in these simple organisms are elegantly tuned to their survival. The important thing to note is that the cells responding to environmental chemical signals are not specialized sensory cells or neurons involved in complex networks; indeed, in unicellular organisms the receptor, transducer, and effector are all the same cell! As we shall see in later sections of this book, there are certain common features in the cell biology of the receptor-transduction systems in olfactory sensory cells and some of these simple organisms. For example, many of these specific chemical signals are active in fairly low concentrations, some involve second messengers, whereas others involve changes in ionic flux across the cell membrane, resulting in transient voltage changes.

1.3 Origins of special chemoreceptor neurons in invertebrates

As animals acquired simple nervous systems and their behavior grew more complex, reception of chemical signals became the function first of simple nerve endings, then of special cell types. For example, nerve endings lying beneath the outer body surface of jellyfish are responsive to chemical stimuli in the environment. Behavioral responses to chemical stimuli in these primitive animals are still essentially limited to feeding or defensive actions.

1.3.1 Nematodes

In the nematode *Caenorhabditis elegans,* the choice of which life cycle it enters as a larva depends on the availability of an ample food supply for the numbers of larvae. When enough food is available, the animals develop to adulthood through four larval stages. Under conditions of crowding and starvation, the development of young animals is arrested, and they differentiate into specialized "dauer larvae" that can survive under severe environmental conditions for several months (Bargmann & Horvitz 1991). If sufficient food becomes available, the dauer larva changes its developmental program to become a fertile adult. The decision of which path to follow is mediated by a competition between two chemical signals, one emanating from the food (bacteria), and the other, a pheromone, emanating from the organisms themselves. The concentration of the pheromone is dependent on the population density of nematodes. High concentrations of the pheromone and scarcity of food will cause them to enter the dauer larva stage. The activities of four classes of chemosensory neurons control the direction of larval development (Bargmann & Horvitz 1991). In this case, then, the chemosensory neurons responsive to the dauer pheromone provide a link between an environmental chemical signal and developmental regulation.

1.3.2 Annelids and mollusks

With the appearance of invertebrate animals that have more complicated nervous systems, such as annelids (e.g., earthworms, leeches) and mollusks (e.g., snails, clams, octopus), chemical communication became more highly specialized. In these animals two kinds of sense organs, each with specialized cells, constitute the receptive apparatus for detecting chemical signals in the environment. The distance chemoreceptors, or olfactory sensory cells, are located in the head region, sometimes on appendages, such as tentacles or antennae (Figure 1.1) (Emery 1975, 1992; Chase 1986; Ache 1987; Hallberg, Eloffson, & Johansson 1992). These cells can respond to low concentrations of stimulus in the air or water, depending on where the animal lives. The contact chemoreceptors or "taste" cells are located around the mouth parts and respond to stimuli in solution when the cells come into direct contact with the solution. The taste cells are usually less sensitive to chemicals (i.e., their threshold for a response is usually a higher concentration of the stimulus molecule than is the case for olfactory responses). In annelids and mollusks, the primary functions of these chemical sense organs are related to feeding and defense behaviors, although some reproductive and social behaviors are also mediated by the chemical senses.

Behavioral studies in snails have shown clearly that many amino acids elicit feeding responses (W. Carr 1967). In these gastropods, olfactory structures have many similarities to those of mammals. Both the primary olfactory sensory cells and the glomeruli in the central nervous system, where they form synapses, are structurally similar to the analogous structures of mammals (Emery 1975; Chase 1986; Chase & Tolloczko 1986). Moreover, new olfactory sensory cells are continually made in snails, as in vertebrates (Chase & Rieling 1986). Chase (1986) suggests that the "fact that similar olfactory structures have evolved in these two animal groups implies that there has been a limited number of solutions to the problems of olfactory perception, and that the relevant biological principles may be universally applicable."

1.3.3 Arthropods

In arthropods, such as insects and crustaceans, the olfactory sense is more versatile. It is used for many different types of behaviors, not only finding food and avoiding danger, as in lower animals, but in reproduction, territorial marking, finding the nest, social behavior, and so

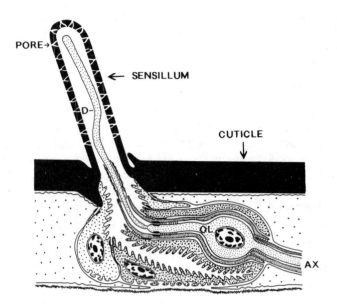

Figure 1.5. Schematic diagram of a prototypic arthropod olfactory sensillum with a single sensory neuron (OL) and three auxiliary cells. The sensory neuron is a bipolar cell, with a dendrite (D) extending into the sensillum, and an axon (AX) projecting to the central nervous system. In this diagram, only one sensory cell is shown for simplicity. Sensilla may contain two, three, or more cells. The sensillum is a hollow, fluid-filled tube with a cuticular wall, and there are tiny pores in the cuticle. The fluid, produced by auxiliary cells, protects the dendrite against desiccation and also contains stimulus-binding molecules and enzymes. Access to the sensory cell dendrite from the environment is by way of the cuticular pores. (Redrawn from Kaissling 1986.)

forth. The sensory cells for the olfactory apparatus are on the antennae. The latter become exquisitely sensitive in some insects. For example, when the female gypsy moth becomes sexually receptive, she secretes a mixture of volatile chemical substances. These act as a sexual attractant, or pheromone. They can be detected by the male at distances of several miles (if the wind is right). In fact, the olfactory apparatus on the male moth is so sensitive that it is responsive to only a few molecules of the pheromone (reviewed by Kaissling 1986). The sex attractant pheromone mixture has been used to bait traps for male gypsy moths in efforts to prevent them from finding mates, and thus control the population of this pest (Beroza & Knipling 1972).

The sensory neuron is a bipolar cell, with its dendritic ending near the tip of the sensillum (Figure 1.5); its axon terminates within a glomerulus in the olfactory lobe of the central nervous system. The basic structure of the primary olfactory pathway, then, is conserved in arthropods, both the water-dwelling crustaceans and land-dwelling insects.

1.4 The olfactory system in vertebrates

The participation of the olfactory sense in several types of behavior, particularly feeding and reproduction, has been preserved in vertebrates. Most vertebrates have well-developed olfactory systems. One measure of the importance of the olfactory system to a species is the relative proportion of the brain devoted to it. In fishes and amphibians, the olfactory region of the brain occupies about one-sixth of its total volume, and in some mammals (e.g., insectivores) the olfactory regions occupy about one-fourth of the surface area of the cerebral cortex (Figure 1.6) (Allison 1953; Negus 1958). However, some vertebrates, notably birds, aquatic mammals, and primates, have come to rely more on other senses, and their olfactory systems are less well developed. In fact, in whales the nasal organ has been profoundly modified into a perpendicular tube, opening on the dorsal aspect of the head, as these animals have adapted to a totally aquatic life. Adult toothed whales (odontocetes) have no peripheral olfactory sense organ at all. It is present during early embryonic development, but is lost in later fetal stages (Oelschläger & Buhl 1985; Oelschläger 1989).

In air-breathing vertebrates the olfactory organ is incorporated into the respiratory system. The nasal cavity forms the most anterior end of the respiratory system. It is divided into three sectors: a vestibule or entry region, a respiratory region, and an olfactory region. In contrast, fish and other aquatic animals with gills have an olfactory organ that usually is not directly incorporated into the respiratory apparatus (Negus 1958).

1.4.1 Fishes

In cyclostomes (jawless fish, e.g., lampreys, hagfish) there is a single opening above the mouth that leads into paired nasal cavities. In elasmobranchs (cartilaginous fish, e.g., sharks, rays) the paired olfactory organs are located in cartilaginous nasal capsules on the underside of the snout, entirely separate from the mouth (Zeiske, Theisen, & Gruber 1987). Each olfactory organ has an incurrent opening and an excurrent opening separated by a nasal flap (Figure 1.2). The olfactory organ is exposed to water flowing through the incurrent opening and out again through the posterior excurrent nasal opening. The pressure for the flow is achieved by forward motion of the animal through the water.

In bony fish (teleosts), the general external and internal form of the paired olfactory organs is similar, but they are located on the dorsal side of the head. Each organ is in a cavity with two openings, anterior and posterior (Figure 1.2). There are, however, many variations in teleosts – some olfactory organs have an anterior naris open to the

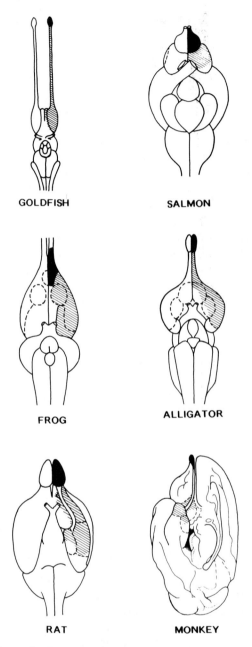

Figure 1.6. Schematic diagrams of the brains of several vertebrates, showing the relative size of the olfactory bulb (solid black) and the distribution of projections from the bulb (hatched). (Redrawn after Allison 1953 and Negus 1958.)

outside and a posterior one opening into the pharynx. Others are nearly completely exposed to the surface, covered only by a small skin flap (reviewed by Yamamoto 1982).

The nasal mucous membranes of the various kinds of fishes, cyclostomes, elasmobranchs, and teleosts, are similar. There are variable numbers of folds or ridges arranged in rows parallel to each other, or in a rosette (Figure 1.7). Olfactory sensory epithelium usually is found in the valleys between the ridges and a short distance up their sides. The remainder of the ridge is lined with nonchemosensory epithelium.

1.4.2 Amphibians

In neotene amphibians (those that retain their gills and tails and do not undergo metamorphosis into air-breathing animals), the paired nasal cavities are divided into an anterior respiratory chamber and a posterior olfactory chamber. The mucous membrane in the olfactory region is raised into several transverse folds connected with a longitudinal fold, very much as is seen in fishes (Farbman & Gesteland 1974). The olfactory sensory cells are found at the bottoms of the folds. The nasal cavity opens posteriorly into the pharynx.

In air-breathing amphibians the olfactory organ is placed within the respiratory pathway. Respiratory movements associated with simple breathing or sniffing draw air into and out of the nasal cavity and over the surface of the olfactory sense organ. Two nostrils in the snout permit air to enter the nasal sacs, which are lined mostly with olfactory epithelium. Exit is by way of an opening into the pharynx. Amphibians have a relatively simple nasal cavity, often containing an eminence under the ventral mucosa (Figure 1.8).

1.4.3 Reptiles

The evolutionary strategy for enlarging the size of the olfactory organ (i.e., increasing the number of sensory cells that can be exposed to stimuli in the respiratory airstream) begins to become evident in reptiles. The lateral wall of the nasal cavity in some of these animals projects a bony or cartilaginous extension or fold toward the nasal septum (Figure 1.9). The medial surface of the extension, called a turbinal (or turbinate), is lined with olfactory epithelium. The nasal cavity and turbinal extensions vary in size and shape in different groups of reptiles. These turbinal folds, which become much more elaborate in mammals, provide the means for augmentation of the olfactory organ by increasing the surface area within the nasal cavity without significantly increasing its volume.

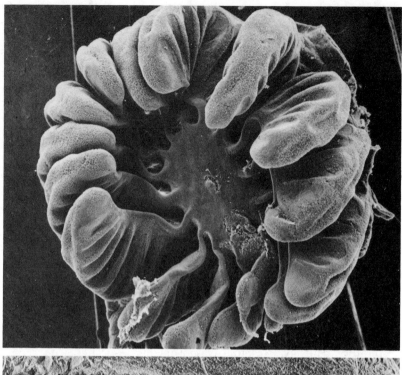

Figure 1.7. Scanning electron micrographs showing the olfactory sensory region in a trout (above) and a catfish (below). In both fishes the sense organ is a highly folded structure. The sensory cells are at the bottom and side walls of the folds. The tops of the folds and the neighboring epithelium are nonsensory. (Upper figure courtesy of Dr. David Moran.)

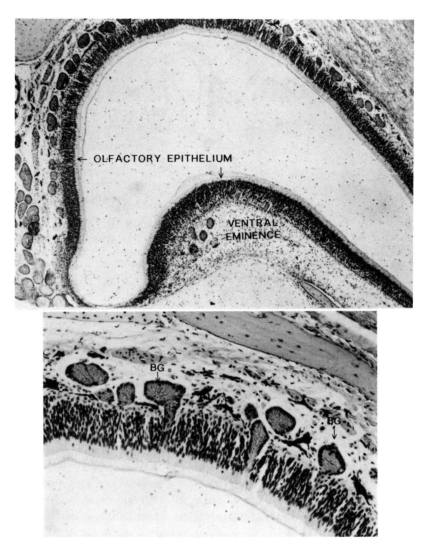

Figure 1.8. Photomicrograph of a frontal section through a frog nasal cavity. The ventral eminence and the dorsal region are lined with olfactory mucous membrane, containing the sensory epithelium and Bowman's glands. In the lower figure, a higher magnification of the frog olfactory mucosa is shown in which Bowman's glands (BG) are prominent.

1.4.4 Birds

In warm-blooded animals (the birds and mammals) the nasal cavity is divided into three compartments: the anterior chamber, lined with stratified squamous epithelium, the middle chamber, lined with respiratory

Figure 1.9. Drawing of a section through a reptilian head (the slow worm) showing the nasal cavities and a single turbinate (T) that increases the surface area of the cavity. A thick line (arrow) is used to show which part of the nasal cavity is lined with sensory epithelium. The remainder of the nasal mucosa is nonsensory. (Adapted from Negus 1958.)

epithelium, and the posterior chamber, lined with olfactory epithelium (Allison 1953). The posterior chamber of the nasal cavity opens into the pharynx. Inspired air, which must be at body temperature and moist when it reaches the lungs, is warmed and humidified primarily in the anterior and respiratory chambers of the nasal cavity, before it reaches the olfactory organ.

Birds have not been used extensively in cell biology studies of olfaction, although many species have well-developed olfactory systems (Bang 1971). Electrophysiological evidence (e.g., Tucker 1965; Wenzel & Sieck 1972) indicates that at least some species of birds are sensitive to olfactory cues. Behavioral evidence (e.g., Walker et al. 1986) and experimental evidence from the field indicate that kiwis and some vultures and procellariforms use olfaction in orientation and food-finding (e.g., Wenzel 1968, 1985; Hutchison et al. 1984; Houston 1987). Pheromones that stimulate the olfactory system are involved in the control of sexual behavior in ducks (Balthazart & Schoffeniels 1979), and the ability to respond to odors undergoes seasonal changes associated with breeding (e.g., starlings have a lower threshold to many odor stimuli in the spring, when they are in breeding condition, than in the fall, when they do not breed) (Clark & Smeraski 1990).

1.4.5 Mammals

In mammals, the nasal cavity is generally relatively large. Its lateral walls are elaborated into a variably complex series of folds by the bony

RABBIT

HUMAN

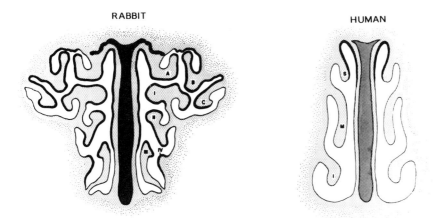

Figure 1.10. On the left is a diagrammatic representation of a frontal section through a rabbit nasal cavity, showing elaborate scrolling of ectoturbinals (A, B, C) and endoturbinals (I, II, III, IV) on the lateral aspect of the nasal cavity. On the right is a diagram of a frontal section through an adult human nasal cavity, showing superior (S), middle (M), and inferior (I) turbinates. In both drawings, the thick line along the surface of the nasal cavity is where olfactory epithelium is found. In the rabbit the olfactory epithelium is much more extensive. (From Farbman 1991, with permission.)

turbinal extensions. In animals with a keen sense of smell (macrosmatics, such as dogs, deer, rodents, etc.), the highly folded maxilloturbinal in the respiratory chamber is covered by a very vascular respiratory mucosa. In this chamber, the surface area is amplified severalfold, and the air passing through is warmed, humidified, and filtered (Figure 1.2).

In the posterior olfactory chamber, the turbinal folds originate from the ethmoid bone and often are arranged in lateral and medial rows, referred to as ectoturbinals and endoturbinals, respectively. Much of the surface area of the ethmoid turbinals is covered with olfactory mucosa (Figure 1.10). In macrosmatic mammals, turbinal folds, lined with olfactory tissue, extend even into some of the paranasal sinuses (Negus 1958).

On the other hand, many primates, including humans (microsmatic mammals, those with a comparatively poor sense of smell), have uncomplicated turbinal folds. The superior and middle turbinals (ethmoid) are simple folds, and only the medial aspect of the superior turbinal has olfactory epithelium. The inferior turbinal is a maxilloturbinal; that is, it originates from the maxillary bone, not the ethmoid (Figure 1.2 and 1.10).

1.5 The olfactory sensory apparatus

Olfactory sensory neurons in higher invertebrates and those in aquatic and terrestrial vertebrates are rather similar. The sensory cell is a bipolar neuron, with a single axon going to the central nervous system, and a dendrite that terminates on a body surface. From the apical end of the dendritic terminal, narrow appendages extend toward the outside world (Figures 1.3 and 1.5) (see Chapter 2). The appendages, which are bathed in fluid or mucus, contain the receptor and transduction apparatus associated with olfaction.

In terrestrial vertebrates an accessory olfactory organ makes its appearance, namely, the vomeronasal organ. The vomeronasal system, as we shall see in Chapter 4, may have a special function in certain reproductive and social behaviors.

Some mammals have an isolated patch of olfactory epithelium near the ventral edge of the septum, in the respiratory region of the nasal cavity (see Figure 2.19). This so-called septal organ (organ of Masera) may have a special role in olfaction, perhaps as an "early warning" monitor of the external environment. Two additional neural components in the nasal cavity, the trigeminal and terminal nerves, may also be associated with the sense of smell in vertebrates (Graziadei 1977).

Of the five neural elements associated with the olfactory region of the nose, three, the olfactory proper, septal, and vomeronasal sensory cells, have a common structural basis (Graziadei 1977): All are bipolar cells, and each has a peripherally located cell body within a sensory epithelium lining a surface at the boundary between the animal's interior and the external world. The other two, the trigeminal and terminal nerves, have free nerve endings in the olfactory mucosa. Their cell bodies are situated closer to the central nervous system, as are most other sensory neurons.

1.6 Vulnerability of olfactory sensory cells

Olfactory sensory cells are located on a body surface where they can advantageously sample the chemical stimuli in the environment. The advantage of the location, however, is offset by an increased vulnerability to physical and chemical injuries. For example, desiccation is a major problem in terrestrial animals. Various means for protecting and preserving the cells have evolved. In insects, for example, the sensory cells are located in sensilla (Figures 1.1 and 1.5) that extend from antennae or antennal branches like the bristles of a brush. The sensilla have a tough, resistant cuticle that encloses the cells. The sensillar wall is perforated by microscopic openings that permit access of volatile molecules, but are small enough to retard loss of sensillar fluids. The sensory cells are further protected from drying by being bathed in secretions produced by neighboring nonsensory cells. Protection against drying in terrestrial vertebrates is by secretion of mucus and fluids from supporting cells within the epithelium and from glands associated with the olfactory mucosa (i.e., Bowman's glands) (Figure 1.8).

The olfactory sensory cells of land-dwelling vertebrates are located deep within the nasal cavity, somewhat protected from the external environment, but they are still susceptible to infection, desiccation, and chemical and physical trauma. Adding to the vulnerability is the fact that vertebrates generally have a longer life span than invertebrates. Consequently, more time is available for potential insults that might compromise the animal's olfactory ability and therefore its survival. The evolutionary strategy to protect vertebrates, particularly those dwelling on land, against loss of olfaction is a mechanism to make new sensory cells continually, not only by a wound-healing response to trauma but also by a mechanism that provides new cells continuously even under physiological conditions. Thus, in the event that sensory cells are damaged, a population of new cells is readily available to replace them (see section 2.7). Similarly, in land-dwelling snails, in which olfactory cells are not well protected by a resistant cuticle, the sensory cells are continually replaced (Chase & Rieling 1986).

1.7 Some unique features of the olfactory system

The olfactory neurons are unique anatomically because they are the only primary sensory neurons (in vertebrates) in which the cell body is located so close to the periphery. Peripherally located sensory cell bodies are common in invertebrates; partly for this reason, the olfactory system is regarded as a primitive part of the vertebrate nervous system.

The trigeminal and terminal nerves, on the other hand, are typical of most vertebrate sensory neurons in that their cell bodies are located within a sensory ganglion, closer to the central nervous system, and their representation at the surface of the body is by way of an axonal process. The cell bodies of these sensory neurons are more protected by virtue of their location within the skull, and therefore are less vulnerable to the kinds of injuries that can afflict olfactory sensory cells.

The location of the cell body in the periphery is only one feature of the olfactory system that characterizes it as unique among sensory systems. As we shall see later, the olfactory axons are all unmyelinated, and they are organized in a way that is typical of axons in an early stage of embryonic development in the peripheral nervous system. Both myelinated and unmyelinated nerve axons in the peripheral nervous system of vertebrates are ensheathed by supporting cells, known as Schwann cells. In unmyelinated nerves of the adult, axons are usually isolated from one another within sheetlike wrappings of Schwann cell cytoplasm;

in contrast, embryonic axons may be grouped together within a single bundle wrapped by the Schwann cell cytoplasm (see Figure 2.18). This is the case in olfactory nerve, both in the embryo and in the adult. In other words, the olfactory nerve is the only nerve in the peripheral nervous system that retains its embryonic relationship to the cells that ensheath it.

Another unique feature of the olfactory system is that it projects olfactory information to a cortical region of the brain without a thalamic relay. Incoming information from other sensory systems is processed in the thalamus before it is projected to the cerebral cortex, whereas in the olfactory system the projection goes from the olfactory bulb to several parts of the olfactory cortex without passing to the thalamus first. However, parts of the olfactory cortex, notably the pyriform cortex, project to the thalamus, and olfactory information is then passed to the neocortex, presumably where conscious perception of odors occurs.

1.8 Some aspects of olfactory behavior

Although the primary focus of this book is cell biology, it is interesting to consider some aspects of animal behavior traceable to olfaction. In several instances observations of functional responses to chemical signals have led investigators to seek the biological mechanisms underlying the behavior. In this section we shall briefly discuss a few examples of olfactory-driven behavior. In Chapter 6, the cellular basis for other examples of behavior is discussed. By no means is this intended to be a complete survey of this important aspect of research.

1.8.1 Olfaction in the newborn mammal: maternal–infant relationships

In most mammals the olfactory system is functional at birth, although their eyes and ears may be closed. When visual and auditory input is thus restricted, the sense of smell is extremely important for survival of newborn mammals, because it is the primary sense utilized in locating the mother's nipple, their major source of nutrition. Several experiments have shown that mammalian pups will be unable to suckle if their olfactory sense is compromised by experimental surgery or if the olfactory cues associated with the mother's nipple are removed by washing the teats (Hofer, Shair, & Singh 1976; Teicher & Blass 1976, 1977; reviewed by Pedersen & Blass 1981). Nipple-search behavior in rabbits is apparently guided by a pheromone in the mother's milk (Hudson & Distel 1983, 1984, 1986).

Within a short time after birth, mammals, humans included, learn to distinguish their own mothers from others by their unique odors. This was shown in humans by presenting an infant with a breast pad from its own mother and one from another lactating woman. The infant consistently turned its head toward the pad from its own mother (MacFarlane 1975; Russell 1976). Remarkably, women are able to recognize their own newborn infants by their odors after only a few minutes to an hour of exposure to them (Kaitz et al. 1987). Thus, olfaction is an extremely important sensory modality in the bonding of mother and infant in many mammalian species. In Chapter 6 we shall return to some aspects of mother–infant bonding that have been studied more extensively.

1.8.2 Olfaction in nest-finding and homing behavior

As the newborn mammal grows, many environmental odor cues are influential in the development of its behavior. For example, the odor of the nest is reinforcing to young pups, as they learn very early to detect the odors associated with it; when displaced from the nest, they will soon orient themselves to return to it (reviewed by Alberts 1981).

Another noteworthy example of behavior related to olfaction is the homing instinct of the Atlantic salmon (reviewed by Hasler, Scholz, & Horrall 1978; Stabell 1984). These fish hatch in freshwater streams and, as juveniles or young adults, swim to the ocean where they may, in some instances, spend a couple of years, often hundreds of miles away from their home streams. When the mature adult is ready to spawn, it returns to its home stream by using olfactory cues. The salmon's ability to distinguish its own kin from closely related families, presumably by the odor of skin secretions in the water, is as remarkable as the male gypsy moth's ability to detect female pheromones. Similarly, female sea turtles, hatched on land, swim out to sea where they may spend several years. As adults ready to lay eggs, they return to the island where they were hatched; this behavior, though not well understood, is also thought to be based on olfactory cues (Koch, Carr, & Ehrenfeld 1969).

Field experiments on homing in pigeons indicate that olfaction is an important sense for long-distance navigation. Long-term olfactory deprivation severely impairs the ability of pigeons to return to their home loft after having been released at distances of 20 km or more from the loft (Papi et al. 1973; Wallraff 1988).

1.8.3 Olfaction in mating

Much of the sexual behavior associated with the olfactory system is mediated by the vomeronasal organ (see Chapter 4). Fish have no vomeronasal organ, but there have been several studies showing that the olfactory system is involved in mating behavior. In goldfish (*Carassius auratus*), for example, a pheromone produced by the ovaries has been identified, namely, $17\alpha,20\beta$-dihydroxy–4-pregnen–3-one. When this 21-carbon steroid pheromone is released into the water by females, it functions as a priming pheromone for males (Dulka et al. 1987). Within a few hours of exposure, males exposed to low concentrations in the water (10^{-10} M) respond by a rapid increase in blood gonadotropic hormone, an increase in sperm and seminal fluid production, and an increased sexual arousal (Stacey & Sorensen 1986). Electrophysiological studies have shown that the male detection threshold for this compound is actually 10^{-12} M, and the response is highly specific when compared with responses to 24 closely related sex steroids (Sorensen et al. 1990). The specificity is thought to be due to a fairly precise ligand–receptor interaction.

A second pheromone mixture of F prostaglandins is later released at ovulation by female goldfish. These pheromones rapidly stimulate male spawning behavior (Sorensen et al. 1988). The male goldfish has at least two classes of olfactory receptors for F prostaglandins, one highly specific and sensitive to 15-keto-prostaglandin $F_{2\alpha}$ (threshold concentration 10^{-12} M), and the other somewhat less specific to prostaglandin $F_{2\alpha}$ (Sorensen et al. 1988). Thus, female goldfish release two pheromones sequentially that have important and different effects on mating behavior in male goldfish.

Examples of the importance of olfaction in vertebrate mating behavior are far too numerous to be covered here. Skin-derived substances are important in mating, or at least in sex recognition, among reptiles (Garstka & Crews 1981; Mason & Gutzke 1990). In mammals, most of the odorants involved in mating behavior originate in glands around the anogenital area, and some can be found in urine. In Chapter 4, some of the chemical substances that have been identified will be discussed.

1.8.4 Olfaction in the predator–prey relationship

The role of olfaction in food-finding is critical in carnivorous animals. For example, chemicals from earthworms elicit prey attack by garter snakes. The attack response is exhibited even in neonatal animals that have been hatched in the laboratory and never fed before (Schell et al. 1990). These responses are mediated by the vomeronasal organ and are at least partially dependent on whether or not the tongue is present. Flicking of the tongue is involved in chemical access to the vomeronasal organ in snakes. Rattlesnakes use chemical cues on the fur of rodents to follow the trails of these prey (Chiszar et al. 1990).

However, the prey preferences of some animals are at least partly determined by what has been called olfactory imprinting, a form of learning that occurs relatively early in postnatal or posthatching development. Laboratory experiments with ferrets have shown that the prey presented within the third postnatal month (i.e., between 60 and 90 days after birth) becomes the preferred prey (Apfelbach 1986). This has been related to developmental changes occurring in the olfactory bulb at that age.

On the other side, the hunted also have the ability to detect the hunter. For example, the odor of a weasel is readily detected by one of its food sources, the short-tailed vole, and the vole avoids or flees from areas marked by weasel odor (Stoddart 1979). The broad-headed skink (a lizard) can detect chemical stimuli from snakes that are its predators (Cooper 1990). Not only can the prey detect the predator, but also they can warn their own kin about imminent danger. For example, in some species of fishes, alarm substances are released that elicit fleeing and hiding behavior in members of their own species (Pfeiffer 1978). Mice release an alarm pheromone when stressed (Rottman & Snowdon 1972). When certain species of deer are alarmed, they discharge an odor from skin glands that elicits alertness and flight in other deer (Müller-Schwarze 1971, 1979).

2
Structure of Olfactory Mucous Membrane

This chapter is devoted to a review of the structure of olfactory epithelium in the main nasal cavity, including the septal organ. It also reviews the innervation of the mucous membrane in the main nasal cavity by the trigeminal and terminal nerves. A separate chapter (Chapter 4) will be devoted to the vomeronasal olfactory system.

2.1 Structure of the olfactory mucosa in the main nasal cavity

The olfactory sense organ is made up of a mucous membrane, or mucosa, consisting of two major layers: an epithelium and a connective tissue, the lamina propria (Figures 2.1 and 2.2A). A thin basement membrane separates these two layers. Like mucous membrane elsewhere in the body, it is maintained in a moist condition by glandular secretions that form a layer of mucus covering the external surface of the epithelium.

The olfactory mucosa in many vertebrates is colored yellow to brown, whereas the adjacent respiratory mucosa is pink. Carotenoids and non-carotenoid pigments associated with lipids in the olfactory epithelium probably account for the yellow–brown color. It was once thought that the pigment played a role in olfactory function, but there is no evidence to support this (Moulton & Beidler 1967), and there are few, if any, who hold that view today.

2.1.1 Epithelium

The epithelium contains the three major cellular components of the sense organ: the sensory cell, the supporting (sustentacular) cell, and the basal cell. In addition, there are a few minor cell types. The histology and fine structure are remarkably similar in all classes of vertebrates (Bloom 1954; Bannister 1965; Andres 1966, 1969; Frisch 1967; Graziadei & Bannister 1967; Gemne & Døving 1969; Graziadei 1971, 1973b, 1977; Kolnberger 1971; Kratzing 1971a,b, 1972, 1975, 1978, 1984; Naessen 1971a,b; Farbman & Gesteland 1974; Jourdan 1975; Graziadei & Monti Graziadei 1976; Breipohl & Fernandez 1977; Ciges et al. 1977; Kerjas-

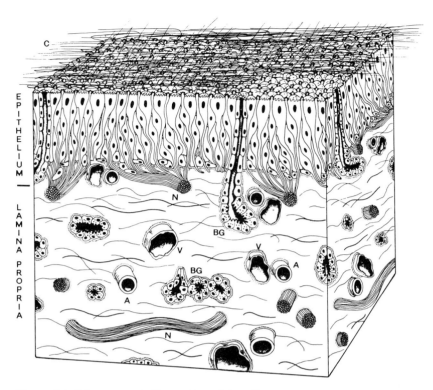

Figure 2.1. Diagrammatic illustration of the olfactory mucous membrane. The epithelium and lamina propria are shown. The long cilia on the surface are matted in a layer of mucus on the epithelial surface and lie parallel to the surface. Within the lamina propria are Bowman's glands (BG), bundles of olfactory nerve processes (N), and blood vessels, both small arteries (A) and veins (V). Ducts from Bowman's glands open onto the surface. For clarity, the numbers of olfactory nerve bundles and the cell bodies from which they originate are understated in this diagram. Compare with Figure 2.2A.

chki 1977; Menco 1977, 1980a,b,c,d, 1983, 1984; Breipohl et al. 1982; Moran et al. 1982b, 1985b; Muller & Marc 1984; Morrison & Costanzo 1990).

In a histological section through the olfactory epithelium, one sees several layers of cell nuclei (Figure 2.2). The oval nuclei of supporting cells form a single layer close to the surface, and the round nuclei of sensory cells usually make up six to eight or more layers in the lower one-half to two-thirds of the epithelium. Some nuclei of basal cells along the basement membrane are flat; others are round or polyhedral (the globose basal cells). The presence of nuclei situated at different levels creates the illusory appearance that the cells are stratified into several

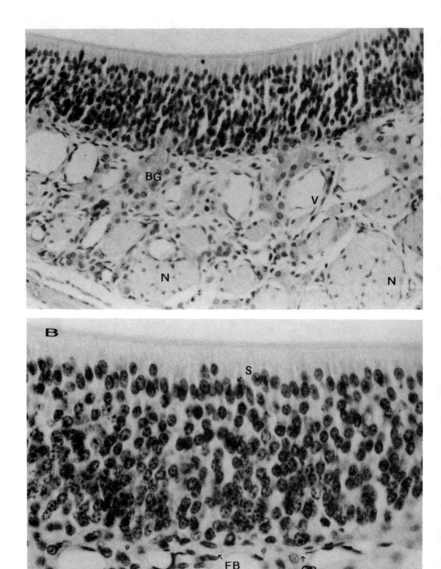

Figure 2.2. A: Histological section of rat olfactory mucosa showing the pseu-dostratified epithelium and lamina propria. Bowman's glands (BG) and ducts (D) are seen. The lamina propria contains large bundles of nerve (N) and some blood vessels (V). B: Histological section of rat olfactory epithelium at a higher magnification. The oval supporting cell nuclei (S) are the most superficial. Eight to 10 layers of sensory neuronal nuclei containing patchy chromatin are present. Globose (GB) and flat (FB) basal cells are seen.

layers; in fact, olfactory epithelium is not truly stratified, because each cell maintains direct contact with the basement membrane, albeit by very slender processes (Figure 2.1). This cellular organization provides the basis for the classification of olfactory epithelium as "pseudostratified."

The total thickness of the epithelium varies from one species to another, from about 30 μm in the mole (Graziadei 1973b) to more than 400 μm in amphibians (cf. Farbman & Gesteland 1974; Mackay-Sim, Breipohl, & Kremer 1988), but it also varies in different regions of a single animal (Graziadei 1973b; Mackay-Sim & Patel 1984). Given that the supporting cell nuclei and basal cell nuclei each form a single layer within the epithelium, the variation in epithelial thickness is directly related to the number of sensory neurons (Mackay-Sim et al. 1988).

In the rat nasal septum, sensory cells constitute approximately 75–80% of all the epithelial cells, supporting cells about 15–20%, and the basal cells most of the remainder (Farbman et al. 1988). There are no quantitative data on the minor cell types. In tiger salamander olfactory epithelium, 61% of cells are sensory neurons, 15% supporting cells, and 24% basal cells (Mackay-Sim & Patel 1984). In the bullfrog, the ratio between sensory neurons and supporting cells is reported to be between 5.2 : 1 and 5.7 : 1 (Takagi & Yajima 1965). These figures are based on counts of the nuclei within the epithelium in histological sections.

In Chapter 1 the olfactory sensory neuron was described as a bipolar cell with a dendritic process reaching the epithelial surface. *It is important to realize that this description applies to mature sensory neurons. At any given time in the life of a vertebrate not all sensory neurons are mature, and the dendritic endings of young or immature cells may not reach the epithelial surface* (see section 2.7). Consequently, the number of sensory neuronal nuclei is not equivalent to the number of mature cells. A better estimate of mature cells would be a count of the number of ciliated dendritic endings on the epithelial surface.

The number of sensory endings per unit area on the epithelial surface has been estimated in several species of animals (reviewed by Menco 1983). The estimates vary somewhat among different animals, depending on the counting method (i.e., some counts were obtained from scanning electron micrographs of olfactory surfaces, and others from light microscopic images of sections through olfactory epithelium). Estimates using the former method are probably more accurate (Menco 1983). In salamanders the dendritic knob density is 4.5×10^4 per square millimeter of epithelial surface (Mackay-Sim et al. 1988); in frogs it is about the same (Menco 1980a, 1983). Estimates in rabbits vary from 12 $\times$ 10^4/mm^2 (Allison & Warwick 1949) to 19–43 $\times$ 10^4/mm^2 (Mulvaney & Heist 1971a); in rats, 5–10 $\times$ 10^4/mm^2 (Menco 1980a; Hinds & McNelly 1981); in the ox, 5–9 $\times$ 10^4/mm^2 (Menco 1977, 1980a). It has been suggested that there is an "optimal" surface density of mature olfactory dendritic knobs. When this density is attained, there may be a physical or chemical mechanism that prevents the dendritic tips of immature neurons from extending to the epithelial surface and undergoing the final stages of maturation. This results in the accumulation of a population of young, almost mature neurons within the epithelium (see Figure 2.17), and further, it may account for changes in the mitotic rate (Mackay-Sim et al. 1988).

2.1.2 Lamina propria

The lamina propria may be two to three times as thick as the epithelium. It contains thin collagen and elastic fibers and most of the cellular components of connective tissue, including fibroblasts, macrophages, mast cells, and leukocytes. The relatively large caliber of the thin-walled blood vessels in the lamina propria permits a substantial blood flow to reach the mucosa. Exocrine glands (glands of Bowman; see section 2.5) and nerve bundles of the olfactory and trigeminal nerves are also found in the lamina propria. The lamina propria of the nasal septum contains, in addition, bundles of the vomeronasal and terminal nerves.

2.2 Cytology of the sensory neuron

2.2.1 Tight junctions

Near the epithelial surface, the sensory cells are isolated from one another by supporting cell cytoplasm; in other words, the distal parts of the sensory cell dendrites and their knobs generally do not make direct contact with each other, except during development. Just below the epithelial surface, supporting and sensory cells form typical tight junctions with one another. Tight junctions are characteristically found joining cells in the upper regions of *all* columnar epithelia. They are not unique to olfactory epithelium.

In electron micrographs of thin sections of olfactory epithelium, the tight junctions appear as fused regions of adjacent lateral cell membranes near the apical pole of the cell (Figure 2.3). The morphology is better appreciated in freeze-fracture preparations that permit visualization of the internal structure of the membrane. In these preparations, the tight junction is seen to consist of a variable number of interweaving strands. The strands form a series of ridges within the membrane that are in register with grooves in the adjoining cell membrane (Figure 2.4). The number of strands is correlated with the tightness of the junction (i.e., junctions with 1–3 rows of strands are leaky, whereas those with 8–12 rows are tighter). The olfactory epithelium falls into the latter category.

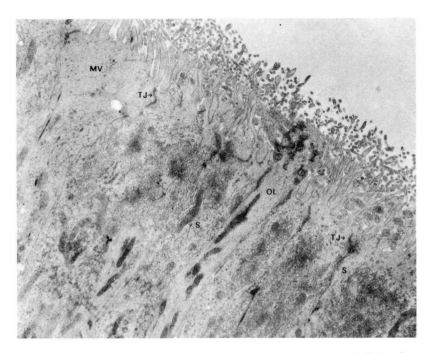

Figure 2.3. Transmission electron micrograph of the olfactory epithelial surface from a rat showing a sensory cell (OL) and its dendritic knob with cilia. Tight junctions (TJ) join the cells with adjacent supporting cells (S) near the surface. The sensory cell dendrite typically contains microtubules, some smooth endoplasmic reticulum, and mitochondria, but few ribosomes. The supporting cell cytoplasm is characterized by an abundance of smooth endoplasmic reticulum (ser) organized into parallel stacks of membranous sacs. A microvillous cell (MV), one of the minor cell types of olfactory epithelium, is seen in this micrograph.

Taken together, the tight junctions form an epithelial barrier that inhibits passage of molecules and ions between cells in either direction, from the external surface into the tissues, and vice versa. In other words, for ions and molecules to move across the epithelium, they must circumvent the tight junction by passing into the epithelial cells and then out again on the other side of the junction. Consequently, the transepithelial movement of ions and molecules is controlled by *cellular activity*. The effectiveness of the tight junction barrier in olfactory and other epithelia is related to the electrical resistance that can be measured across epithelium (i.e., a low resistance is associated with a leaky barrier, one that permits some leakage of ions between cells of the epithelium). Epithelia exhibiting a high resistance are associated with a tight barrier (Claude & Goodenough 1973). The integrity of the tight junctions, then,

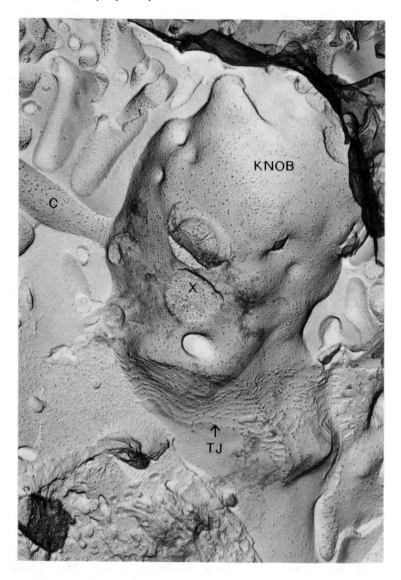

Figure 2.4. Freeze-fracture electron microscopic preparation of an olfactory dendrite knob. This image shows the many strands of intramembranous particles making up the tight junction (TJ), below the knob. An olfactory cilium (C) projects from the knob. The fractured bases (X) of other cilia are also seen. (Courtesy of Dr. B. Menco.)

is an important influence determining the electrophysiological properties of the olfactory epithelium.

2.2.2 Dendrite

The dendrite contains many parallel microtubules running lengthwise in its shaft (Figures 2.3 and 2.6). Smooth endoplasmic reticulum, both vesicular and irregular, and elongated mitochondria are found in the dendritic stalk, but few ribosomes are present. The mature dendrite terminates in a slight expansion rising a short distance above the epithelial surface. This expansion, the dendritic knob, typically has several narrow appendages projecting from it into the overlying mucus. The shape of the dendritic knob is variable (Figure 2.5). For example, in the dog, the knob may be up to 4 μm long and about 1.5 μm thick, whereas in the frog it is spherical and about 2 μm in diameter (Steinbrecht 1969). In some fishes and aquatic amphibians (Figure 2.6), and in the vomeronasal organ in mammals and lizards, it does not rise much above the epithelial surface.

The dendritic knob, including its ciliary or microvillar extensions, is the part of the cell directly exposed to odors entering the nasal cavity, that is, the site where stimuli can interact with the cell to elicit a response. There are two kinds of dendritic extensions or appendages. In the main nasal cavity in terrestrial animals the appendages usually are cilia that sprout from basal bodies in the apex of the dendritic knob, whereas, in the vomeronasal organ they are typically microvilli. Curiously, in most animals the vomeronasal sensory cells have the basal bodies, but usually do not sprout cilia. In air-breathing vertebrates, some cells in the main nasal cavity, thought to be sensory, have neither cilia nor basal bodies, but do have several short microvilli (Jourdan 1975; Moran et al. 1982a,b; Rowley Moran, & Jafek 1989). In bony fishes and gilled amphibians, neither of which have a vomeronasal organ, the sensory neurons in the olfactory organ are a mixture of ciliated and microvillous cells (Figure 2.6). Elasmobranch olfactory sensory cells usually have microvilli and no cilia (Zeiske et al. 1987).

2.2.3 Cilia

The number of cilia on a dendritic knob varies from species to species, from as few as 4 or 5 to as many as 25 or 30 (reviewed by Menco 1983). The average number of cilia in many animals is from 10 to 15. The cilia vary in length from about 50 μm in mammals to more than 200 μm in frogs. Menco (1980a) estimated that the ciliary membranes increase the receptive surface of the cell by a factor of 25 to 40.

Each cilium is about 0.3 μm in diameter at its base. The interior

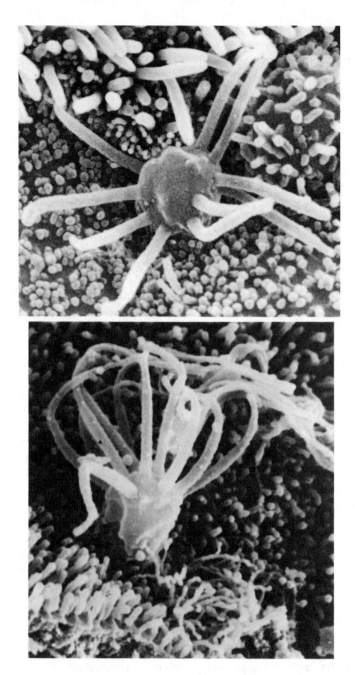

Figure 2.5. Top: Scanning electron micrograph of a dendritic knob from olfactory epithelium of a fetal rat in the 20th day of gestation. (Courtesy of Dr. B. Menco.) Bottom: Scanning electron micrograph of a dendritic knob from adult human olfactory epithelium. (Courtesy of Dr. E. Morrison.)

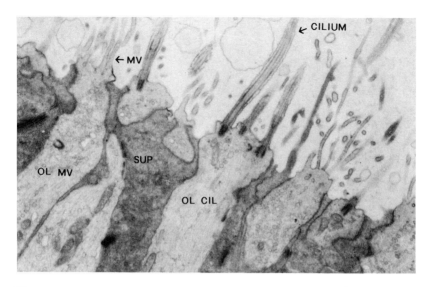

Figure 2.6. Transmission electron micrograph of the olfactory epithelial surface of an amphibian, the mudpuppy, *Necturus maculosus*. The dendritic knobs are only slightly above the level of the epithelial surface. Ciliated (OL CIL) and microvillous (OL MV) sensory neurons are typically mixed together in olfactory epithelium of aquatic vertebrates. The dendrite contains elongated microtubules and mitochondria.

contains a structure known as an axoneme. As in cilia of other cells, the axoneme is constituted of a 9(2) + 2 pattern of microtubules (9 peripheral attached pairs, or doublets, of microtubules and two single ones in the center) (Figure 2.7). In mammals, the olfactory cilium becomes narrower about 2–3 μm from its base and tapers toward its tip, where it is about 0.1 μm in diameter. In frogs, the proximal, wide segment of the cilium is about 20% of the total length (i.e., 20–40 μm long) (Reese 1965). In the narrow distal segment, some microtubules are gradually lost, so that cross sections of the cilium approaching its tip reveal fewer and fewer microtubules (Figure 2.7). The microtubules in the olfactory ciliary axoneme in many vertebrates, mammals in particular, have no arms, spokes, or other microtubule-attached structures associated with motility and are therefore immotile (Bannister 1965; Lidow & Menco 1984). Cilia in other vertebrate classes, however, do have arms and are motile (Lidow & Menco 1984).

From a functional standpoint it should be emphasized that the cilia are the parts of the sensory cell first encountered by an odor stimulus in the nasal cavity. By virtue of their length and the multiple cilia on each dendritic knob, the cilia collectively form a densely packed mat

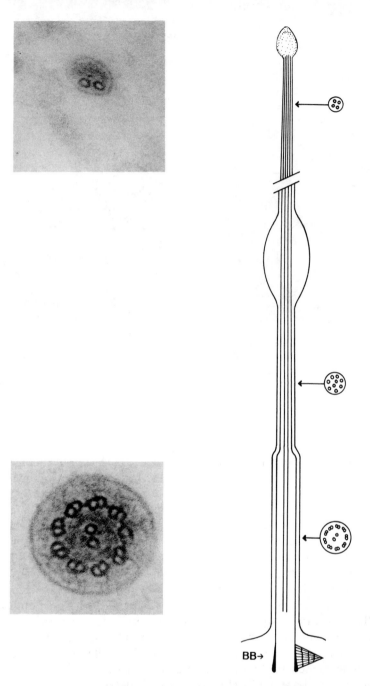

Figure 2.7. Diagram of an olfactory cilium. The cilium extends from a specialized structure, the basal body (BB) in the apex of the cell. Typical cross sections at the different levels are shown diagrammatically and in the two electron micrographs. (Electron micrographs courtesy of Dr. B. Menco.)

within the mucus on the epithelial surface. An appreciation of the density of the ciliary mat can be gained by considering that in most mammals each cilium extends in a plane parallel to the epithelial surface (Figure 2.1) and may cover a linear path of 50 μm or more, intertwining with cilia of other cells. In that path, it may pass over as many as 15 other dendritic knobs, each of which is radiating its own cilia (Menco 1977). In effect, the ciliary mat, covered by a thin film of mucus, shields the epithelial surface. From a purely physical standpoint, then, the odor stimulus reaching the nasal cavity must interact with ciliary membranes.

Because odor stimuli are thought to interact with cilia, there has been a great deal of interest in the molecular structure of ciliary membranes. One can predict, at the very least, that these membranes should contain the molecules representing the olfactory receptive sites, the transducing molecules, ion channels, some enzymes, and perhaps other molecules. Within the past few years, adenylate cyclase and G-proteins, thought to participate in signal transduction, have been localized in olfactory ciliary membranes (Pace et al. 1985; Pace & Lancet 1986; Pfeuffer et al. 1989; Mania-Farnell & Farbman 1990) (see section 3.3). Further, it has become clear that these participants in signal transduction are linked to ion channels by second messengers (Nakamura & Gold 1987) that can directly activate the channel and elicit a current (Firestein & Shepherd 1989; Firestein, Darrow, & Shepherd 1991) (see Chapter 3).

The freeze-fracture method has been used to advantage in the study of molecular domains in membranes of many cell types. In olfactory ciliary membranes this technique reveals large numbers of intramembranous particles thought to represent protein molecules (Kerjaschki & Hörandner 1976; Menco et al. 1976; Masson et al. 1978; Menco 1977, 1980b,d, 1983, 1984, 1987, 1988a,b). Near the ciliary base, the intramembrane particles are organized into several circumferential strands, referred to as the ciliary necklace (Figure 2.8). Most mammalian olfactory cilia have seven necklace strands; frogs have six (Menco 1980d). Several suggestions have been made about the specific function of the ciliary necklace (e.g., it may be involved in mechanical attachment of cilia, or it may provide a barrier to lateral diffusion of molecules within the membrane) (Menco 1983). To date, however, there is insufficient evidence to determine the function.

Above the necklace, the olfactory ciliary membrane contains particles at an average density of 1,000–1,500/μm^2, compared with a particle density of only 200–300/μm^2 in ciliary membranes of nonsensory nasal respiratory epithelial cells. In immature olfactory cilia of fetal mice and rats, the particle density is much lower, only about 20% of the adult value. As the cilium matures, the number of particles inserted into the membrane increases until adult levels are reached (Kerjaschki 1977; Menco 1988a). The increase in particle density has been correlated with

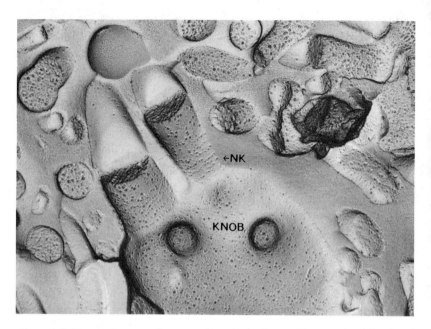

Figure 2.8. Freeze-fracture electron micrograph of a rat olfactory dendrite show-ing two cilia (C) and their necklaces (NK) within the membrane near the base. The necklaces consist of several strands of intramembranous particles. Above the necklace and on the knob itself are other intramembranous particles, or-ganized more randomly. (Courtesy of Dr. B. Menco.)

the onset of specific electrophysiological responses to odors (Menco 1988a).

The intramembranous particles range in diameter from 8 to 11 nm. With very high resolution techniques, Menco, Minner, and Farbman (1988) have shown preliminary evidence that the particles have different substructures. Interestingly, some of them seem to have pores, which may represent transmembranous channels. These new, high-resolution freeze-fracture methods, combined with specific molecular markers, may allow identification of each of the particles within the ciliary membrane, as well as membranes of other cell types.

2.2.4 Cell body

The cell bodies of sensory cells are located in the middle to lower third of the epithelium, where they are often in direct membrane-to-membrane contact with one another (i.e., there may be no supporting cell cytoplasm between them as is the case near the apical end of the cell). The nucleus is generally round to oval and contains a nucleolus

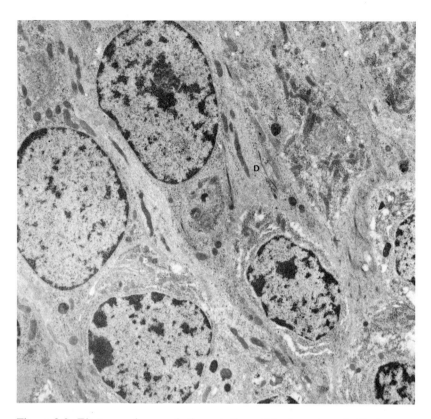

Figure 2.9. Electron micrograph through the cell body region of olfactory epithelium, showing several sensory cell nuclei containing patchy, dense chromatin. At this level, the cells are in direct contact with one another. Around the nucleus there is rough endoplasmic reticulum and often one or more Golgi bodies. The cytoplasm on the apical pole of the nucleus tapers into a dendrite (D) with few ribosomes, many microtubules, and elongated mitochondria.

and clumped chromatin (Figure 2.9). Ribosomes are present in abundance, both free and in association with the rough endoplasmic reticulum around the nucleus. A Golgi apparatus and dense bodies, probably representing lysosomes, are commonly present, but mitochondria and microtubules are sparse in the cell body region.

2.2.5 Axon

The proximal pole of the cell body narrows into an axon, about 0.2 μm in diameter, which joins with other axons to form small nerve bundles or fascicles (Figure 2.1). These pass out of the epithelium to join other

# 38 Structure of olfactory mucous membrane

fascicles, the so-called fila olfactoria, which project to the olfactory bulb, where the axons terminate. In fishes, amphibia, reptiles, and birds the fascicles are collected into a single olfactory nerve that projects to the bulb. In mammals, however, the bulb is situated relatively close to the mucosa, and the fascicles do not collect into a single nerve, but project individually through a sievelike bony plate, the cribriform plate of the ethmoid bone, to reach the bulb. In some mammals there may be up to 200–300 individual olfactory nerve bundles; in humans there are about 44, with 22 on each side (Negus 1958).

2.2.6 Biochemistry of the sensory neuron

2.2.6.1 Common neuronal proteins Most of the criteria used to characterize the olfactory sensory neuron are morphological, but in recent years there have been efforts to determine the similarities and differences between these sensory cells and other neurons. Olfactory sensory neurons contain some of the molecules common to other neurons, such as neuron-specific enolase, a glycolytic enzyme (Takahashi et al. 1984; Yamagishi et al. 1989), and the neuron cell adhesion molecule (Carr et al. 1989; Miragall, Kadmon, & Schachner 1989; Key & Akeson 1990a,b). Young neurons contain the growth-associated protein B–50, otherwise known as GAP 43 (Verhaagen et al. 1989) (see section 2.7). On the other hand, some disagreement exists concerning whether or not other neuron-specific molecules exist in sensory neurons (e.g., the three neurofilament protein subunits, heavy, medium, and light; the heavy subunit has a molecular mass of about 200 kDa, the medium about 160 kDa, and the light about 68 kDa). In an immunological study on rat olfactory epithelium, Schwob, Farber, and Gottlieb (1986) used antibodies against all three types of neurofilament protein and found little evidence for their presence. Takahashi et al. (1984) used an antibody against the medium subunit (from rat brain) and found it within some axons in human fetuses. Yamagishi et al. (1989) used a similar antibody and found neurofilament immunoreactivity in cell bodies and dendrites of human and guinea pig sensory neurons, but not in axons, where one might expect it to be expressed abundantly. Bruch and Carr (1991) found a 200-kDa neurofilament in rat sensory neurons by immunoblotting and immunohistochemistry, but the 68- and 160-kDa polypeptides were absent. Talamo et al. (1989) found only the dephosphorylated form of the medium subunit in axons of human sensory neurons, whereas the phosphorylated form was not demonstrable, nor were the other subunits. In the same study, autopsied olfactory epithelium from patients with Alzheimer's disease had increased reactivity to neurofilament antibodies against the dephosphorylated medium subunit, and to heavy and light subunits as well. The differences in ability to demonstrate neurofilament protein may be related to the relative immaturity of the sensory neurons (see section 2.7) or to species differences.

2.2.6.2 Olfactory marker protein Because olfactory sensory neurons are unusual in several ways, there have been efforts to determine the biochemical basis of their uniqueness. The most notable of these has been a series of papers from the laboratory of Dr. Frank Margolis, who discovered and characterized a specific protein found almost exclusively in olfactory sensory neurons (Margolis 1972, 1980c, 1988). Known as olfactory marker protein (OMP), this is a soluble, acidic molecule with a molecular mass of about 19,000 Da and an isoelectric point of pH 5. OMP is found in the cytosol of mature sensory neurons and is distributed essentially throughout the entire cell (Figure 2.10), from the proximal part of the cilia (Menco 1989) to the axon terminal (Farbman & Margolis 1980; Monti Graziadei, Stanley, & Graziadei 1980b). It is also found in vomeronasal sensory cells. It is not found in immature neurons (i.e., those located in the deeper regions of the epithelium). Consequently, it is often used as a marker to identify mature neurons and distinguish them from immature ones (Farbman & Margolis 1980; Miragall & Monti Graziadei 1982). Antibodies to rat or mouse OMP cross-react with olfactory epithelium from many mammalian species, including humans (Nakashima, Kimmelman, & Snow 1984) and garfish, but not with that from frogs, birds, or invertebrates (Margolis 1988).

The amino acid sequence of OMP is known (Sydor et al. 1986), and its gene has been cloned (Rogers et al. 1987). It is a unique protein in that its amino acid sequence and three-dimensional structure do not closely resemble those of any known protein; in other words, it cannot be linked with any general family or superfamily of known proteins. It constitutes about 0.1–1.0% of the total protein in the tissue. The function of OMP remains to be discovered, although a great deal of effort has been expended in many laboratories to do so.

Other proteins suggested to be unique to olfactory sensory cells have been identified. Most of these are glycoproteins, isolated from membrane fractions enriched in olfactory cilia (Chen & Lancet 1984; Chen et al. 1986a,b; Anholt, Petro, & Rivers 1990). The functions of these proteins have not been ascertained, although at least some are thought to be receptor molecules (Fesenko, Novoselov, & Krapivinskaya 1979; Fesenko et al. 1983; Chen & Lancet 1984; Chen et al. 1986a,b; Lancet 1986; Fesenko, Novoselov, & Bystrova 1987, 1988) (see Chapter 3).

2.2.6.3 Carnosine Although olfactory sensory neurons have been extensively studied, none of the well-known neurotransmitters of other neurons has been shown to be present in their axon terminals (Macrides

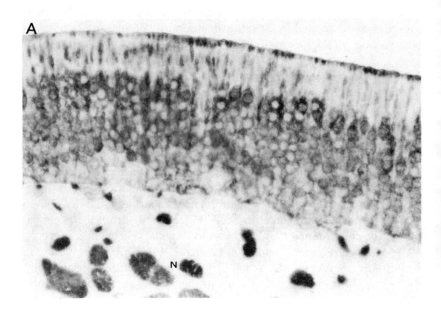

Figure 2.10. A: Histological section through olfactory epithelium of rat showing the distribution of olfactory marker protein (OMP) in the cytoplasm in most layers of sensory neurons. The protein is absent from supporting cells and many of the younger, deeply situated sensory neurons. OMP is also present in axons (N). B: Section through the olfactory bulb showing OMP-positive terminals of olfactory nerves from the main nasal cavity (NC) in bulbar glomeruli (G) and OMP-positive terminals of vomeronasal nerves in the accessory olfactory bulb (AOB).

& Davis 1983; Halász & Shepherd 1983). It was therefore of great interest when a dipeptide, carnosine (β-alanyl-L-histidine), was found in abundance in olfactory sensory neurons of several vertebrate species (Margolis 1974, 1980b; Neidle & Kandera 1974; Burd et al. 1982b; Sakai et al. 1987). This dipeptide is synthesized directly from its precursor amino acids by the action of an enzyme, carnosine synthetase (Harding & Margolis 1976; Crowe & Pixley 1991), and is transported to the axon terminals (Burd et al. 1982b). The lytic enzyme, carnosinase, is found in supporting cells (Margolis et al. 1983). Carnosine appears at the end of the second trimester in rat fetuses, indicating that carnosine synthetase is present relatively early in development (Margolis et al. 1985).

A discussion of the possible role of carnosine as the neurotransmitter of olfactory sensory neurons is included in section 5.5.1.

2.2.6.4 Pharmacological studies of olfactory sensory neurons Membrane fractions of olfactory epithelium have been studied with radioactive

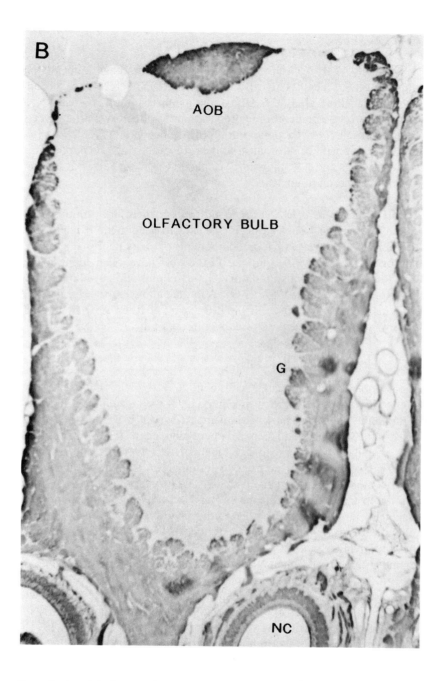

B

AOB

OLFACTORY BULB

G

NC

ligands that bind to various neurotransmitter and drug receptors. The binding assay methods were essentially identical with those used on tissues from the central nervous system. Acetylcholine receptors of the muscarinic type were found (Hirsch & Margolis 1981; Hedlund & Shepherd 1983), as were binding sites for diazepam, a benzodiazepine (Hirsch

& Margolis 1981; Anholt et al. 1984). Recent findings indicate that dopamine D2 receptors are also located in olfactory receptor cells and axons (Nickell et al. 1991). Dopamine agonists inhibit the activity of adenylate cyclase in membrane preparations of olfactory epithelium (B. Mania-Farnell & A. Farbman unpublished observations).

Although the pharmacological studies are interesting, the implications of the findings and how they may be related to the function of olfactory sensory cells remain unclear.

2.2.6.5 Proteins associated with signal transduction The proteins associated with signal transduction in the olfactory epithelium include G-proteins and adenylate cyclase, components of many neuronal and non-neuronal cell types. Two unique G-proteins have been identified in olfactory epithelium, a stimulatory $G_{olf} \alpha$ subunit and an inhibitory G-protein (Jones & Reed 1987, 1989). A unique adenylate cyclase has also been identified (Pfeuffer et al. 1989; Bakalyar & Reed 1990). The functions of these molecules are discussed in detail in section 3.3.

2.3 Cytology of supporting cells

The supporting (sustentacular) cells contain oval nuclei that usually are aligned in a single row, the most superficial of the nuclei in the olfactory epithelium. The distal part of the supporting cell is cylindrical, and it reaches the surface to end in apical microvilli (Figure 2.3). The proximal part or stalk of the supporting cell is a narrow sliver of cytoplasm that winds between receptor neurons to terminate in a slight expansion on the basement membrane (Figure 2.11). The expansion is usually in close relation to acinar cells of Bowman's gland or capillaries in the connective tissue. Because of this, one suggested function of supporting cells is to regulate the passage of substances between the epithelial surface and the underlying connective tissue (Rafols & Getchell 1983). In some animals (e.g., frogs, salamanders, mudpuppies) the supporting cells of the olfactory epithelium contain secretory granules and are thought to be the source of some of the mucus on the epithelial surface. In fact, stimulation of olfactory epithelium with some odorants, particularly irritants, evokes secretion by supporting cells (Okano & Takagi 1974). In mammals, there is no morphological evidence for a secretory function. Mammalian supporting cells often contain a striking abundance of smooth endoplasmic reticulum in their supranuclear cytoplasm (Figure 2.3). Abundance of this organelle in other cellular systems is associated with high levels of ionic flux and with detoxifying enzymes, both probable functions of supporting cells, as discussed later.

Supporting cells can be classified into at least two types based on morphological differences (Rafols & Getchell 1983; Chuah, Farbman,

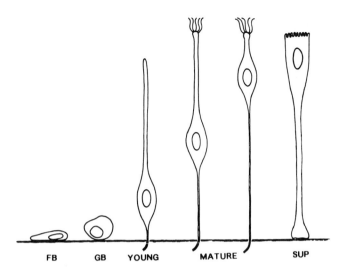

Figure 2.11. Diagrammatic representation of the major cell types in olfactory epithelium. From left to right: flat basal cell (FB); globose basal cell (GB); young sensory neuron with an axon; mature sensory neurons with axons and cilia; supporting cell (SUP) with an expanded basal process.

& Menco 1985), but nothing is known about how these structural differences may relate to function. In salamanders, a distinguishing morphological feature between the two types of sustentacular cells is that the so-called Type II cell stalk gives rise to many processes from which cytoplasmic veils extend around the cell bodies of sensory neurons, whereas the Type I stalk is unbranched (Rafols & Getchell 1983). In rat olfactory epithelium, length of the microvilli is a distinguishing feature between two types of supporting cells (Chuah et al. 1985; Menco & Farbman 1985a). No comparison has been made between the types seen in salamander and those seen in rat.

Freeze-fracture studies on olfactory epithelium have shown that the particle density within membranes of supporting cell microvilli is in the range of 1,600–2,000/μm^2, even higher than in ciliary membranes of sensory neurons (Menco 1980b, 1988a). It is also interesting that the supporting cell apical surfaces and, to a lesser extent, their microvilli contain rod-shaped intramembranous particles (Figure 2.12) (Kerjaschki & Hörandner 1976; Menco 1980a,b, 1984, 1988a). Membranes of Bowman's gland duct cells and those of some acinar cells also contain rod- or dumbbell-shaped particles, each of which consists of a pair of contiguous globular particles. In early development, supporting cell apical surface membranes contain only single globular particles, but during maturation the rod-shaped particles increase in number until they are

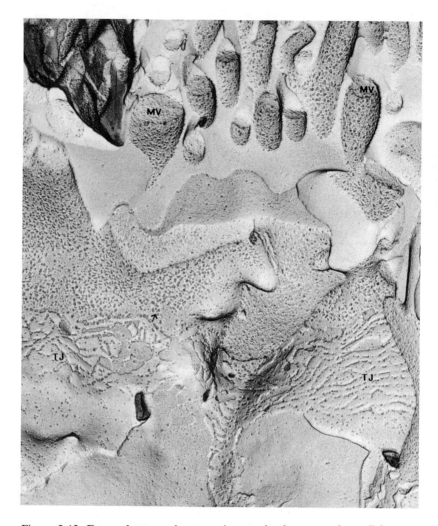

Figure 2.12. Freeze-fracture electron micrograph of a supporting cell from rat olfactory epithelium. Tight junctions (TJ) are seen. The intramembranous particles (arrows) on the apical surface and microvilli (MV) are rod-shaped. (Courtesy of Dr. B. Menco.)

the prevailing type (Menco 1988a). The significance of these unusual intramembranous particles is not understood.

Supporting cell cytoplasm in rats contains a specific cytoplasmic antigen that is immunologically reactive with a monoclonal antibody, SUS–1 (Figure 2.13). The antibody does not react with basal or sensory cells within the epithelium, but does react with some cells in Bowman's glands

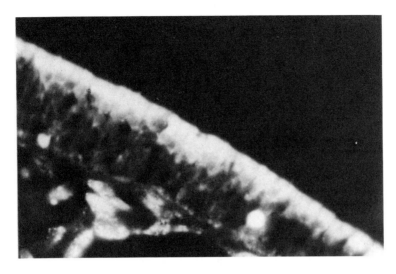

Figure 2.13. Histological preparation of rat olfactory epithelium showing immunoreactivity of supporting cells with a monoclonal antibody, SUS-1. The positive reaction is in the cytoplasm of supporting cells and is revealed by the presence of a fluorescent marker in the supporting cell cytoplasm. (Courtesy of Dr. V. Carr.)

(Hempstead & Morgan 1983). The identity of the SUS–1 reactive molecule is not known. But the fact that sustentacular cells are replenished relatively slowly (Graziadei 1973a) and the observation that SUS–1 reactive cells exist in Bowman's glands suggest the possibility that the two cell types share common embryonic origins (Hempstead & Morgan 1983). It is possible, for example, that migration of cells from Bowman's glands or its ducts may give rise to supporting cell precursors (Mulvaney & Heist 1971b).

Recently, a nonneuronal cell was identified in rat olfactory epithelium that, in almost every way, resembles a supporting cell, but it does not react with the SUS–1 antibody. In fact, it specifically reacts with another monoclonal antibody, 1A–6, and its reactivity is located mostly on its microvilli (Figure 2.14) (Carr et al. 1991). This cell type constitutes less than 1% of the total epithelial cell population (probably less than 5% of the supporting cell population). It occurs more frequently on the turbinates of the nasal cavity than on the septum. The function of 1A–6 immunoreactive cells is not known, nor do we know if its function differs from that of other supporting cells. Microvilli of possibly another minor supporting cell type had a lectin binding pattern different from the usual pattern of most supporting cells (Menco 1992).

2.3.1 Function of supporting cells

Mammalian supporting cells are thought to participate in regulation of the ionic composition of the mucus layer (T. Getchell, Margolis, & M. Getchell 1984). The evidence for this is suggested primarily by two morphological features, namely, the presence in supporting cells of some animals of an abundant array of smooth endoplasmic reticulum (Figure 2.3) (Frisch 1967; Graziadei 1971, 1973b) and the intramembranous rod-shaped particles discussed earlier. Both of these structural elements are associated with cells engaged in fluid and ion transport, and immunocytochemical studies provide some evidence for this function (Menco & Benos 1989).

Supporting cells probably participate in at least one other function, that of detoxification. An enzyme system associated with this function, the cytochrome P–450 system, is associated with the smooth endoplasmic reticulum, where it is thought to function in the enzymatic degradation of olfactory stimuli and/or other substances in the nasal cavity in preparation for their disposal (Dahl et al. 1982; Voigt, Guengerich, & Baron 1985; Dahl 1988; Lazard et al. 1989). In fact, the enzyme activity of the cytochrome P–450 enzyme system in olfactory tissue is higher than in the liver (Reed, Lock, & De Matteis 1986).

Other roles, if any, for supporting cells remain to be discovered. One recent finding of interest is that within 2–6 hr following administration

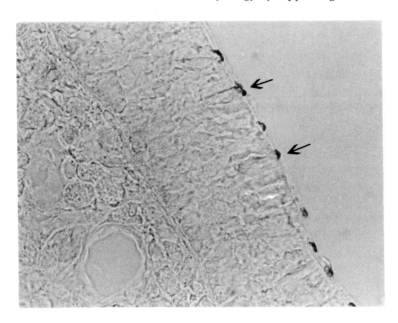

Figure 2.14. Histological preparation of rat olfactory epithelium showing immunoreactivity of some nonneuronal, supporting-like cells with a monoclonal antibody, 1A-6. In this micrograph the positive reaction is revealed by precipitation of a brown reaction product, mostly on the microvillous surface of the supporting cell (arrows). (Courtesy of Dr. V. Carr.)

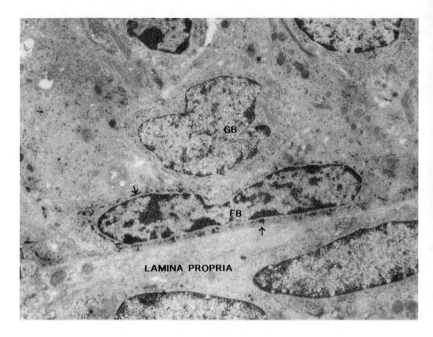

Figure 2.15. Electron micrograph showing a flat (FB) and a globose (GB) basal cell in the olfactory epithelium, just above the lamina propria. The flat basal cell contains bundles of tonofilaments (arrows) that are positive for the protein keratin.

of general anesthesia, the supporting cells in olfactory epithelium express at least two heat shock proteins (HSPs): ubiquitin and HSP–70 (Carr & Farbman 1991). HSPs, also known as stress proteins, are expressed in all organisms, from single-cell organisms to humans, in response to sudden and excessive elevation of ambient temperature or to other stresses placed on the organism. They are expressed in glia in many parts of the central nervous system in mammals in response to elevation of body temperature from 37°C to 43°C. These proteins are thought to serve a protective function.

The anatomy of the supporting cell suggests other possible functions. This cell stretches from the epithelial surface to the basal lamina. Along the extent of its lateral cell membrane it comes into contact with basal cells and sensory neurons of all ages, from immediately postmitotic to mature. One might ask, for example, Do the supporting cell interactions with sensory neurons vary with respect to the age of the latter? Do the supporting cells have anything to do with continued neurogenesis or the regulation of the rate of neurogenesis in olfactory epithelium?

2.4 Basal cells

Basal cells are more or less polyhedral. They rest on the basement membrane, and some of them ensheath small bundles of axons. Two types of basal cells have been described in electron microscope studies, a flattened cell with dark cytoplasm, resting directly on the basement membrane, and a globose basal cell, a rounded cell containing a light cytoplasm (Figure 2.15) (Andres 1966; Graziadei & Monti Graziadei 1979).

The darker basal cells, termed "basal cells proper" (Graziadei & Monti Graziadei 1979), contain many free cytoplasmic ribosomes and tonofilaments and are immunoreactive with antikeratin antibodies (Vollrath et al. 1985; Yamagishi et al. 1989). It is likely that the darker cell is the true stem cell, and it divides asymmetrically to form one dark daughter cell and one globose cell (Calof & Chikaraishi 1989). The globose basal cells usually are superficial to the flattened cells, but they, too, retain the ability to undergo cell division even in adult animals (Andres 1966; Graziadei & Monti Graziadei 1978, 1979). They presumably give rise to the precursors of new neuronal cells (Figure 2.11) (reviewed by Moulton 1975; Graziadei & Monti Graziadei 1978). The globose basal cell does not stain with antikeratin antibodies. It has fewer tonofilaments, a less dense cytoplasmic matrix, significant amounts of rough endoplasmic reticulum, and a prominent Golgi apparatus. This cell type (not the dark, flat basal cell proper) is likely the one that is stimulated to divide rapidly when the olfactory epithelium is stressed, as it is after axotomy or bulbectomy (Schwartz Levey, Chikaraishi, & Kauer 1991; Suzuki & Takeda 1991). The products of its division differentiate into sensory neurons. Data from in vitro studies lend support to the notion that the globose basal cells, or perhaps a subpopulation of them, are the immediate precursors of neuronal cells (Calof & Chikaraishi 1989).

2.5 Bowman's glands

Bowman's glands, located in the lamina propria of the olfactory mucosa, are usually described as branched tubuloalveolar glands that secrete a serous (i.e., serumlike, or watery and protein-rich) product. Secretions are produced in the alveolar (bottom) part of the gland and conveyed into ducts, lined with flattened cells, that pass through the olfactory epithelium and empty onto the surface (see Figures 1.8 and 2.1). Secretions from Bowman's glands provide most of the mucus covering the olfactory epithelial surface. The mucus protects the epithelial surface from drying and temperature extremes and helps to prevent damage caused by infectious agents and noninfectious particles. Further, because the secretions bathe the receptive parts of the sensory cells, they probably influence stimulus access to the cells and possibly affect interaction between stimulus and receptor sites on the cilia.

The pyramidal secretory cells of Bowman's glands contain secretory granules that in histological preparations stain readily with basic stains (basophilic). In amphibians and snakes, these granules contain neutral mucopolysaccharides, but in most other reptiles, and in birds and mammals, acidic mucopolysaccharides are also present in the gland cells (T. Getchell et al. 1984). With the electron microscope, two secretory cell types can be distinguished: a "dark cell" that contains an extensive ribosome-laden granular endoplasmic reticulum and abundant secretory granules, and a "light cell" with a vesicular-appearing cytoplasm and no secretory granules (Frisch 1967). The latter appearance may be due to depletion of their granules after secretion.

Secretion from Bowman's glands can be stimulated by B-adrenergic (M. Getchell & T. Getchell 1984; T. Getchell et al. 1984), α-adrenergic, and cholinergic agonists (M. Getchell, Zielinski, & T. Getchell 1988). This suggests that secretion is regulated by both sympathetic and parasympathetic nerves that terminate on secretory cells (Zielinski et al. 1989b). Substance P-like immunoreactivity has been found in nerve fibers associated with Bowman's glands in mice (Papka & Matulionis 1983), salamanders, and frogs (M. Getchell et al. 1988), and peptidergic fibers have been demonstrated with the electron microscope (Zielinski, Getchell, & Getchell 1989a). Application of substance P to the olfactory mucosa results in depletion of granules in Bowman's gland cells, suggesting this peptide may participate in eliciting gland secretion. It is clear that odor stimulation can also invoke gland secretion (M. Getchell et al. 1988).

In some animals, most notably amphibians, the olfactory region of the nasal cavity contains other, more deeply situated glands in addition to Bowman's glands (see Figure 1.8). Ducts from these glands have not been found to go to the olfactory epithelial surface. The function of these glands is unclear.

2.6 Does response specificity exist in the olfactory system?

Because olfactory cells can respond to so many different kinds of stimuli, a frequently asked question is, Are there different subtypes of sensory cells responsive to various classes of odorants? It seems an inescapable conclusion that the sensory neurons in the olfactory epithelium must function in some way as an analyzer of the chemical substances borne into the nasal cavity. Early studies by Gesteland, Lettvin, & Pitts (1965) on responses of single sensory neurons in frog olfactory epithelium showed that individual neurons are not very specific in their responsiveness to stimuli; that is, most sensory cells respond to several stimuli. The response spectrum of a given sensory cell could overlap partially or completely with that of other sensory cells. Later studies confirmed these observations and suggested that receptor cells may fall into several general groups, based on statistical similarity of their response profiles (Revial, Duchamp, & Holley 1978a; Revial et al. 1978b, 1982, 1983). Thus, odorants do not activate highly specific receptor cells that respond only to a single compound. Rather, stimulation with a given odor activates a subset of the sensory neuron population, each one of which has a broad range of sensitivities overlapping with one another.

Even though single sensory neurons are not odor-specific, we can still ask to what extent the sensory epithelium itself participates in the analysis of the stimulus. There are at least two aspects of this question. One can be framed as follows: Are the various regions of the nasal cavity differentially sensitive to certain classes of odor stimuli? In other words, is there a "spatial map" within the nasal cavity organized such that olfactory sensory cells in one region are more responsive to certain odors and less responsive to others? If this is true, the second aspect concerns whether or not the different regions are populated by sensory cells that can be distinguished structurally or biochemically from one another.

2.6.1 A spatial olfactory map in the nasal cavity

Evidence from several laboratories suggests that there is spatial differentiation of responses to different odorants within the olfactory epithelium (Mustaparta 1971; Kauer & Moulton 1974; Døving & Thommesen 1977; Thommesen & Døving 1977; Kubie, Mackay-Sim, & Moulton 1980; Mackay-Sim & Kubie 1981; Mackay-Sim, Shaman, & Moulton 1982; Thommesen 1982, 1983; Mackay-Sim & Shaman 1984). In most of these experiments, different odorant stimuli were puffed from a very narrow outlet tube onto a tiny area of olfactory mucosa. It was found that a given odor elicited its greatest response amplitude (measured electrophysiologically) over a particular mucosal region, whereas other stimuli elicited their maximal responses in other regions. Although there is much overlap in regions of submaximal sensitivity, the variability in responsiveness suggests that sensory cells having different response spectra are not uniformly or randomly distributed in the nasal cavity; rather, cells with similar spectra are grouped together, at least to some degree.

The methodology used in these experiments deliberately circumvented the influence of another factor that may contribute to the spatial map of responsivity in the olfactory region, namely, the mucus on the epithelial surface. The odorant-sensitive parts of the cells (i.e., the dendritic knob and its cilia and/or microvilli) are bathed in a layer of mucus, approximately 35 μm thick in frogs (Reese 1965), and 5–11 μm thick in mammals (Menco 1983). Odor molecules must enter the mucus and diffuse through it before reaching the responsive part of a sensory cell. When odorant stimuli are taken into the nasal cavity, the stimulus molecules are spatially distributed across the epithelial surface of the nose by differential sorption to the mucosa. In effect, the mucus may concentrate odorants in certain regions of the nasal cavity, or at least influence the passage of different molecules along the sheet (Mozell 1964, 1966, 1970; Mozell & Jagodowicz 1973; Hornung, Lansing, & Mozell 1975; Hornung & Mozell 1977, 1980, 1981; Mozell et al. 1987). The molecular distribution of different odorants along the olfactory receptor sheet seems to depend on the partitioning of the molecules between the air phase and the mucus phase in air-breathing animals. The more this partition favors the mucus, the slower the molecules will travel through the nasal cavity, and within the time interval of a single sniff, the steeper will be their concentration gradient from the external naris to the internal naris (Hornung & Mozell 1981).

It seems reasonable to suggest, then, that the activity pattern in a particular region of the olfactory epithelium is influenced by differential sorption of odorants along the mucosa. Consequently, the presence of a higher concentration of stimulus in a region of high sorption could be responsible for a greater response. Moulton (1976) used the term "imposed" to describe the activity patterns stemming from differential sorption of odorants along the mucosa, to distinguish them from "inherent" patterns that are based upon differences in cellular sensitivity (T. Getchell et al. 1984). The available evidence suggests that both the imposed and inherent response patterns must be taken into account in considering the local variations in sensitivity to odorants exhibited by the olfactory mucosa.

In addition to the regional specificity of responsiveness within the olfactory epithelium, there is evidence for some in the olfactory bulb. The existence of a spatial "olfactory map" on the olfactory bulb had been predicted as early as 1920 by Holmgren's anatomical studies and was substantiated by Adrian (1950), who found regional variations in response to stimuli when he made electrophysiological recordings from the olfactory bulb. Adrian's work has been confirmed by several authors using different techniques, including simultaneous recordings of multiunit activity (Moulton 1965), recordings from single glomeruli in the bulb (Leveteau & MacLeod 1966), spatial mapping of the odor-induced surface potential changes (Døving & Thommesen 1977), differences in amounts of acetylcholinesterase-positive nerve fiber input from the central nervous system (Zheng, Ravel, & Jourdan 1987), selective morphological changes in populations of bulbar mitral cells caused by continuous long-term exposure to specific odorant stimuli (Pinching & Døving 1974), and autoradiographic demonstrations that continuous olfactory stimulation results in spatial differences in glucose metabolism among regions of the bulb (Sharp, Kauer, & Shepherd 1975, 1977; Skeen 1977; Stewart, Kauer, & Shepherd 1979; Lancet et al. 1982). We shall return to a discussion of the topographic code of odor information within the bulb in section 2.9.

2.6.2 Do sensory neuron subtypes exist?

We turn now to the second aspect of the question posed earlier, namely, Are there any specific structural or biochemical differences among the cells themselves that can be associated with regional variations in responsivity? A suggestion that such differences exist came from anatomical studies showing subtle differences in staining properties of sensory neurons (Le Gros Clark 1957). From studies done with the electron microscope, it is now well known that in some mammals, including humans, the main nasal cavity contains primarily ciliated neurons. There are some recent studies suggesting the existence of nonciliated sensory cells (Jourdan 1975; Moran et al. 1982a,b; Rowley et al. 1989). Ciliated and nonciliated sensory cells have been described in nasal cavities of fishes (Bannister 1965; Hara 1975; Yamamoto & Ueda 1977; Breucker, Zeiske, & Melinkat 1979; Cancalon 1983e; Muller & Marc 1984) and water-dwelling amphibians (Figure 2.6) (Farbman & Gesteland 1974; Graziadei & Monti Graziadei 1976).

The existence of only two or three distinct cell types may or may not account for the organism's ability to discriminate among hundreds of different odorants. In the visual system, after all, there are only three types of cones that account for the ability to discriminate among hundreds of hues. In the olfactory system of salmonid fishes there is some evidence suggesting that odor specificity may be correlated with different types of dendritic appendages on sensory epithelial cells (Thommesen 1982, 1983). In the central part of the olfactory region in these fish, the relative numbers of microvillous sensory cells are higher, whereas in the periphery there are relatively more ciliated sensory cells. There is an apparent correspondence between the density of ciliated sensory cells and the sensitivity to bile salts in the peripheral region, as determined electrophysiologically. Moreover, the increased density of microvillous sensory cells in the central region of the organ is related to a higher responsivity to amino acids.

On the other hand, the spatial distribution of ciliated and microvillous olfactory sensory neurons in the channel catfish is not matched by differential specificity to amino acid and bile salt stimuli (Erickson & Caprio 1984). In catfish, the density of microvillous cells is highest in the dorsomedial region of the olfactory lamellae, and ciliated cells are relatively more numerous in the ventrolateral region. However, no significant differences were found between these two regions in physiological responses to amino acids and bile salts. On the basis of the two sets of experiments in fishes, it is not possible to make any conclusive generalizations about functional differences between ciliated and microvillous cells.

Carbonic anhydrase is present in a subpopulation of olfactory sensory neurons (Brown, Garcia-Segura, & Orc 1984). This enzyme is a zinc-dependent, metalloenzyme that catalyzes the reversible hydration of carbon dioxide to produce hydrogen ion and bicarbonate. A histochemical technique was used to show that the enzyme was present in cell bodies, dendrites, and axons of some olfactory neurons, but not in supporting cells. Whether or not these differences are important in olfactory function is not known.

More recent evidence for subtypes among sensory cells has come from studies employing lectin binding that have shown selective binding of some lectins to specific sensory neurons or to specific glomeruli in the bulb (Key & Giorgi 1986; Plendl & Schmahl 1988; Barber 1989; Menco 1992). Others have done immunological investigations using monoclonal antibodies (Mabs) as probes. These studies have revealed that sensory cells in the olfactory epithelium express antigenic differences, thus indicating that they may have different molecular "fingerprints" and therefore may not be a homogeneous population of cells. In some instances a Mab labels a population of sensory cells that line a particular region of the olfactory epithelium within the nasal cavity. For example, one Mab, R4B12, raised against rabbit olfactory bulb, labeled sensory cells in the ventrolateral and caudal regions of the olfactory epithelium in the main nasal cavity, but not in the dorsomedial region (Mori et al. 1985; Onoda & Fujita 1988). The same antibody recognizes a subclass of vomeronasal nerve fibers projecting to glomeruli in the rostrolateral part of the accessory olfactory bulb, whereas another Mab (R5A10) recognizes a complementary subclass projecting to glomeruli in the caudomedial part (Mori et al. 1987). Other monoclonals also recognize subsets of vomeronasal sensory cells in rabbits (Mori 1987a). Schwob and Gottlieb (1986) developed a Mab that stains the sensory cells and their axons in the ventrolateral part of the olfactory sheet in rats, but not those in the dorsomedial part. Others have shown differential immunological staining between the sensory neurons with cell bodies in the more superficial regions of the epithelium and those in the deeper regions (Onoda & Fujita 1988). This staining is very similar to that shown for OMP (Farbman & Margolis 1980); that is, it stains the more mature sensory cells in the epithelium; however, the tissue antigen is different from OMP (Onoda & Fujita 1988).

In other studies, subsets of cells throughout the epithelium (i.e., not restricted to a particular region of the nasal cavity) exhibit labeling with an antibody. A monoclonal antibody known as 2B8 stains a subpopulation of olfactory neurons, about 25% of those stained by antibody to OMP (Allen & Akeson 1985a,b). The reactive antigens in olfactory epithelium show up as two bands on immunoblots of olfactory mucosa, with apparent molecular masses of 215,000 and 160,000 Da. One monoclonal antibody stains only basal cells in olfactory epithelium and may be a useful marker for proliferative components of the epithelium (Akeson & Haines 1989). Blood group antigens H and B are detectable on subgroups of olfactory sensory neurons (Mollicone, Trojan, & Oriol 1985; Astic et al. 1989). In frog olfactory epithelium a monoclonal antibody against carnosine synthetase stains a subset of sensory neurons (Crowe & Pixley 1991). At this time, however, although some of the immunohistochemical data clearly show that there are major subtypes of sensory cells, it cannot be said that there is enough individuality in the molecular fingerprints of sensory cells to account for their ability to detect a large variety of odors. The probes currently available may not be subtle enough to determine whether the discriminating ability of individual sensory cells is to a small or large number of stimuli.

A recent review summarizes the current knowledge about olfactory neuron subclasses (Akeson 1988). For the time being, we are left with the suggestion that there is anatomical, physiological, and biochemical evidence for the existence of subclasses among sensory neurons. Physiological studies have not uncovered any one-to-one relationship between a given stimulus and either the sensory neuron or the glomerulus; rather, there is considerable overlapping of responsivities. Nor can we correlate the results from physiological studies with those from immunological experiments. Such correlations would enable us to say, for example, that cells expressing a certain glycoprotein would be highly responsive to certain classes of stimuli.

2.7 Neurogenesis in olfactory epithelium

A complicating factor in the analysis of response specificity in individual olfactory sensory cells is that these cells are usually short-lived. In fact, they are almost unique among neurons in that they can be replaced, even in the adult warm-blooded vertebrate. In some cold-blooded animals, new neurons in many regions of the nervous system are made as the animal grows (e.g., retinal neurons in the goldfish eye are added as the posthatching fish increases in size). Concomitantly, the tectum in the brain increases in size to accommodate the increased synaptic input from the retina. In warm-blooded animals, however, most neurons in other parts of the nervous system are formed before birth; a few types, including granule cells of the olfactory bulb, are formed postnatally, during infancy. With very few known exceptions (Paton & Nottebohm 1984; Kaplan & Hinds 1977; Kaplan et al. 1985), new neurons are not made in the adult warm-blooded vertebrate. The olfactory epithelium is one of these exceptions. Neurogenesis in this tissue continues throughout life, thus endowing the animal with the ability to replace neurons that have been lost as a result of physical, chemical, or infectious insults.

The work of Nagahara (1940) is usually cited as the first to suggest strongly that olfactory sensory neurons could be replaced. He classified sensory neurons into two groups: "functionally active" and "resting" cells. The functionally active ones were those in which the cell bodies were located more superficially in the epithelium; cell bodies of resting neurons were situated more deeply. When olfactory axons were severed, the functionally active neurons degenerated and disappeared within a few days, but the resting neurons were undisturbed. However, the population of resting cells, which contained a few mitotic cells even under normal conditions, began to show much more active mitosis after nerve section. The resulting daughter cells soon extended proximal and distal processes and differentiated into normal-appearing olfactory sensory neurons, so that about 90 days after nerve section the epithelium appeared more or less as it did before surgical intervention.

Subsequent work by others confirmed that after experimental destruction of olfactory epithelium, by nasal lavage with a zinc sulfate solution or by axotomy, the olfactory epithelium could recover (e.g., Schultz 1941, 1960; Smith 1951). Later studies suggested that even under physiological conditions olfactory cells died and were replaced by new cells. For example, Andres (1966, 1969) noticed that mitotic cells, young sensory cells, mature sensory cells, and dying cells coexisted within the olfactory epithelia of cats and dogs, and he proposed that cells were continually replaced.

A few years later, several workers, using autoradiographic methods, provided conclusive evidence that new neurons were produced by mitosis in the basal cell population and that the daughter cells differentiated into normal, functional sensory neurons (Moulton, Celebi, & Fink 1970; Thornhill 1970; Graziadei & Metcalf 1971; Graziadei 1973a; Moulton 1974, 1975; Graziadei & Monti Graziadei 1978). These studies took advantage of the fact that thymidine, radioactively labeled with tritium ([^{3}H]thymidine), is specifically incorporated into the nuclei of cells that are replicating their DNA in preparation for cell division. The cells that take up the radioactive thymidine can be identified by placing histological sections of tissue in direct contact with a photographic emulsion for an appropriate period of time, usually 1–4 weeks. After the emulsion is developed, the sites where [^{3}H]thymidine was taken up are revealed by the presence of silver grains, which can be observed with a microscope, lying over the nuclei that incorporated the radioactive isotope (Figure 2.16).

This technique enabled investigators to show that the basal cells were the first to incorporate [^{3}H]thymidine. Autoradiographs taken from animals 1–2 weeks after injection of the radioactive label exhibited labeled sensory neurons above the basal layer, and at later time intervals labeled cell nuclei were seen higher in the epithelium (Figure 2.17). These results

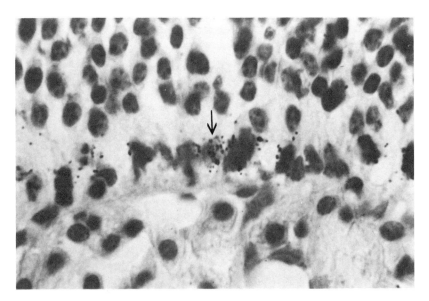

Figure 2.16. Typical autoradiograph of olfactory epithelium from a rat that had been injected with [³H]thymidine 1 day earlier. Black silver grains are seen overlying nuclei of cells in the basal region (arrow). (Courtesy of Dr. V. Carr.)

indicated that those neurons with nuclei more superficially located within the epithelium were both older than those located more deeply and had shorter dendritic processes. These older, presumably more mature cells, as defined by thymidine incorporation, corresponded to the "functionally active" cells of Nagahara (1940), those that had been damaged by axotomy. Nagahara's "resting cells" were the younger cells, not damaged by axotomy, presumably because their axonal processes had not grown to the point where the nerves were surgically severed.

The results from regeneration and autoradiographic experiments gave rise to the notion that olfactory neurons were "disposable" (i.e., that new neurons were made continuously to replace the old ones) and that the neurons had a finite life span of approximately a month in mammals (Moulton 1975; Graziadei & Monti Graziadei 1979), and somewhat longer in amphibians (Simmons & Getchell 1981a). In terms of their disposability, sensory neurons were thought to fall into the same category as epidermal cells, intestinal epithelial cells, and red blood cells, all of which are continually replaced during the lifetime of the organism. The idea grew that daughter cells resulting from mitoses in the basal cell layer of olfactory epithelium differentiated, matured into functional sensory cells as they made synaptic connections with the bulb, and

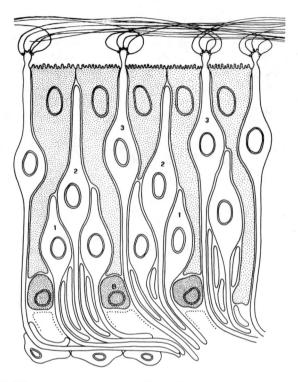

Figure 2.17. Diagrammatic reconstruction of stages in sensory neuron maturation from the basal (B) to the mature cell. In the earliest stages (1) the
postmitotic cell is deeply situated; it grows an axon and begins to grow a dendrite.
In the next stage (2), the cell body rises within the epithelium, and a dendrite
grows out, but does not reach the epithelial surface. In the mature stage (3),
the dendritic knob has reached the surface and has grown cilia.

subsequently died and were removed, that is, that olfactory cells, like
the other cell types (but unlike other neurons), underwent a continual
turnover.

The view that olfactory neurogenesis was an ongoing process was
strengthened by evidence for differential expression of proteins associated with either growing or mature olfactory neurons. B–50 or GAP
43 (also known as F1 or pp46) is a neuron-specific phosphoprotein that
is known to be expressed in growing, developing, and regenerating
neuronal processes (reviewed by Benowitz & Routtenberg 1987; Coggins & Zwiers 1991). In the olfactory epithelium, the expression of B–
50/GAP 43, as revealed by immunohistochemistry, is confined to the
younger sensory neurons (Verhaagen et al. 1989). On the other hand,

OMP is associated with mature olfactory sensory neurons (Farbman & Margolis 1980; Miragall & Monti Graziadei 1982). These two proteins are expressed in a nearly reciprocal fashion, as the cells mature (Verhaagen et al. 1989). Basal cells are not immunoreactive with either antibody.

The idea that true neurons could be replaced in adults upset the preexisting dogma that all neurons, once dead, were irreplaceable (Leblond & Walker 1956). Graziadei and Monti Graziadei (1979), the champions of this novel hypothesis, took the position that the regenerative capacity of the olfactory system was not a response of the system to environmental injury, but a genetic characteristic.

Some experiments in the 1980s raised questions about the concept that the olfactory epithelium was constantly replaced at a genetically predetermined rate. Hinds, Hinds, and McNelly (1984) injected mice with [^{3}H]thymidine and raised them in the filtered-air environment of a laminar-flow hood, to prevent rhinitis resulting from airborne particles or infectious organisms. They showed that some olfactory sensory cells survived for as long as 12 months, considerably longer than the life span of 1 month postulated in earlier experiments (Graziadei & Monti Graziadei 1979). Hinds and his co-workers suggested that in the absence of disease-related destruction of the olfactory epithelium, most or all sensory cells that turned over were *newly formed or not fully mature cells that failed to establish synapses with the olfactory bulb.* The population of cells that died, then, was not necessarily composed of worn-out mature cells, but of those *nearly mature* ones that did not establish synaptic contact in the olfactory bulb. This would mean that functionally mature, synaptically connected cells could remain alive for an indeterminate (i.e., not programmed) length of time and that their life span would be determined not by intrinsic, genetic factors, but extrinsically, by factors related to nutrition, disease, age, hormonal state, injury, and so forth.

Other workers argued that if sensory neurons are replaced according to a genetic program, the ratio between progenitor cells (the basal cells) and mature olfactory cells should be more or less constant (Breipohl et al. 1986; Walker et al. 1990); that is, to maintain equilibrium, each mature cell that died should be replaced by the differentiation product of a cell division in the progenitor cell pool. This does not appear to be the case, however, for the following reasons: First, the ratio between progenitor and mature cells changes with age (Breipohl et al. 1986; Walker et al. 1990). Second, the ratio is different in some parts of olfactory epithelium than in others, even within a given animal (Mackay-Sim & Patel 1984; Mackay-Sim et al. 1988). Third, the number of mitoses in the progenitor cell population can be regulated without changing the number of mature cells. For example, it can be decreased or down-regulated (1) by raising animals in a filtered-air environment (to prevent rhinitis) (Hinds et al. 1984) or (2) by unilateral occlusion of a naris, which protects the olfactory epithelium from potentially harmful environmental influences (Farbman et al. 1988). It can be up-regulated in response to bulbectomy (Carr & Farbman 1992).

Breipohl et al. (1986) agreed with the suggestions of Hinds et al. (1984) that mature sensory neurons, those that are synaptically connected with the olfactory bulb, may live much longer than originally thought and that cell death may occur in the population of neurons that are *almost mature*, or not yet synaptically connected with the bulb. The concept that a population of almost mature sensory neurons exists, available and ready to occupy synaptic spaces vacated by dying mature cells, is important. There is no question that neurogenesis continues in the olfactory epithelium throughout adult life. This new concept, however, suggests that the neurogenic mechanism may really be regulated to maintain a supply of almost mature cells, any one of which, if necessary, can replace a dying, synaptically connected, mature neuron (Walker et al. 1990). This concept has been supported by recent data showing that after intrabulbar injection of a marker (colloidal gold conjugated with concanavalin A), it was retrogradely transported to the epithelium and remained within epithelial cells for at least 90 days, considerably longer than the regular 1-month life span proposed by others (Mackay-Sim & Kittel 1991).

Implicit in this model is the suggestion that an olfactory sensory neuron cannot become fully mature unless it forms an appropriate synaptic connection with its target. In other words, without this contact, the cell may be deprived of a factor required for the final stages of maturation. Consequently, it remains suspended in the almost mature state. One criterion for full maturity in olfactory sensory neurons is a full complement of cilia. Developmental (Cuschieri & Bannister 1975b), organ culture (Chuah et al. 1985), and bulbectomy (Schwob & Szumowski 1989) studies indicate that ciliogenesis is incomplete in the absence of direct contact with the bulb. Furthermore, the results of degeneration/reconstitution studies have supported the contention that the bulb may provide something required for full maturation of the sensory cells (Simmons & Getchell 1981b; Doucette, Kiernan, & Flumerfelt 1983a; Monti Graziadei 1983; Hempstead & Morgan 1985; Farbman 1986; Schwob & Szumowski 1989).

Although this new model is attractive from an evolutionary standpoint, particularly for animals that are heavily dependent on olfaction for their survival, it does not completely refute all the evidence suggesting that there is a genetic component to the short life span of the sensory cell. For example, the anatomical relationship between olfactory axons and their ensheathing (Schwann) cells never progresses beyond what might be considered the embryonic state (Figure 2.18) (Gasser

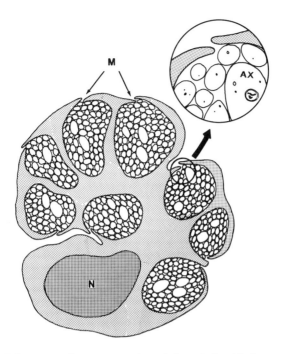

Figure 2.18. Diagrammatic representation of the relationship between olfactory axons (AX) and their ensheathing cells; M, mesaxon.

1956). The fact that the ensheathing cell–axon relationship remains immature is consistent with the idea that the cells are not programmed to live a long time. Moreover, only very small numbers of sensory neurons engage in synthesis of proteins typically found in the intermediate filaments (neurofilaments) of mature neurons (Vollrath et al. 1985; Schwob et al. 1986; Yamagishi et al. 1989). Rather, they continue to synthesize vimentin (Schwob et al. 1986; Ophir & Lancet 1988), an intermediate filament protein usually made transiently by embryonic neurons during their early development (Cochard & Paulin 1984). Further, the adult olfactory neuron continues to express juvenile forms of microtubule-associated proteins (Viereck, Tucker, & Matus 1989).

It is likely that in the everyday life of most or all vertebrate animals, olfactory sensory neurons rarely live for extended periods of time because their location renders them so vulnerable to various kinds of injury. Most vertebrates are heavily dependent on the olfactory sense for their basic needs. Without the ability to make new neurons continually, the survival of these animals might be seriously jeopardized. Consequently, the evolution of a means by which olfactory neurons could be replaced must have been essential for the survival of many vertebrate species. We have already seen that the rate of neurogenesis can be modulated experimentally. It is also clear that the life span of the cells can be modulated (Hinds et al. 1984; Mackay-Sim & Beard 1987; Carr & Farbman 1992). New data suggest that cell death can occur at any stage of olfactory cell development (Breipohl et al. 1986; Carr & Farbman 1990, 1992; Farbman 1990). Indeed, along with the up-regulation of the mitotic rate after bulbectomy (as discussed earlier) there is an increase in the number of cells that undergo early cell death; that is, many newly formed cells die well before they reach maturity (Carr & Farbman 1992). The phenomenon of overproduction of cells during nervous system development is well known (reviewed by Oppenheim 1991). In other neuronal systems, cells can undergo degeneration as early as 2 hr after mitosis (Carr & Simpson 1982). Although the tissue factors regulating mitosis and cell death in the olfactory sensory neuron population are unknown, it is likely that both exist and will ultimately be discovered.

The important thing to keep in mind, from the standpoint of olfactory response specificity, is that at any given time the olfactory sensory cells are a mixed population of young, almost mature, and fully mature cells. Because only the fully mature cells are synaptically connected, they alone carry meaningful information to the brain. Any electrophysiological data obtained exclusively by recording from single olfactory neurons in the epithelium must be interpreted in this context.

One curious phenomenon in reconstitution of olfactory epithelium should be noted. When the olfactory epithelium is experimentally destroyed by a chemical agent (the one commonly used is a solution of zinc sulfate), it is not reconstituted directly. The denuded surface is first covered with a respiratory epithelium (Schultz 1960; Lidow et al. 1987), the precursor cells of which are thought to originate from the epithelial cells of Bowman's glands or their ducts (Mulvaney & Heist 1971b). After the respiratory epithelium is established, it is replaced by olfactory epithelium. This sequence of reconstitution is not identical with what happens in development or in reconstitution of olfactory epithelium after axotomy or bulbectomy. In the latter situations, the original basal cells survive the damage inflicted by experimental manipulation, and they give rise directly to replacement sensory neurons. The reason that wound healing after chemical damage requires an intermediate step is not understood. The experiments do indicate, however, that in adults precursor cells for all cell types in both respiratory and olfactory epithelium may reside in the epithelium of Bowman's glands or their ducts. Given an appropriate stimulus, these precursor cells can give rise, directly or indirectly, to several cell lineages.

2.8 Structure of the olfactory nerve

The olfactory nerve in most vertebrate species is made up of the un-myelinated axons originating from the proximal poles of sensory neu-rons. Its structure differs from that of other sensory nerves and from other unmyelinated nerves in a few ways. The axons are of very fine caliber and are remarkably uniform in diameter, about 0.1–0.3 μm. In other unmyelinated nerves, individual axons are ensheathed by Schwann cells, one axon to one mesaxon. Olfactory axons are ensheathed several axons to a single mesaxon (Figure 2.18).

The ensheathing cells of olfactory nerve differ from typical Schwann cells of peripheral nerves (Kreutzberg & Gross 1977; Fraher 1982; Dou-cette 1984, 1989, 1990). An olfactory nerve bundle may contain more than a single ensheathing cell, unlike other unmyelinated nerves (Fraher 1982). Furthermore, in the development of the nerve, there is a pro-gressive increase in the number of axons per bundle (Fraher 1982), whereas in other peripheral nerves there is a progressive decrease until there is a single nerve in relation to each mesaxon. This means that olfactory axons in a given bundle are essentially in contact with one another, not individually insulated by ensheathing cell cytoplasm as in other peripheral nerves. Moreover, they travel parallel to each other for extended distances. A possible consequence of this arrangement is that changes in ionic concentration resulting from activity in one axon (i.e., an increase in the concentration of extracellular potassium) could reduce the excitability of its neighbors (Daston, Adamek, & Gesteland 1990).

The organization of axons suggests the possibility that new axons growing to the central nervous system are guided in the right direction by cues on preexisting axons. In addition, the ensheathing cells of ol-factory nerve differ from peripheral nerve nonmyelinating Schwann cells in that some contain an intermediate filament protein, glial filament acidic protein (GFAP), typically seen in central nervous system glial cells (Barber & Lindsay 1982).

The unusual structure of the olfactory nerve has made it a useful model for study of axonal membranes, particularly in some fishes that contain an exceptionally long nerve (Chacko et al. 1974; Kracke & Chacko 1979; Kreutzberg & Gross 1977). In garfish, an olfactory nerve may be 20–30 cm long, depending on the size of the fish (Easton 1971). The volume of axon cytoplasm in the nerve is about five times greater than that of ensheathing cell cytoplasm, and the axon surface is about 30 times the ensheathing cell surface; the ratio of surface to volume for axons of a typical olfactory nerve is about 5,400 times that for the squid axon of the same diameter (Easton 1971). Thus, this nerve represents an extreme in high-density axonal packing and is exceptionally well suited for studies on the molecular composition of axonal membrane. It has been shown, for example, that membranes of garfish olfactory axons have a relatively low protein : lipid ratio, about 0.35, and about 75% of the total lipid is phospholipid. The axonal membranes also have a high $Na^+ - K^+$ ATPase activity (Chacko et al. 1974; Kracke & Chacko 1979).

2.8.1 Axonal transport in the olfactory nerve

The length of the olfactory nerve and its relatively homogeneous axonal content have made it a suitable model for other types of studies, both morphological and functional. For example, microtubules are the most prominent cytoskeletal elements of axons and are of particular interest because they form the substrate for rapid transport of elements within the axoplasm (reviewed by Vale 1987). In the frog, most axons contain two to three microtubules (Burton 1987), whereas in the pike the average number is four (Kreutzberg & Gross 1977); each microtubule is more than 400 μm long (Burton 1987) and often contains some dense material within its lumen (Burton 1984). Axonal microtubules are polarized, with the "plus" end of the microtubule located distal to the cell body (Burton & Paige 1981); the plus end is the locus of nucleation sites, where growth (or reduction) in length of the microtubule occurs most rapidly.

Other studies on olfactory axons have shown that the smooth endoplasmic reticulum is branched and irregular and appears to be continuous over long distances (Kreutzberg & Gross 1977; Burton & Laveri 1985). This organelle is thought to play a role in the regulation of axonal calcium, probably to facilitate fast axonal transport (Burton & Laveri 1985).

The relatively high degree of homogeneity of olfactory axons has made the olfactory nerve a useful model in the study of axonal transport and the relationships between axonal transport and regeneration (Gross & Beidler 1973, 1975; Cancalon & Beidler 1975; Gross & Kreutzberg 1978; Cancalon 1979, 1982, 1983a,b,c,d, 1987, 1988; Elam & Peterson 1979; Cancalon & Elam 1980a,b; Elam 1982; Cancalon, Brady, & Lasek 1988). Axons themselves do not contain ribosomes, the organelles required for synthesis of proteins and other macromolecules. These macromolecules are synthesized in the cell body and are carried by specific transport mechanisms along the axon to its termination. In order to generate and maintain the volume of cytoplasm in the axons, neurons must produce large amounts of proteins and be able to transport them, sometimes for long distances. Axonal transport, then, is initiated when the axon first develops and continues throughout the life of the mature neuron.

Two kinds of axonal transport have been described: fast (100–400 mm/day) and slow (0.5–10 mm/day) (reviewed by Hammerschlag & Brady 1989). Both occur in the olfactory nerve, and they occur simultaneously (Buchner et al. 1987). Membrane- and vesicle-associated components travel mostly by fast transport (small vesicles, elements of the smooth endoplasmic reticulum, membrane proteins, neurotransmitters, lipids, etc.). Fast transport in the anterograde direction (i.e., from the cell body toward the axon terminal) is dependent on energy derived from hydrolysis of adenosine triphosphate (ATP) by an ATPase, kinesin. Kinesin binds the membranous elements to microtubules and provides the "motor" for transporting them anterogradely through the axon (reviewed by Vale 1987). Fast axonal transport can occur in the retrograde direction as well, from the axon terminal toward the cell body. This, too, is an energy-dependent activity, but it is dependent on a different ATPase, an enzyme thought to be similar to dynein, the "motor" in motile cilia (reviewed by Vallee, Shpetner, & Paschal 1989). Kinesin and the dynein-like ATPase, then, are each involved in unidirectional fast transport, the one in the anterograde direction, the other in the retrograde direction.

In the olfactory system, an illustration of how fast transport may be important is the following: During growth of the axon, products required for expansion of the membrane are transported rapidly to the growth cone at the axon terminal and are inserted as new membrane. When the growing axon of a sensory neuron reaches its appropriate synaptic target, the growth cone at its terminal somehow senses this, and a signal is sent in the retrograde direction to the cell body, which responds by initiating synthesis of the macromolecules required for the formation of a functional synapse. Subsequently, the newly synthesized synaptic molecules are transported anterogradely to the axonal terminal. Although the precise mechanism for this is not known, its existence is implicit.

Examples of the kinds of molecules transported by slow flow include soluble cytoskeletal and structural elements, such as actin, subunits of neurofilaments and microtubules, metabolic enzymes, and so forth. In fact, most (75–80%) of the substances transported along axons are carried by slow flow. However, the specific mechanisms for slow flow are not yet understood.

2.8.2 Transport into the central nervous system via the olfactory nerve

Because of its anatomical location, the olfactory sensory neuron provides a possible point of entry for foreign substances, such as proteins, particles, viruses, and so forth, to the central nervous system. For example, in the early part of the twentieth century it was thought that encephalitis viruses entered the nasal cavity and invaded the brain along the olfactory nerve. To prevent such infection during encephalitis epidemics, the nasal cavity in persons at risk was irrigated with zinc sulfate to destroy the olfactory epithelium, and thus prevent viruses from entering the brain (Takagi 1989, p. 121).

Experimental verification for the transport of viruses within axoplasm of nerves came from studies of intraneural spread of poliomyelitis virus in monkey sciatic nerve (Bodian & Howe 1941a,b). Subsequently, experimental studies were done on the olfactory system using *Herpes simplex* (DeLorenzo 1970; Twomey et al. 1979, Tomlinson & Esiri 1983; Esiri & Tomlinson 1984; Merkel & Maibach 1984; Stroop, Rock, & Fraser 1984; McLean, Shipley, & Bernstein 1989a), vesicular stomatitis (Lundh, Kristensson, & Norrby 1987), and other viruses (Jackson, Tigges, & Arnold 1979; Monath, Cropp, & Harrison 1983; Morales et al. 1988). Viruses and other foreign proteins, such as wheat germ agglutinin conjugated to horseradish peroxidase (Shipley 1985; Baker & Spencer 1986; Itaya 1987), other proteins (Kristensson & Olsson 1971), and gold particles (DeLorenzo 1970), as well as dyes (Holl 1981; Suzuki 1984), can be transported from the nasal cavity by way of olfactory axons into the bulb. There is a degree of specificity to the uptake and transport mechanism. Some viruses affect the olfactory sensory neurons only minimally (e.g., Sendai virus), although this particular virus does cause extensive infection of the nearby respiratory epithelium (Lundh et al. 1987).

Importantly, these particles can be transported transneuronally into higher centers of the central nervous system via bulbar output neurons. For example, tracers and viruses placed into the nasal cavity have been found in the anterior olfactory nucleus, olfactory tubercle, and pyriform cortex, all regions to which bulbar output cells project (Shipley 1985; Baker & Spencer 1986). Further, transport to higher centers can occur retrogradely from the bulb via nerve fibers projecting to the bulb from other parts of the brain. Tracers from the nasal cavity have been found in regions of the central nervous system that project to the bulb, but do not receive direct connections from the bulb. These regions include (1) the basal forebrain, especially the horizontal limb of the diagonal band, a major source of cholinergic fibers to the olfactory bulb (Shipley 1985; Baker & Spencer 1986; Lundh et al. 1987), (2) the locus coeruleus, a major source of noradrenergic fibers, and (3) the raphé, a major source of serotonergic fibers (Shipley, Halloran, & DeLaTorre 1985; Lundh et al. 1987). Neurons from these centers also project widely to many other parts of the brain.

It is at least theoretically possible, then, that the olfactory system is a portal of entry to the central nervous system for viruses, toxins, and the like and provides a pathway by which these particles can bypass the blood–brain barrier. This pathway has been implicated in the causation of certain neurological diseases, most notably Alzheimer's disease (Roberts 1986) and Parkinson's disease (Hornykiewicz 1983). Even viruses that enter the blood circulation via routes other than the nasal cavities can be transported to the head and can be released into the nasal cavity in secretions. From there they can be taken up by olfactory cells and transported to the central nervous system (Monath et al. 1983).

2.8.3 Regeneration of olfactory axons

It is well known that olfactory axotomy results in retrograde degeneration of almost the entire sensory neuronal population. However, in studies on nerve degeneration in other parts of the nervous system it has been found that the extent of neuronal cell death is inversely proportional to the distance between the lesion and the cell body. In most vertebrates the olfactory nerve is relatively short. Do olfactory neurons not survive axotomy simply because their axons are short and the lesion is necessarily placed close to the cell body, or are these neurons particularly sensitive to axotomy? Pike and garfish have been the animals of choice in studies to answer these questions, because the length of the olfactory nerve in these species has made it possible to crush it at various distances up to several centimeters away from the cell body. When such experiments were done, it was found that olfactory neurons behaved like other neurons; that is, most cell bodies could survive lesions made more than 10 cm away and would regenerate a new axon (Cancalon 1987).

Cancalon and Elam (1980a,b) have studied regeneration in the garfish olfactory nerve after a nerve crush close to the olfactory mucosa. After the injury, the mature olfactory neurons degenerate. Curiously, after a nerve crush, slow axonal transport persists along the detached axonal stumps (Cancalon 1982). Degeneration spreads distally from the site of injury, but it may take several days, depending on the temperature at which the fish are kept (Cancalon 1983d).

Following the injury to the olfactory nerve, three new groups of axons enter the olfactory nerve in succession. The first group, about 3–5% of the original axon population, represents the immature neurons that were growing at the time of injury and survived the axonal crush, probably because their axons had not yet grown to the site of the injury. These cross the crush site about 1 week after injury (when the fish are kept at 21°C) and advance along the nerve at a rate of about 5 mm/day. The second group, about 3–5% of the original axon population, reaches the crush site 2 weeks after injury and advances at a rate of about 2 mm/day. These probably represent a population that was maturing at the time of injury but had not yet started to grow axons. The third group of fibers, representing about 50–70% of the original number of axons, crosses the injury site about 4 weeks after the injury and grows at about 0.8 mm/day. These represent cells generated after the injury.

2.8.4 Glia of olfactory axons

We have already discussed several features of the olfactory system that are unique. Added to these is the fact that the primary olfactory (and vomeronasal) axons are essentially the only ones able to grow into the adult central nervous system. Other primary sensory axons are incapable of doing this. Furthermore, if the appropriate targets are removed, olfactory axons can grow into other parts of the central nervous system, induce glomerular formation, and even form synapses (Graziadei, Levine, & Monti Graziadei 1978; Graziadei & Kaplan 1980; Barber 1981b).

Raisman (1985) has suggested that a specialized neuroglial arrangement may be responsible for this unusual property of olfactory and vomeronasal axons. We have already noted that ensheathing cells of axons in the olfactory nerve differ from ordinary Schwann cells in that they contain a protein more typical of neuroglial cells, namely, glial fibrillary acidic protein (GFAP) (Barber & Lindsay 1982; Vollrath et al. 1985).

The olfactory nerve layer of the bulb contains two types of glial cells. One is an astrocytic cell type found between nerve bundles in this layer of the bulb. The astrocytes do not ensheath the axons (Doucette 1984, 1989, 1990; Raisman 1985). The second glial cell type in the olfactory nerve layer actually ensheaths the axons. These cells, probably derived from olfactory placodal epithelium (Doucette 1989, 1990; Marin-Padilla & Amieva B. 1989; Chuah & Au 1991), ensheath axons entering the bulb (Doucette 1984; Raisman 1985) and accompany them until they reach the glomerulus. This glial cell does not enter the glomerulus, but surrounds it and comes in contact with astrocytes that accompany mitral cell dendrites to the glomerulus. This type of glial cell, then, may provide a favorable substrate for the growth of newly generated sensory axons and direct them into the adult olfactory bulb (Raisman 1985; Doucette 1989). Both glial cell types in the nerve layer of the bulb are morphologically different from those that are seen in deeper bulb layers and in other regions of the central nervous system (Raisman 1985; Mack & Wolburg 1986).

2.9 Are there topographical projections from epithelium to bulb?

If, as noted previously, certain areas of epithelium respond better to some stimuli than to others, is this selectivity projected to the first synaptic station in the glomerulus of the bulb? In other words, is there any sort of topographical organization of the sensory neuronal projections to the bulb that is spatially related to the epithelium, or perhaps functionally specific? In the visual system, the topographical relationship between sensory ganglion cells in the retina and their synaptic targets in the thalamus is well known and is fairly precise.

All of the evidence from several studies suggests that precise point-to-point projections are not present in the primary olfactory pathway (Kauer 1974, 1981, 1991). However, some general statements can be made. For example, sensory neurons in dorsal regions of the nasal cavity project largely but not exclusively to the dorsal part of the bulb, and similarly for ventral, medial, and lateral regions of the olfactory epithelium onto corresponding parts of the bulb (Le Gros Clark 1951, 1957; Land 1973; Land & Shepherd 1974; Costanzo & Mozell 1976; Jastreboff et al. 1984; Astic & Saucier 1986; Saucier & Astic 1986; Astic, Saucier, & Holley 1987; Duncan et al. 1990). The important point to keep in mind is that a given cluster of sensory cells in any particular part of the olfactory epithelium may project to more than one region of the bulb.

2.10 The septal olfactory organ

The septal organ (organ of Masera) is an isolated patch of sensory epithelium on the ventral region of the nasal septum in the main nasal cavity. This patch of epithelium is separated from the main olfactory epithelium by a region of respiratory (nonolfactory) epithelium and lies caudal and slightly dorsal to the level of the vomeronasal organ, near the entrance to the nasopharynx (Figure 2.19).

Most of what has been said about the cellular composition of the main olfactory organ pertains also to the septal organ. The neurons are bipolar cells, similar to those in the main olfactory organ, but there are some subtle differences. The Golgi apparatus is generally more prominent than in sensory cells of the main nasal cavity, and the organization of the perinuclear smooth endoplasmic reticulum is described as having an "onion ring" appearance (Graziadei 1977). The numbers of intramembranous particles in the sensory cilia may be slightly higher in the septal organ (Miragall et al. 1984), but this requires verification.

Rodolfo-Masera (1943) and others (e.g., Adams & McFarland 1971; Katz & Merzel 1977; Kratzing 1978; Breipohl, Naguro, & Miragall 1983) examined this specialized area of olfactory epithelium in several mammalian species. They suggested that by virtue of its location, its function may be to monitor the airstream for odor stimuli during quiet respiration, when only limited amounts of air might reach the main olfactory epithelium. In other words, this organ might function to alert the animal to odor stimuli in its surroundings, and thus play a role in eliciting a behavioral change resulting in more directed sniffing. This notion is

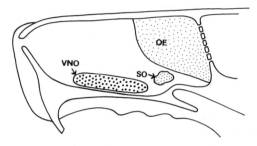

Figure 2.19. Diagram of a sagittal section through a rodent head showing the distribution of the various types of olfactory sensory epithelium; OE, the olfactory epithelium proper; SO, the septal organ; VNO, the region of the vomeronasal organ. The VNO is actually a tubular organ *within* the septum, not on the surface as are the other two.

supported by physiological evidence showing that the septal organ responds to a broad range of odor stimuli and may be more sensitive to lower concentrations of odorants than are the sensory cells in the main nasal cavity (Marshall & Maruniak 1986). Another possible function has been suggested by the work of Wysocki, Wellington, and Beauchamp (1980), who suggested that the location of the septal organ in the ventralmost region of the nasal cavity might play a role, along with the vomeronasal organ, in sensing compounds of low volatility.

The sensory neurons of the septal organ project to the main olfactory bulb, as do those of the main olfactory organ. Axons from septal organ sensory cells form two fascicles, independent of the vomeronasal and other olfactory fascicles, and pass to the main olfactory bulb, where they terminate (Bojsen-Møller 1975). Tracing studies using retrograde injections of horseradish peroxidase have shown that in rats the axons terminate in the medial side of the posterior half of the olfactory bulb (Pedersen & Benson 1986; Astic & Saucier 1988). Astic and Saucier (1988) have claimed that the ventromedial aspect of the posterior part of the bulb is the prime target of projecting olfactory neurons from the septal organ, but Pedersen and Benson (1986) have shown that projection also goes to the so-called modified glomerular complex on the dorsomedial aspect of the main olfactory bulb, near the boundary of the accessory bulb. As a consequence of the separate projections of septal and main olfactory organs, the central nervous system receives parallel, possibly reinforcing, inputs carrying sensory information.

2.11 Trigeminal innervation of the nasal cavity

The nasal mucosa, both olfactory and respiratory, receives sensory innervation via two branches of the trigeminal (V cranial) nerve. Trigeminal nerve fibers originate in the cells of the trigeminal ganglion, in the floor of the cranial cavity. The anterior ethmoid nerve, a branch of the ophthalmic division of the trigeminal, innervates the mucosa in the anterior region of the nasal cavity. The nasopalatine nerve, a branch of the maxillary division, innervates the mucosa in the posterior region of the nasal cavity (reviewed by Silver 1987). Both of these nerves apparently respond to chemical stimuli in the nose. The finely myelinated A_δ and unmyelinated C fibers are thought to carry the chemosensory information. In addition to chemical information, trigeminal sensory fibers bring somatosensory information (touch, temperature, pain, etc.) to the central nervous system, specifically the trigeminal nucleus of the hindbrain and upper spinal cord.

Although several workers in the late nineteenth and early twentieth centuries had described the presence of extrinsic (nonolfactory) innervation in the olfactory mucosa (reviewed by Graziadei & Gagne 1973), there was only suggestive evidence that these nerve fibers originate from the trigeminal nerve. Recently it has been shown that capsaicin treatment eliminates or severely reduces trigeminal chemosensitivity in the nasal cavity without affecting olfaction or taste significantly (Silver et al. 1985; Mason, Greenspon, & Silver 1987). Capsaicin is an irritant phenolic amide component of pepper (capsicum) seeds. Trigeminal responsivity to volatile chemical stimuli disappears in adult animals chronically treated with large doses of capsaicin (Silver et al. 1985). Further, when neonatal animals are given capsaicin, the peptidergic fibers disappear, and there is no trigeminal response to odor stimuli (Silver et al. 1989). One major class of nasal mucosa nerve fibers sensitive to capsaicin is the population exhibiting immunoreactivity when treated with an antibody against calcitonin-gene-related peptide (CGRP) (Ånggard et al. 1979; Silver et al. 1989).

The knowledge that CGRP-immunoreactive fibers are sensitive to capsaicin strongly suggests that these fibers mediate nasal trigeminal chemoreception, although the possibility that other fibers are capsaicin-sensitive cannot be entirely ruled out. It is interesting that, at least in rodents, CGRP-immunoreactive fibers usually are also immunoreactive when treated with the antibody against another peptide, substance P. Bouvet et al. (1987b) showed that substance P immunoreactivity in frog olfactory epithelium was completely abolished after the trigeminal nerve was sectioned, thus confirming the fibers' origin from the trigeminal nerve.

Antibodies to both peptides, CGRP and substance P, have been used as probes to identify and trace trigeminal fibers in olfactory and respiratory mucosa. In rodents and amphibians, immunoreactive fibers are found in the lamina propria around blood vessels and Bowman's glands, as well as in the epithelium, although in one study on mouse olfactory mucosa, substance P fibers were found only in the lamina propria (Papka & Matulionis 1983). Most intraepithelial fibers are in the basal region, but some reach toward the surface (Lundblad et al. 1983; Bouvet et al. 1987b; M. Getchell et al. 1989; Silverman & Kruger 1989). Electron microscope studies have confirmed the identification of peptide-immunoreactive nerve fibers that terminate between cells just below the epithelial tight junction (Finger et al. 1990).

It is interesting to speculate on the possible roles of peptides in the olfactory mucosa. Substance P is thought to act as a transmitter or neuromodulator in several parts of the nervous system. In the nasal cavity, it may also affect the activity of nonsensory epithelial cells, blood vessels, and secretory elements. When substance P is applied directly to the surface of frog olfactory mucosa, it elicits low-threshold electrical potentials (Bouvet, Delaleu, & Holley 1987a). Studies on single olfactory sensory neurons indicated that application of substance P affects their spontaneous activity: In most cases (21 of 33) there was an increase in the spontaneous activity, and in some (9 of 33) there was a decrease (Bouvet, Delaleu, & Holley 1988).

2.12 The terminal nerve

The nervus terminalis, or terminal nerve, is found in all vertebrates except cartilaginous fishes. It innervates the mucous membrane on the anteroventral septum of the nasal cavity, and in fishes its branches go also to the retina (Demski & Northcutt 1983; Stell et al. 1984). In most vertebrates the terminal nerve forms a diffuse system of neurons within a plexus on the medial aspect of the olfactory bulb and within the nasal cavity. The intracranial neurons send fibers peripherally along the medial surface of the bulb, through the cribriform plate, to innervate the nasal septum. In rodents, terminal nerve fibers are incorporated within the vomeronasal nerve or the anterior ethmoid nerve (Bojsen-Møller 1975). The fibers can be distinguished from other nerves in the nasal or olfactory mucosa because they are positive for the presence of acetylcholinesterase activity (Bojsen-Møller 1975; Wirsig & Getchell 1986; Wirsig & Leonard 1986b).

Some of the terminal nerve ganglion cells and fibers are immuno-reactive for luteinizing hormone-releasing hormone (LHRH), whereas trigeminal, olfactory, and vomeronasal nerves are not (Schwanzel-Fukuda & Silverman 1980; Witkin & Silverman 1983; Stell et al. 1984; Wirsig & Getchell 1986; Wirsig & Leonard 1986b; Witkin 1987). LHRH is an important peptide in reproductive physiology. It is produced in the hypothalamus and prompts the release of luteinizing hormone (LH) from cells in the anterior lobe of the pituitary gland. LHRH could also be a peptide neurotransmitter, particularly in the olfactoretinal pathway

(Stell et al. 1984). The terminal nerve ganglion cells, however, are a morphologically heterogeneous population of unipolar, bipolar, and multipolar neurons (Wirsig & Leonard 1986b; Zheng, Pfaff, & Schwanzel-Fukuda 1990). Neurons that are immunoreactive with an antibody for LHRH usually do not contain acetylcholinesterase, and vice versa, but in a small population these are co-localized (Wirsig & Leonard 1986b). Within the population of cells in the nerve plexus and ganglia, there are various synapses between LHRH-immunoreactive and non-immunoreactive elements (Zheng et al. 1990).

The data from experiments using histochemical and immunohistochemical probes suggest that many of the peripheral endings of the terminal nerve are around blood vessels and mucosal glands. There is conflicting evidence concerning whether or not these nerves enter the olfactory epithelium in the main nasal cavity, although it seems clear that they do enter the vomeronasal epithelium (Wirsig & Leonard 1986a). The latter nerve fibers originate from cell bodies that are located along the terminal nerve outside of the cranial cavity. Immunohistochemical staining with antibodies to LHRH or histochemical staining for acetylcholinesterase has shown no evidence for nerve terminals in epithelium of the main nasal cavity (Wirsig & Getchell 1986; Wirsig & Leonard 1986b). On the other hand, after placement of a tracer molecule, horseradish peroxidase, in the nasal cavity, the tracer can be detected in terminal nerve ganglion cells (Demski & Northcutt 1983; Jennes 1986). Because the nerve is heterogeneous, it may be that fibers that are neither LHRH- nor acetylcholinesterase-reactive do project to the olfactory epithelium in the main nasal cavity. In fishes, some of the terminal nerve ganglion cells project via the optic nerve to the retina (Münz et al. 1982; Springer 1983). The cells projecting from the retina apparently do not go to the olfactory region (Springer 1983).

Is the terminal nerve a sensory nerve? In particular, does it play a role in chemical sensation? Some argue that because the cell bodies of the ganglion cells have peripheral and central projections, it is likely to be a sensory neuron (Bojsen-Møller 1975; cf. Wirsig & Leonard 1986a). The morphology of at least some of the ganglion cells is similar to that of typical sensory neurons (i.e., they are unipolar or bipolar, although many are multipolar) (Fujita, Satou, & Ueda 1985; Wirsig & Leonard 1986b). Others argue that because it is anatomically and developmentally associated with the vomeronasal and olfactory nerves (Pearson 1941; Bojsen-Møller 1975; Fujita et al. 1985; Buhl & Oelschläger 1986; Jennes 1986; Schmidt, Naujoks-Manteuffel, & Roth 1988; Hofmann & Meyer 1989), it could be functionally related.

Experiments in goldfish have led to the suggestion that the terminal nerve is a chemosensory nerve involved in sexual behavior, possibly sensitive to pheromones. In male goldfish, stimulation of either the optic nerve or the medial olfactory tract (which carries terminal nerve fibers to the central nervous system) resulted in release of sperm (Demski & Northcutt 1983). Support for a sexual behavior function is derived from hamster experiments in which damage of the terminal nerve was shown to impair male mating behavior (Wirsig & Leonard 1987). Others take the position that because the terminal nerve ends on blood vessels and glands in the nasal septum mucous membrane, it is likely part of the autonomic nervous system and controls these end organs (Buhl & Oelschläger 1986). The acetylcholinesterase fibers ending in the vicinity of the vomeronasal organ may be related to the autonomic control of the vomeronasal organ (i.e., the vomeronasal pump) (Wirsig & Getchell 1986) (see section 4.2). The nerve may subserve both sensory and autonomic functions, and perhaps others yet undiscovered.

Most of the central fibers of the terminal nerve in mammals end in or near the medial septal nucleus of the basal forebrain (Jennes 1986) and the accessory olfactory bulb (Wirsig & Leonard 1986a); in amphibians they end in the preoptic region, anterior commissure, olfactory bulb, medial septum, and hypothalamus (Wirsig & Getchell 1986; Schmidt et al. 1988; Hofmann & Meyer 1989); in fish they end in the same regions of the brain as in amphibians (Springer 1983).

The embryological origin of the terminal nerve is from the medial part of the olfactory placode (Pearson 1941; Buhl & Oelschläger 1986). Cells migrate out of the placode toward the forebrain (see section 6.1.3). Some of these cells become LHRH-positive during their migration and have been traced ultimately to the hypothalamus, where they become part of the hypothalamopituitary axis involved in growth, development, and maintenance of the reproductive system (Schwanzel-Fukuda & Pfaff 1989; Wray, Nieburgs, & Elkabes 1989). The path of these LHRH-positive cells from the nasal region to the diencephalon is marked in the adult by the terminal nerve. It is likely that the terminal nerve ganglion cells originate from the olfactory placode (Farbman & Squinto 1985), but there is an indication that some of them may migrate out of the telencephalon (Pearson 1941).

3

Functional Aspects of Olfaction

3.1 Characteristics of an olfactory stimulus molecule

In terrestrial animals the surface of the olfactory sensory epithelium is covered by a thin film of watery mucus. In amphibians, this film, which may be as much as 35 μm thick, is described as a superficial watery layer overlying a deep, more viscous component (Reese 1965; T. Getchell et al. 1984). The division of the olfactory mucus into only two regions may be an oversimplification, in light of recent studies suggesting that there may be several domains within the mucus compartment (B. Menco & A. Farbman unpublished data). In mammals, the film of mucus is usually thinner, about 5–10 μm deep. Olfactory stimuli must pass through this complex mucus to reach the receptor sites on the sensory cell membrane. Obviously, any molecule that stimulates the olfactory sense must be soluble in the mucus overlying the olfactory epithelium surface. Although many odorants are hydrophobic, they are soluble in the aqueous mucus, as discussed later. A theoretical analysis of diffusion times through the mucus is available (T. Getchell et al. 1984).

A second requirement for distance chemoreception (in air) is that the stimulus molecules be volatile (i.e., they must have a relatively high vapor pressure. This obviously is not a requirement for olfaction in aquatic animals, which respond to different types of stimuli; water solubility is more important than volatility.) Studies of the relationship between odor-stimulating ability and molecular structure indicate that most odorous substances have molecular masses less than 400 Da (Laffort & Dravnieks 1974; Meredith 1983). The volatility of substances at the upper end of this molecular mass range is so low that most cannot easily be detected. A few steroid molecules are odorous, suggesting that there are receptors with high affinities for these. Certain nonvolatile molecules are stimuli for the vomeronasal system in some species. In these cases, the sensory organ receives the stimulus in the liquid phase by physical contact with the stimulus source (see Chapter 4).

The sensory quality and the perceived intensity of the stimulus molecule are influenced by several of its physicochemical properties, in addition to volatility, solubility, and size. The concentration of the stimulus may be a factor in determining how many sensory neurons respond; cells with a low threshold will respond at lower concentrations, while others with higher thresholds will be recruited at increased stimulus concentrations. The flexibility or rigidity of the odorant molecular structure may determine how it interacts with a specific receptor molecule (Beets & Theimer 1970; Wright 1982; Boelens 1983). The functional groups on the molecule can influence how it is perceived (Beets 1978). For example, it is known that about 2–3% of tested humans cannot detect isovaleric acid, but these same individuals can detect the corresponding aldehyde (Amoore 1977, 1982). Further, the D- and L-isomers of the same compound can elicit different odor sensations. For example, L-carvone elicits a spearmint sensation, and D-carvone elicits a caraway sensation (Friedman & Miller 1971; Russell & Hills 1971; cf. Polak et al. 1989a). Structure–activity relationships of odorant molecules have been reviewed by Beets (1978) and Boelens (1976, 1983).

3.2 The initial events in the response of the sensory cell to the odorant stimulus

Once again we return to a fundamental problem in our understanding of the olfactory system: If we are able to (1) distinguish among thousands of different odorant chemicals and (2) evaluate the intensities of the stimuli, where and how are the determinations made? Is there some degree of specificity in the interaction between stimulus and receptor site on the sensory cell, a specificity that determines the nature of the response by the cell? What role is played by the mucus surrounding the parts of the sensory cell that presumably contain the odor detection apparatus? We have already discussed the evidence for spatial mapping or regional variations in response specificity within the nasal cavity and the differential sorption of odor stimuli across the entire receptor sheet (section 2.6.1). These inherent and imposed factors no doubt influence the olfactory response.

In this section we focus on the initial events in olfaction, the interaction between the stimulus and the sensory cell that results in the generation of a nerve impulse.

3.2.1 Odorant-binding proteins

Many airborne odorants are hydrophobic molecules. How do they partition into an aqueous medium, such as olfactory mucus or the fluid surrounding the olfactory dendrite in the arthropod sensillum on the antenna? An active area of investigation has focused on agents in the mucus that may modify the stimulus, or perhaps that bind the stimulus in a way that enables it to be presented to the receptor (e.g., an odorant-binding protein) (Vogt & Riddiford 1981; Pelosi, Baldaccini, & Pisanelli 1982; T. Getchell et al. 1984; Bignetti et al. 1985a,b; Pevsner, Sklar, & Snyder 1986; Snyder, Sklar, & Pevsner 1988a,b; Vogt, Prestwich, & Riddiford 1988; Snyder et al. 1989; Vogt et al. 1989; Vogt, Prestwich, & Lerner 1991a; Vogt, Rybczynski, & Lerner 1991b).

The growth of this area of research has in one sense been serendipitous, because the driving force in the initial investigations was the search for an olfactory receptor molecule. The ligand binding techniques used in these early experiments disclosed the existence of certain binding proteins in olfactory mucus of vertebrates and in the fluid of insect sensilla. In vertebrates, a soluble odorant-binding protein has been discovered. The protein originates from glands in the nasal cavity (Pelosi et al. 1982; Pevsner et al. 1985). Odorant-binding proteins are small, water-soluble, extracellular proteins located in the fluid surrounding the sensory dendritic endings. Because they are soluble and extracellular, it was recognized that they could not be the olfactory receptor molecules; receptors would have to be membrane-associated (see section 3.2.3.1). However, these soluble binding proteins are located only in the nasal cavity. Their presumed function is to augment the capture rate of hydrophobic odorants by (1) helping to solubilize the odor molecules, thus enhancing their partitioning into the aqueous fluid surrounding the dendrites, and (2) aiding in the transport of the odorants to receptor proteins located in the dendrite membrane (Vogt 1987; Gyorgyi, Roby-Shemkowitz, & Lerner 1988; Pevsner et al. 1988, 1990; Vogt et al. 1989).

The vertebrate olfactory binding protein binds to a wide range of structurally diverse odorants with variable affinities that roughly parallel the detection thresholds in humans (Pelosi et al. 1982; Bignetti et al. 1985a,b; Pevsner et al. 1985, 1986; Snyder et al. 1989). Molecular cloning techniques have been used to deduce the amino acid sequence of the binding protein. It has a molecular mass of about 18,000 Da, and its structure places it within a family of proteins all of which serve as carriers for small lipophilic molecules (Pevsner et al. 1988; Snyder et al. 1989). It seems, then, to meet the criteria for an odorant-binding protein noted earlier.

Stimulus-binding proteins have been identified in insects as well. Pheromone-binding proteins have been identified in the antennae of males in several species of moths (Vogt & Riddiford 1981; Gyorgyi et al. 1988; Vogt et al. 1989). These proteins (1) are unique to the antennae, (2) usually are not seen in significant amounts in females, (3) have a molecular mass of about 15,000 Da, and (4) bind the sex pheromone emitted by the female of the species. In the gypsy moth, two genetically distinct pheromone-binding proteins are expressed; both bind the sex pheromone disparlure, but with different affinities (Vogt et al. 1989).

In addition, insect antennae contain general olfactory binding proteins, found in both sexes and associated with olfactory sensory neurons thought to respond to general odorants. Two subclasses of general olfactory binding proteins have been identified. Molecular cloning and sequencing of both the pheromone-binding proteins and general olfactory binding proteins from several insect species have been carried out (Vogt et al. 1991a,b). The amino acid sequence of pheromone-binding proteins appears to be highly variable from one species to another, but the sequences of each of the two classes of general olfactory binding proteins seem to be highly conserved from species to species.

Vogt et al. (1991b) pointed out some implications of the conservation of interspecies general olfactory binding proteins in view of the fact that various moth species they studied utilize very different habitats. For example, *Manduca sexta* feed and deposit eggs only on tobacco plants, whereas *Bombyx mori* feed specifically on mulberry leaves. The finding that the two known classes of general olfactory binding proteins are highly conserved between these two and other moth species suggests that the various species are equally receptive to the same odorants irrespective of odorant habitat. However, the fact that the binding proteins are nearly identical does not necessarily imply that the receptor molecules are as well.

3.2.2 Location of the receptor site

It is generally assumed that the odorant stimulus interacts with receptor sites in the membrane of the sensory cell, and the sites for odorant detection are located within the dendritic knob and its appendages: the cilia or (in nonciliated sensory cells) microvilli (reviewed by T. Getchell et al. 1984). This conclusion is based on several kinds of evidence, the most convincing of which comes from physiological, morphological, and biochemical studies. Olfactory cilia are required for the generation of a physiological response, as determined by recording the electro-olfactogram (EOG). When cilia are removed, no EOG can be recorded in response to a stimulus; when they are regenerated, the EOG returns (Bronshtein & Minor 1977; Burns et al. 1981; Mair, Gesteland, & Blank 1982; Adamek et al. 1984; Lidow et al. 1987). A biophysical analysis of the onset latencies of excitatory responses showed that the sites where responses were initiated included the dendritic knob and the proximal part of the cilium (T. Getchell et al. 1980). Recent electron microscopic immunocytochemical data suggest, however, that the specific molecules involved in olfactory transduction, namely, G-proteins and adenylate cyclase (see section 3.3), are localized in the special modified distal parts of the cilia (Menco 1991).

The biochemical data are based on the findings, from several laboratories, that isolated cilia preparations from the olfactory mucosa bind odorant stimuli with high affinity (Rhein & Cagan 1980; Brown & Hara 1981). Although these data cannot be considered definitive, for reasons discussed later, taken together with the morphological, electrophysiological, and degeneration–reconstitution studies, they make a strong case in favor of the argument that the receptor sites for olfactory detection are located primarily in the ciliary membrane.

3.2.3 Identification of the receptor molecule

What is the nature of these receptor sites? Are they specific molecular detectors? If so, are the receptor sites constituted of proteins, as is assumed by many, or can the lipid in the membrane be perturbed by odor stimuli in such a way that a response is elicited?

In this section we review (1) the evidence in favor of a protein receptor initiating the response of the sensory cell and (2) that favoring the notion that the responsivity of the olfactory cell can be explained simply on the basis of the lipid in the membrane.

3.2.3.1 Binding studies Some of the earliest attempts by biochemists to search for olfactory receptor proteins were based on electrophysiological and behavioral data showing that amino acids, in relatively low concentrations (10^{-6}–10^{-9} M), were effective olfactory stimuli for fishes (Sutterlin & Sutterlin 1971; Suzuki & Tucker 1971; Hara 1973; Døving & Holmberg 1974; Caprio 1977, 1978). The rationale for the biochemical experiments was based on receptor binding assays that had been developed in pharmacological studies on other regions of the nervous system, and in hormone binding studies in endocrine organs. Application of these techniques to the olfactory system resulted in the development of an assay based on the binding of radioactively labeled amino acids to sedimentable fractions isolated from olfactory tissues of fishes. The trout *Salmo gairdneri* (Cagan & Zeiger 1978; Brown & Hara 1981), coho salmon *Oncorhynchus kisutch* (Rehnberg & Schreck 1986), carp and skate (Fesenko et al. 1983), and catfish (Kalinoski, Bruch, & Brand 1987; Bruch & Rulli 1988) were the animal models usually employed for these studies. Although there was some low-affinity binding of radioactively labeled amino acids to both olfactory and nonolfactory tissues, most of the high-affinity binding was associated with olfactory tissue.

In the trout studies, most of the high-affinity binding of labeled amino acids was found in a fraction of olfactory mucosa enriched in cilia (Rhein & Cagan 1980). The binding was tissue-specific (i.e., it did not occur in nonolfactory tissues, including brain and gills). Moreover, high-affinity accumulation of amino acids on the olfactory tissue fraction was dependent on the presence of sensory cells, because denervation experiments, in which these cells were destroyed, resulted in a considerable reduction of high-affinity binding (Brown & Hara 1981). Results from these studies on trout olfactory tissues suggested that there were multiple receptor sites for amino acids: one that seemed to bind L-threonine, L-serine, and L-alanine, the TSA site; a site that bound L-lysine and L-arginine, the L site; and a site that bound β-alanine, site A_b (Cagan & Zeiger 1978; Rhein & Cagan 1983).

3.2.3.2 Physiological studies Excellent correlation between function, measured electrophysiologically, and binding, measured biochemically, has been found in experiments with channel catfish (*Ictalurus punctatus*), in which many more amino acids were tested. Cross-adaptation studies have shown that olfactory cells contain different receptor sites for acidic (A), basic (B), and neutral L-amino acids; at least two partially interacting neutral sites exist, one for hydrophilic neutral amino acids containing short side chains (SCN), and the second for the hydrophobic amino acids containing long side chains (LCN) (Caprio & Byrd 1984). The SCN binding site is analogous to the threonine-serine-alanine site of trout described in the previous section, but it also exhibits binding with L-valine, L-histidine, glycine, and L-glutamine (Caprio & Byrd 1984); the basic amino acid-binding site in catfish presumably corresponds to the L site described for the trout. In biochemical experiments on catfish, the four relatively independent binding sites for L-α-amino acids do not compete with one another (Kalinoski et al. 1987; Bruch & Rulli 1988). The agreement between the binding studies and results from cross-adaptation experiments supports the notion that there are distinct olfactory receptor sites in the catfish.

The catfish studies are in agreement with those in trout and salmon because they suggest that there may be a limited number of amino acid receptors in aquatic species with broad specificity for groups of structurally related ligands, rather than a large number of highly selective sites. The data on fish present the strongest evidence for the notion that there may not be specific detectors for each of hundreds of odorants. Rather, there may be classes of detectors that can be stimulated by more than one stimulus.

3.2.3.3 Are receptor molecules proteins? The stereospecificity hypothesis Over the years there have been several theories purporting to explain the basis of the stimulus–receptor interaction in the sensory neuron. In this section we discuss only two of these theories, those most recently proposed, for which there are abundant supporting data. The two are not mutually exclusive; indeed, they build on one another.

If there were a unique receptor molecule for each of hundreds of thousands of stimuli, a large amount of genetic information would be required for the synthesis of these receptors. Because nature tends to be reasonably economical, the notion that genetic information would be used in this way is intuitively unappealing. Nevertheless, in a recent paper it has been suggested that there may indeed be hundreds of genes devoted to relatively large numbers of specific receptor molecules in the olfactory sensory cell (Buck & Axel 1991). This will be discussed in section 3.2.3.5.

The two theories pertaining to stimulus–receptor interactions in the olfactory system are modeled on the way nature has dealt with color vision, on the one hand, and the immune system, on the other.

In color vision, hundreds of hues can be discriminated, but there are only three different color detectors in the vertebrate retina for primary colors: the red, green, and blue cone cells. The ensemble response of large numbers of these primary color detectors is decoded by the brain to enable the organism to perceive so many hues.

Amoore (1963a,b, 1967b, 1970, 1977, 1982) has compared smell with color vision and proposed that there are primary odors, as there are primary colors, and that our perceptions of odors depend on an ensemble response of a relatively small number of different primary receptors. In this model, several (as few as seven or eight) specific receptors may exist on olfactory sensory cells, each one related to a so-called primary odor. Odorant molecules of similar size, shape, and structure would fit into or onto receptor sites of a complementary size and shape, as a key fits a lock. Each site would accept only molecules that had a matching three-dimensional configuration (Figure 3.1). The correct stimulus then would elicit a response in the matching receptor. This does not exclude the possibility that some stimuli might interact with more than one receptor site, perhaps with varying degrees of affinity. [Wright (1982) has suggested that the "vibrational" qualities of one or both molecules, i.e., the stimulus and receptor, might be changed from their intrinsic values by a collision between the two, thus leading to an interaction.] This model predicts that structurally similar molecules would elicit similar responses. In this instance, the three-dimensional shapes of the molecules are more important in determining the response than is their chemical makeup. But the model also permits stimulus–receptor interaction with varying degrees of affinity (i.e., the receptor–odorant interaction is not absolutely specific) (Margolis & Getchell 1991). Thus, a smaller number of receptor sites could be coded in some combinatorial way to discriminate among thousands of different stimuli. This would limit the number of receptor molecules and would not place a heavy burden on the cellular genome.

For purposes of this discussion, an important aspect of the Amoore theory is that it suggests there are specific molecules in the sensory cell

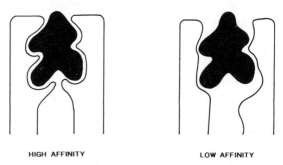

HIGH AFFINITY   LOW AFFINITY

Figure 3.1. Sketch depicting how the shape of a molecule (in black) might fit closely with that of a high-affinity receptor, on the left, or a low-affinity receptor, on the right.

membrane, most likely proteins, that account for olfactory specificity. Supporting evidence for this notion is derived from the several descriptions of specific olfactory deficits, presumably genetically determined, in humans (e.g., Amoore, Forrester, & Buttery 1975; Amoore & Forrester 1976; Amoore, Forrester, & Pelosi 1976; Amoore 1977, 1982; Wysocki & Beauchamp 1984) and in mice (Price 1977; Wysocki, Whitney, & Tucker 1977). Individuals with these deficits are unable, or much less able, to detect certain specific odor molecules or classes of molecules, but have normal olfactory sensitivity to most other odorants. The implied reason for the specific deficit in an otherwise normally functioning olfactory system is a genetically based absence of or defect of a high-affinity receptor (protein) molecule.

Further support for the Amoore theory comes from the observation that olfactory responses are stereospecific. This may come about because the chemical structures of some compounds have two variations that may be mirror images of one another. An analogy: The right and left hands are similarly constructed, but the right hand will not fit properly into a left-handed glove (Wright 1982). Thus, two compounds may have identical physicochemical properties, but if their chemical structures are mirror images of one another, they may elicit different olfactory responses because only one may fit into or onto a particular receptor site (Amoore 1963a,b). The two versions of the molecule are referred to as isomers. The only difference between them is their ability to rotate the plane of polarized light; thus, they are called optical isomers. For example, as mentioned earlier, L-carvone smells like spearmint, but the D-isomer smells like caraway. The two isomers of α-ionone also smell differently (Polak et al. 1989a). Moreover, isomers can elicit different electrophysiological responses (Caprio & Byrd 1984) and bind to olfactory membrane preparations with different affinities (Bruch & Rulli

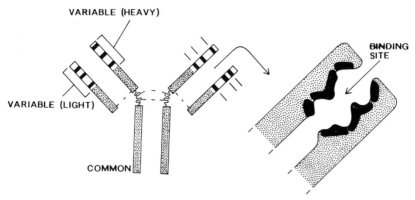

Figure 3.2. Diagram of an immunoglobulin molecule showing the common and variable regions. The molecule is made up of a heavy chain, composed of a common (Fc) and a variable region, and a light chain, made up of a variable region. The two variable regions together make up the Fab fragment. The amino acid sequence in the Fab region of the protein is variable, and this will affect its three-dimensional shape. (In this diagram, the bands in the variable regions are spaced differently to indicate the variations in their amino acid composition.) The putative receptor region is shown enlarged at the right. Compare the receptor region with the sketch in Figure 3.1.

1988). If the receptor molecules are proteins, this should not be surprising, because all protein molecules in animal cell membranes are constituted of L-amino acids, not the D-isomers. Thus, it is not a major leap of the imagination to assume that the three-dimensional configuration of a membrane receptor protein could determine whether it is complementary to one or the other of two stimuli that are optical isomers.

3.2.3.4 Are receptor molecules analogous to immunoglobulins? Another model, first suggested by Thomas (1974), proposes that the olfactory system in vertebrates is analogous to the immune system in the sense that both have evolved large receptor repertoires to recognize an enormous variety of extrinsic chemical substances (Boyse et al. 1982). In the immune system, thousands of molecules can interact with receptors on lymphocytes, the major cellular participants in the immune system, to elicit antibody formation. The antibody molecules themselves (immunoglobulins) and the receptor molecules on lymphocytes are constituted of a large common region, known as the Fc region, common to all immunoglobulins, and a small variable region, the Fab region, that differs from molecule to molecule. The entire molecule is put together by splicing a common region and a variable region (Figure 3.2). It has been suggested that the receptor molecules in the olfactory system may be constructed in a similar way (Lancet 1984).

The olfactory system, according to this model, can accommodate its large repertoire with genetic frugality by using some of the same tricks used by the immune system. For example, if olfactory receptor molecules were analogous to immunoglobulins, they all might have a constant region identical in amino acid sequence within the lipid bilayer of the sensory cell membrane, and a variable region with amino acid sequence heterogeneity in the recognition part of the molecule on the external surface of the membrane (Lancet 1984, 1988). Consistent with this idea is the evidence from experiments with rodents that are genetically identical except at the major histocompatibility complex (MHC). Both mice and rats can distinguish the characteristic odors of individuals of their own identical genetic makeup from those that differ genetically only in the MHC locus (Yamazaki et al. 1979, 1983a,b, 1990; Boyse et al. 1982; Beauchamp, Yamazaki, & Boyse 1985; Brown, Singh, & Roser 1987; Brown 1988; Brown, Roser, & Singh 1989). The MHC is a linked cluster of genes that code for glycoproteins in cell membranes. These membrane glycoproteins carry the histocompatibility antigens that are responsible for the rejection of tissues transplanted from one individual to another in nearly every species of animal. These glycoprotein molecules and their breakdown products are also found in serum and in urine. The metabolites of these glycoproteins excreted in the urine, or perhaps characteristic mixtures in particular ratios, are thought to be the prime sources of odors that identify individuals of a given species (Yamazaki et al. 1990).

If we continue to follow the analogy of the immune system, it is interesting to consider what might happen when a single olfactory cell is stimulated. In the immune system, a single clone of lymphocytes or plasma cells presumably makes a single molecular species of antibody. The ability of the immune system to make a large number of different antibodies is dependent on its ability to generate a clone of immune cells for each antibody. If we extrapolate this to the olfactory system, it suggests that the individual olfactory sensory neuron makes a single type of receptor molecule. Thus, many olfactory cell types would be distinguishable on the basis of the molecular species of receptor they make. This reasoning is not supported by the evidence that single receptor cells respond to more than one stimulus (Gesteland 1964). On the other hand, if a given receptor molecule in a cell membrane has different degrees of affinity for stimuli, then concentration-dependent phenomena can confound the analysis (Sicard & Holley 1984).

3.2.3.5 Experimental evidence indicating that olfactory receptor molecules are proteins The search for olfactory receptors has been an area of intensive research, especially during the past 10–15 years. Most investigators believe, indeed assume, that olfactory receptors are proteins and that the response begins with the binding of the stimulus to the receptor protein in the membrane of the cilium.

There are several good reasons to believe that receptor sites in the sensory epithelium are proteins (Lancet 1986). We have already noted two, namely, several specific genetically based olfactory deficits have been described, and olfactory responses are stereospecific. A third reason is the evidence that saturable and reversible odor-specific binding occurs with high affinity in preparations of membrane fractions from olfactory tissues of fishes (Cagan & Zeiger 1978; Rhein & Cagan 1980, 1983; Brown & Hara 1981; Fesenko et al. 1983; Rehnberg & Schreck 1986; Kalinoski et al. 1987; Bruch & Rulli 1988) and rats (Fesenko et al. 1979). Fractions from control tissues do not exhibit the same degree of binding affinity to the tested substances.

A fourth line of evidence, particularly pertinent to the question, comes from a study in which antibodies were made to putative receptor proteins for two stimuli: anisole (Price 1977; Goldberg, Turpin, & Price 1979) and benzaldehyde (Price & Willey 1987). These antibodies block the physiological response to the odorants, though not specifically. These earlier studies were done with polyclonal antibodies. When monoclonal antibodies against the benzaldehyde and anisole receptors were used, however, some were fairly specific in blocking only the responses (measured by EOG) to the corresponding stimulus, but not to others (Price & Willey 1988).

Other studies have described the effects of less specific blocking agents for a putative receptor molecule (e.g., agents that specifically bind to functional groups of proteins or glycoproteins, rather than to specific protein molecules) (M. Getchell & Gesteland 1972; T. Getchell & M. Getchell 1974; Dodd & Persaud 1981; Shirley, Polak, & Dodd 1983; Shirley et al. 1987). When concanavalin A, a lectin that binds to the glycan of glycoproteins, was applied to the rat olfactory mucosa, responsivity, as measured by the EOG, was diminished to 60% of 112 odorants tested (Shirley et al. 1987). The concanavalin A-induced reduction in amplitude of the EOG, at least among alkane odorants, depends on the molecular size of the stimulus and is maximal with *n*-pentane (Polak et al. 1989b). In vitro binding studies have shown that lectins inhibit receptor binding (Kalinoski et al. 1987).

N-ethylmaleimide, a sulfhydryl blocking agent, irreversibly blocks the electrical response of the olfactory receptor organ of the frog to odorous stimuli (M. Getchell & Gesteland 1972; T. Getchell & M. Getchell 1974). If the odorous substance ethyl *n*-butyrate is applied to the receptor in high enough concentration before exposure to the blocking agent, it protects the response to itself and odorous substances closely related to it, while responses to other compounds are abolished. A procedure for covalent modification of Schiff base-forming binding sites in proteins has been used in salamanders to produce a selective odor blindness to cyclohexanone (Mason, Clark, & Morton 1984).

Fifth, olfactory cilia have a high density of intramembranous particles, as shown by freeze-fracture morphological studies (see Figure 2.8) (Kerjaschki & Hörandner 1976; Menco et al. 1976; Kerjaschki 1977; Menco 1977, 1980a,b,d, 1983, 1984, 1987; Masson et al. 1978), and the size of the particles corresponds to the expected size of putative receptors (Menco et al. 1976). More recent evidence indicates that these intramembranous particles constitute a heterogeneous population of differing sizes and shapes (B. Menco personal communication).

Finally, cyclic nucleotide-processing enzymes, which in other cells have been found in conjunction with protein receptors, are present in high concentrations in olfactory cilia and can be activated by odor stimuli (e.g., Pace et al. 1985; Shirley et al. 1986; Sklar, Anholt, & Snyder 1986; Bruch & Teeter 1990).

Although no one of these observations, in itself, proves the existence of protein receptors in membranes of sensory cells in the olfactory apparatus of vertebrates, taken together they make a compelling argument for their existence. However, the evidence is all indirect, and direct proof for the isolation of a receptor molecule remains elusive.

Lancet (1986) proposed criteria for any protein that might be an olfactory receptor molecule:

1. tissue specificity
2. enrichment in the olfactory cilia
3. glycosylation
4. transmembrane disposition (integral membrane protein)
5. correct bilayer concentration (major component of membrane)
6. diversity (sequence heterogeneity)
7. specific recognition by function-modulating reagents (antibodies, lectins)
8. interaction with transductory proteins
9. reconstitution of odorant modulation of enzymatic activities

Some putative protein receptors have been isolated that meet most of the criteria. A 95-kDa glycoprotein, gp95, has been isolated from cilia of frog olfactory mucosa (Chen & Lancet 1984; Chen et al. 1986a,b; Lancet 1986). Vogt et al. (1988) have isolated a 69-kDa pheromone-binding protein from male moth antennae. Fesenko et al. (1979, 1987, 1988) isolated an odor-binding glycoprotein from rat olfactory epithelium. The molecular mass of the glycoprotein that bound camphor and decanal was 140 kDa, and it consisted of two subunits, 88 and 55 kDa,

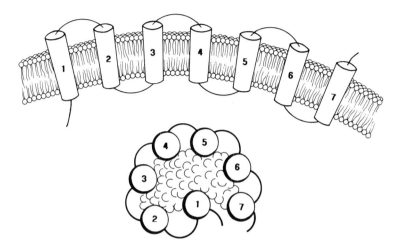

Figure 3.3. Top: Hypothetical model of a protein with seven transmembrane domains, linearly arranged along a membrane of lipid bilayers. Bottom: Extracellular view of the membrane surface showing how the molecule might be organized. The receptor region would be enclosed by the extracellular peptide loops between adjacent transmembrane domains and would contain the stimulus-binding site.

of which the larger was capable of binding odorants. The same authors have described some characteristics of a possible receptor for amino acids from the olfactory epithelium of the skate and the carp. The L-α-amino acid receptors of catfish, which have not been isolated (Bruch & Kalinoski 1987; Kalinoski et al. 1987; Bruch & Rulli 1988), also meet most of the criteria.

The application of molecular biological techniques to the problem of identifying odorant receptor molecules is extremely promising. In a recent paper, Buck and Axel (1991) based their approach on the rationale that there are commonalities in structure among most or all known receptor molecules (in other systems) that are linked to a transduction apparatus involving G-proteins and second messenger systems, as the olfactory system apparently is (see section 3.3). For example, each of these proteins had seven transmembrane domains (Figure 3.3), and in most of them some of the transmembrane domains were highly conserved (i.e., there were few variations in the amino acid sequence). Because the amino acid sequences of many receptor molecules were known, it was possible to synthesize corresponding oligonucleotides coding for the conserved intramembranous sequences and isolate the DNA matching these oligonucleotides. By using these oligonucleotides as probes, these investigators identified and characterized 18 clones coding for proteins with seven transmembrane domains. The family of cloned proteins was found in olfactory tissue, but not in brain, retina, or several nonneural tissues. The amino acid variability in the identified proteins was in regions of the molecule thought to be important in ligand binding in other cellular proteins that traverse the membrane seven times. Their data led them to suggest that there may be a considerable diversity of genes, perhaps hundreds, in a large multigene family that code for olfactory receptor molecules. They also felt it unlikely that DNA rearrangement, as occurs in the immune system, could account for the diversity of the proteins. At this writing, these results appear very exciting, but they are preliminary, as the authors were careful to note. The identified proteins have not yet been shown to meet all of the criteria for odorant receptor molecules. When the necessary control experiments are done and functional data are obtained, we can better evaluate the suggestion that hundreds of different genes are required for coding of olfactory receptors.

A few receptors for external molecules have been identified in membranes of unicellular organisms. The gene for the cAMP receptor in *Dictyostelium discoideum* amoebae has been cloned (Klein et al. 1988), as have the genes for the α-pheromone (Burkholder & Hartwell 1985) and the a-pheromone (Hagen, McCaffrey, & Sprague 1986) in yeast (see section 1.2). Each of these protein receptor molecules has seven hydrophobic regions, indicating that there are seven transmembrane domains.

3.2.3.6 Support for the argument that receptor molecules are lipids

One school argues that the olfactory response in a sensory cell is really dependent on the lipid in the membrane. The evidence for this comes from experiments showing that (1) odorants can generate membrane currents by directly interacting with lipid bilayers in artificial membranes (Koyama & Kurihara 1972; Nomura & Kurihara 1987a), and (2) modification of the lipid in the bilayer can affect the threshold of odorants that generate membrane currents (Kurihara, Miyake, & Yoshii 1981; Kashiwayanagi & Kurihara 1985; Nomura & Kurihara 1987b).

These data, together with the fact that most odorants are lipophilic molecules, have led some investigators to conclude that perturbations of the lipid components of the sensory cell membrane contribute to the cellular response to odorants (Davies 1971; Koyama & Kurihara 1972; Dodd & Persaud 1981; Kashiwayanagi & Kurihara 1984, 1985; Kurihara, Yoshii, & Kashiwayanagi 1986; Nomura & Kurihara 1987a,b). In other words, these investigators are proposing that the response to an olfactory stimulus may *not* be mediated by a protein receptor (Lerner et al. 1988; Bruch 1990a,b; Bruch & Gold 1990).

Much of the basis of this work comes from experiments in the laboratory of Professor Kenzo Kurihara and colleagues, who worked with lipid monolayers from olfactory epithelium (Koyama & Kurihara 1972; Kurihara et al. 1981; Nomura & Kurihara 1987a,b) and with the N–18 clone of mouse neuroblastoma cells (Kashiwayanagi & Kurihara 1984, 1985). They claim that the neuroblastoma cell is a good model for olfactory cells because it responds to 20 different stimuli in the same stimulus concentration range as those that trigger responses in olfactory sensory cells, and the response is dose-dependent. Responses in their studies were monitored by intracellular electrodes and by detection of changes in fluorescence of a voltage-sensitive dye. Changes in the lipid composition of the neuroblastoma cell membrane were made by exogenous application of stearic acid and cholesterol, and some of these membrane changes resulted in response alterations to certain of the odorant stimuli. The authors argued that the N–18 neuroblastoma cell is unlikely to carry any specific receptor proteins for odorants. The fact that it is depolarized by odorants and that alterations in lipid composition of the cell membrane can enhance or diminish responses to some stimuli suggest that specific proteins are not required for odor reception. The implication is that various odorants are adsorbed at different sites in the membrane. The membrane composition of the olfactory cell is postulated to vary from cell to cell (i.e., different combinations of lipids and proteins will be present, and this may account for the different sensitivities to odorants, although the hydrophobic regions of proteins may also contribute to the response) (Kashiwayanagi & Kurihara 1984, 1985; Nomura & Kurihara 1987b). In more recent studies, these investigators used liposomes of complex composition, including some proteins, and measured the changes in membrane fluidity, using fluorescent compounds. They found that the variety of changes in membrane fluidity was greater in the more complex proteoliposomes than in simple liposomes. They concluded that odorants may be adsorbed on a variety of sites formed by different combinations of lipids and proteins in olfactory membranes (Kashiwayanagi et al. 1990).

Punter, Menco, and Boelens (1981) also have argued that the hydrophobic membrane regions are important in olfaction and taste, as well as in other chemoreceptive phenomena of cells. Their study focused on the relative amount of energy required for transfer of a molecule from the water phase outside the cell to the lipid phase of its membrane. Using a series of n-aliphatic alcohols and n-aliphatic fatty acids, they showed that the effect on cell membranes involved in olfaction and taste was similar to that for membranes involved in chemotaxis and in anesthesia. They concluded that the phospholipids in the hydrophobic domain of the membrane provide the microenvironment of intramembranous proteins and that the mobility of the proteins is partly dependent on the types of lipid molecules around them.

However, the proponents of the argument that protein receptors need not be invoked to explain olfactory cell responses to odorant stimuli concede that incorporation of certain species of proteins into the artificial membrane system does lead to enhancement of the response to odorants (Kurihara et al. 1986). Moreover, they point out that the lipid-based mechanism cannot explain the reception of pheromones, which is more likely based on specific receptor proteins (Koyama & Kurihara 1972; Nomura & Kurihara 1987a).

3.2.3.7 A possible resolution of the protein-versus-lipid argument How do we reconcile the two bodies of data showing, on the one hand, that perturbation of lipids in artificial membranes and in nonolfactory cells results in a depolarization in response to odorant stimuli and, on the other hand, overwhelming evidence that receptor sites are constituted of protein molecules? The notion that proteins are most important and can account for specificity in odor detection better explains the enormous diversity of responses of the olfactory system to hundreds of different stimuli. Whereas it may be true that odorants can elicit electrical activity in artificial membranes in the absence of protein, it is less likely that perturbation of the lipid component of an olfactory sensory neuron membrane by an odorant might alone be specific to a class of stimuli.

One clue that might help to understand the contribution of lipids to the olfactory response is suggested by some experiments of Gesteland, Yancey, and Farbman (1982), who examined the developmental changes in responsivity of single olfactory sensory cells to a set of 12 odorants. These experiments, done in rat fetuses, showed that at certain stages in development (i.e., before the 16th fetal day), each olfactory cell was responsive to all 12 stimuli in the set (i.e., there was no *selective* responsivity). At subsequent stages of development, an increasing proportion of cells exhibited selective responsivity (i.e., the cells responded to one or a few, but not all, odorants). In other words, the sensory cell population matured from a group of generally irritable cells that showed no discrimination, but did exhibit electrical activity, to become specialized cells that responded to only a subset of stimuli. It is well known that a single cell in a mature animal may respond to more than one odorant at concentrations well over threshold (Gesteland et al. 1965; Sicard & Holley 1984; Sicard 1985; Gesteland 1986), but the response specificity of a given cell to a group of odorants varies from cell to cell. It is also known that the ciliary membrane in the fetal olfactory cell contains fewer intramembranous particles than that in the adult (Kerjaschki 1977; Menco & Farbman 1985a; Menco 1988a), suggesting a lower protein content. It is possible, then, that the nonspecific electrical response in fetal or immature olfactory cells is based at least partly on perturbations of lipid (i.e., may not be receptor-mediated). However, the specific responses in mature cells may be modified by, and may be more dependent upon, the specific proteins in the membrane that act as receptors.

3.3 The role of second messengers in olfactory transduction

3.3.1 Adenylate cyclase

The mechanism by which the receptor cell in the epithelium transduces the chemical stimulus into a nerve impulse has received a great deal of attention during the past few years and is, at this writing, a very active area of research. It is now generally thought that transduction occurs largely in the olfactory cilia projecting from the apical (dendritic) end of the cell into the mat of mucus on the surface of the nasal lining (Ottoson 1956; T. Getchell 1977; Adamek et al. 1984; Lancet 1986; Anholt 1987, 1989; Anholt et al. 1987; Gold & Nakamura 1987; Lidow et al. 1987; Nakamura & Gold 1987; Kurahashi 1989, 1990; Kurahashi & Kaneko 1991). Biochemical studies on cell fractions enriched in cilia have shown that these structures contain a relatively high activity of the enzyme adenylate cyclase, more than 10 times as high as that in brain, which itself has one of the highest levels of activity for this enzyme (Pace et al. 1985; Pace & Lancet 1986). Adenylate cyclase catalyzes the formation of cyclic AMP, which participates in many cellular processes.

The activity of adenylate cyclase in cilia preparations can be stimulated, in a dose-dependent manner, by certain kinds of odorant stimuli. For example, in catfish, amino acids elicit an increase in GTP-dependent cyclic AMP formation, hence indicating a stimulation of adenylate cyclase activity (Bruch & Teeter 1990). In rat cilia preparations, hydrophobic odorants (floral, fruity, herbaceous, etc.) stimulate the activity of adenylate cyclase, but the enzyme is refractory to stimulation by other odorants (Pace et al. 1985; Pace & Lancet 1986; Sklar et al. 1986; O'Connell, Costanzo, & Hildebrandt 1990). Lowe, Nakamura, and Gold (1989) argue that those odorants that do not stimulate adenylate cyclase activity may simply not be good stimuli in the test animal. Another, more likely possibility is that other transduction mechanisms may be involved in addition to that mediated by adenylate cyclase, such as the phosphoinositide system, as discussed later.

The presence of elevated levels of adenylate cyclase activity implicated the participation of cyclic nucleotides as second messengers in the transduction process. Indeed, it has been shown that the adenylate cyclase of olfactory cilia is activated by odorants with a potency that correlates with their effectiveness in EOG recordings (Lowe et al. 1989). The cyclic AMP formed by the activity of adenylate cyclase may directly gate (open) conductance channels in the membrane, thus initiating the nerve impulse (Menevse, Dodd, & Poynder 1977; Gold & Nakamura 1987; Nakamura & Gold 1987; Trotier & MacLeod 1987; Vodyanoy & Vodyanoy 1987; Bruch & Teeter 1989; Kurahashi 1989, 1990; Bruch & Gold 1990). In isolated patch-clamped frog cilia, both cyclic AMP and cyclic GMP participate directly in the gating of conductance channels in the ciliary membrane (Gold & Nakamura 1987; Nakamura & Gold 1987; Kurahashi 1990; Kurahashi & Kaneko 1991). Indeed, the gene for a cAMP/cGMP-gated channel has been cloned from mammalian olfactory epithelium and has properties similar to the cGMP-gated channel of photoreceptors (Dhallan et al. 1990). The number of cAMP-gated channels has been estimated in olfactory cilia of toads and newts. The mean number of channels in an excised patch of ciliary membrane from toad was $2,400/\mu m^2$, and from newt, $920/\mu m^2$ (Kurahashi & Kaneko 1991). The density of channels in membrane patches excised from the dendrite or cell body was much lower than in those excised from cilia. In newts, the average was about $2/\mu m^2$, and in toads about $6/\mu m^2$. Cyclic AMP and cGMP also participate in the gating of channels in isolated ciliary membranes incorporated in artificial phospholipid layers (Bruch & Teeter 1990).

To a large extent, signal transduction in olfactory receptor cells is thought to mimic that in other systems, such as visual reception and neurotransmitter or hormone action, where cyclic nucleotides act as second messengers (Pace et al. 1985; Lancet 1986; Pace & Lancet 1986; Anholt 1987, 1989; Lancet & Pace 1987). It begins with the binding of an odor stimulus to a receptor (e.g., Adamek et al. 1984; T. Getchell 1986; Lancet 1986; Anholt 1987; Gold & Nakamura 1987; Lancet & Pace 1987; Nakamura & Gold 1987). Binding of the stimulus activates the receptor, which in turn activates a guanine nucleotide-binding protein (GTP-binding protein, or simply G-protein). The activated G-protein presumably stimulates the increase in adenylate cyclase activity. This results in the breakdown of adenosine triphosphate (ATP) into cAMP, which then directly (or indirectly) alters the conductance channels in the membrane (Figure 3.4). Amplification of the process may occur at the level of G-protein activation, as many molecules of G-protein may be activated by a single activated receptor molecule. A much greater degree of amplification of the signal occurs at the second messenger level. The system is recycled as cytoplasmic phosphodiesterase transforms cAMP into AMP.

However, odor stimuli can activate adenylate cyclase in nonolfactory cells, which presumably contain no odor receptors (Kashiwayanagi & Kurihara 1984, 1985; Lerner et al. 1988). This has led to further study of the possible role of Ca^{2+} and calmodulin in the regulation of adenylate cyclase activity. On the basis of evidence that olfactory cilia contain calmodulin, Anholt and Rivers (1990) speculate that activation of olfactory sensory cells by odorants may be initiated by the influx of calcium

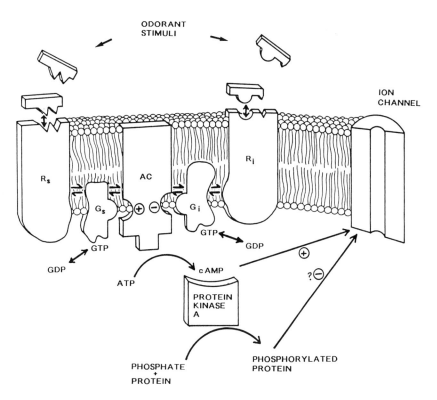

Figure 3.4. Model of the adenylate cyclase system. Odorant stimuli bind to protein receptor sites (R_s and R_i) within the membrane. The receptors, activated by binding to stimuli, activate G-proteins (G_s and G_i) to exchange GDP for GTP. The activated G-proteins either stimulate (+) or inhibit (−) the activity of adenylate cyclase (AC) to produce cyclic AMP (cAMP) from adenosine triphosphate (ATP). The cAMP can gate open (+) the ion channel directly, and/or perhaps can work indirectly through a protein phosphorylation step to close it (−) after the cell fires.

by odorant-gated calcium channels. The increase in intracellular calcium could lead to the formation of a Ca^{2+}/calmodulin complex, which then would activate olfactory adenylate cyclase. Higher concentrations of odorant stimuli might bypass the Ca^{2+}/calmodulin pathway, and the adenylate cyclase would be activated by the receptor–G-protein pathway.

Results on whether or not this truly occurs are conflicting. On the one hand, it is claimed that calmodulin can stimulate olfactory adenylate cyclase activity in preparations of bullfrog cilia (Anholt & Rivers 1990). On the other hand, the olfactory-specific adenylate cyclase isolated from rat olfactory cilia is said to be insensitive to calmodulin (Pfeuffer et al.

1989). Thus, the role of calmodulin in transduction is not clear. The conflicting results may be due to species differences.

A further word should be said about the role of G-protein in the transduction cascade of events. The olfactory epithelium contains three well-characterized types of G-proteins: G_s (stimulatory), G_i (inhibitory), and G_o (other). The effector protein for the stimulatory and inhibitory G-proteins (G_s and G_i) is adenylate cyclase. G-proteins generally are heterotrimers, composed of a guanyl nucleotide-binding α subunit, a β subunit, and a γ subunit. The β and γ subunits of most or all G-proteins are not significantly different from each other; the α subunits are distinctive at or near the binding sites for receptor and effector proteins. Six $G\alpha$ subunits have been identified in rat olfactory epithelium by gene cloning techniques, namely, a G_{olf}, specific for the sensory neurons, a G_s, another stimulatory G-protein, a G_o, and three G_i clones; one of the G_i subunits is a novel species (Jones & Reed 1987, 1989; Jones, Barbosa, & Reed 1989). Messenger RNA corresponding to each of the six $G\alpha$ subunits has been detected. The stimulatory $G\alpha_{olf}$ is strongly implicated in the signal transduction cascade because it is the predominant stimulatory G-protein in the cilia (Jones & Reed 1989; Jones 1990; cf. Pace & Lancet 1986). It should be kept in mind, however, that more than one G-protein may participate in olfactory transduction. In *Dictyostelium discoideum,* the slime mold, the surface receptors to two different stimuli, folic acid and cAMP (recall that cAMP in this organism is an external chemical stimulus), are coupled to different G-proteins (Kesbeke et al. 1990).

Cloning studies on the gene for adenylate cyclase in olfactory epithelium have shown that it differs from that in brain (Pfeuffer et al. 1989; Bakalyar & Reed 1990). Activation of olfactory adenylate cyclase occurs only in the presence of guanosine triphosphate (GTP) (Pace et al. 1985; Pace & Lancet 1986; Shirley et al. 1986; Sklar et al. 1986; Vodyanoy & Vodyanoy 1987). The activation of G-protein occurs when the guanosine diphosphate (GDP), to which it is bound in its inactive state, is exchanged for GTP. Coincident with this exchange is the dissociation of the β and γ subunits from the α subunit (leaving the latter to effect the GDP–GTP exchange and reach its "activated" state).

The evidence for participation of the cyclic nucleotide-mediated transduction in olfactory cilia comes from several sources. First, there is an unusually high adenylate cyclase activity in cilia (Pace et al. 1985; Shirley et al. 1986). High levels of the α_s and β subunits of G-protein are also present in cilia and make their appearance in the developing embryo's cilia at approximately the same time that the cell acquires the ability to generate an action potential (Mania-Farnell & Farbman 1990). Immunohistochemical studies have shown that adenylate cyclase is also demonstrable at this time, although G_{olf} is not demonstrable by this technique until a day or two later in embryonic life (Dau et al. 1991). Second, adenylate cyclase activity is enhanced by certain odorants in the presence of GTP (Pace et al. 1985; Pace & Lancet 1986; Shirley et al. 1986; Sklar et al. 1986; Bruch & Teeter 1990). The activity of this enzyme in the cilia preparations can be stimulated, in a dose-dependent manner, by certain kinds of odor stimuli, in particular hydrophobic odorants that are floral, fruity, herbaceous, and so forth (Pace et al. 1985; Pace & Lancet 1986; Sklar et al. 1986; O'Connell et al. 1990). Third, in patch-clamp studies on frog, newt, and salamander olfactory cells and/or cilia, cAMP and cGMP both directly affect the gating of conductance channels (Nakamura & Gold 1987; Firestein & Shepherd 1989; Kurahashi 1989, 1990; Firestein, Shepherd, & Werblin 1990; Firestein et al. 1991; Frings & Lindemann 1991; Kurahashi & Kaneko 1991). Although cGMP has been shown to have a strong in vitro effect on cilia, its role in physiological sensory transduction is not yet clear. There is conflicting evidence for the presence of guanylate cyclase in olfactory cilia preparations of mammals. Some claim that it is present in rat and pig (Steinlen, Klumpp, & Schultz 1990); however, others have been unable to show the existence of guanylate cyclase activity in olfactory tissue (Pace et al. 1985; Pace & Lancet 1986; Shirley et al. 1986; Sklar et al. 1986). It seems to be present in silkmoths, but a detailed study of these insects has suggested that cyclic GMP is not directly involved in the transduction process generating the receptor potential (Ziegelberger et al. 1990).

3.3.2 Phosphoinositide system

The case for participation of the cyclic nucleotide cascade in olfactory transduction has been made by several laboratories and is now generally accepted. However, it seems clear that this mechanism is not the only one involved in transduction of the stimulus into a nerve response. Another transduction mechanism, involving the phosphoinositide system, is apparently also operative in these cells (Figure 3.5). One of the first clues came from experiments using isolated olfactory cilia of catfish. In this preparation, the amino acid stimulus L-alanine stimulates the activity of the enzyme phosphatidylinositol–4,5-bisphosphate phosphodiesterase (phospholipase C). Stimulation is rapid and dose-dependent (Huque & Bruch 1986). The enzyme activity was also stimulated in the presence of GTP, thus suggesting the participation of G-proteins in stimulation of phosphoinositide turnover in olfactory receptor neurons. The hydrolysis of phosphatidylinositol–4,5-bisphosphate by phospholipase C results in the formation of diacylglycerol and inositol–1,4,5-trisphosphate (IP_3), both of which act as second messengers in other cellular systems.

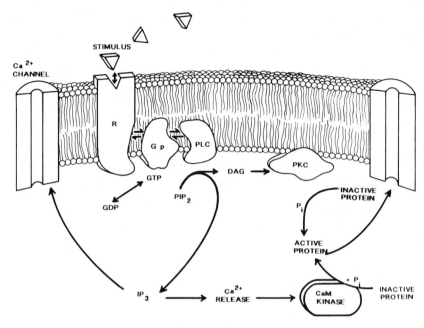

Figure 3.5. Model of the phosphoinositide system. Odorant stimuli bind to a protein receptor site (R) in the membrane. The stimulus-activated receptor activates a G-protein (G_p) to exchange GDP for GTP. The activated G-protein stimulates the activity of the enzyme phospholipase C (PLC), which acts on phosphatidylinositol-4,5-bisphosphate (PIP_2) to produce inositol trisphosphate (IP_3) and diacylglycerol (DAG). DAG stimulates the activity of protein kinase C (PKC) to phosphorylate proteins (inactive to active protein), which may then affect gating of ion channels. IP_3 may directly gate calcium channels in the membrane; it also causes calcium release from membrane-bound calcium stores in the cytoplasm, and the calcium release may be involved in the activation of the enzyme calcium-calmodulin (CaM) kinase, which phosphorylates proteins that may affect gating of ion channels.

Inositol–1,4,5-trisphosphate has been linked to an odorant-induced influx of calcium in isolated catfish olfactory cells (Restrepo & Teeter 1990; Restrepo et al. 1990). The IP_3 apparently directly gates calcium channels in the membrane, leading to depolarization in the cells. Calcium channel blocking agents inhibited the electrical activity. Thus, at least in catfish olfactory sensory cells, calcium is an important participant in the IP_3-mediated signal transduction cascade. The suggestion that calcium may participate in signal transduction by formation of a calcium/calmodulin complex that can directly stimulate the activity of adenylate cyclase (Anholt & Rivers 1990) has already been discussed (section 3.3.1).

Other evidence supports a role for the IP_3 system in the transduction of olfactory stimuli. Odor-induced formation of second messengers in rat olfactory cilia was monitored by a rapid kinetic device in recent studies (Boekhoff et al. 1990b; Breer & Boekhoff 1991). Application of micromolar concentrations of one stimulus, citralva, induced a rapid (in about 50–100 msec), transient elevation of the cAMP level, while the concentration of IP_3 was not affected. In contrast, another stimulus, pyrazine, caused a rise in the concentration of IP_3 without affecting the level of cAMP. In these studies, the odorant-induced second messengers in olfactory cilia are very transient (Boekhoff et al. 1990b; Breer & Boekhoff 1991). At low stimulus concentrations the increased second messenger level decayed within 100 msec. A transient molecular signal in the transduction process is essential for olfactory sensory neurons that can respond repeatedly. A careful analysis of the kinetics for the odorant-induced response indicated that the two transduction systems, involving formation of cAMP and IP_3 apparently are operating simultaneously, and they appear to be mediated by different G-proteins (Boekhoff et al. 1990b; Breer & Boekhoff 1991; cf. Restrepo et al. 1990).

By means of these rapid kinetic techniques, several odorant compounds in the floral, fruity, herbaceous, and putrid families of stimuli were assayed to determine if compounds with similar odors activated the same second messenger system. If this were true (i.e., if compounds that had similar sensory qualities used the same second messenger system), they might be linked to the same or similar receptors. It was found that the compounds with similar odor qualities did not consistently induce the formation of one second messenger to the exclusion of the other (Breer & Boekhoff 1991).

The studies just discussed were all performed in vertebrates, in which at least two signal transduction systems seem to operate, using the second messengers cAMP and IP_3. In insects, there is no evidence that cAMP is involved in olfactory transduction. All the evidence points to IP_3 as the second messenger (Ziegelberger 1989; Boekhoff et al. 1990a; Breer, Boekhoff, & Tareilus 1990). In lobsters, cAMP apparently is not involved in transduction (McClintock, Schütte, & Ache 1989), although other evidence implies the existence of more than one transduction mechanism. For example, whole-cell patch-clamp studies indicate that a single cell can respond either by depolarization or hyperpolarization, depending on the stimulus (McClintock & Ache 1989b; Schmiedel-Jakob et al. 1990; Michel, McClintock, & Ache 1991). It seems inconceivable that a single transduction mechanism would be involved in both types of conductance change (Ache et al. 1989). If there are olfactory transduction mechanisms other than the cAMP- and IP_3-mediated ones, they remain to be discovered. For example, results of experiments in artificial membrane systems suggest the possibility that stimulus-gated conductance changes can occur directly (i.e., without the intervention of second messengers) (Labarca, Simon, & Anholt 1988). These experiments have yet to be replicated in other laboratories.

3.3.3 Other comments on second messenger systems

One of the questions arising from these studies is whether a given sensory cell possesses both second messenger systems or whether in the process of cell differentiation only one of them is expressed. The fact that single neurons can respond both by depolarization and hyperpolarization suggests that more than one transduction mechanism indeed can exist in a given sensory neuron. It is probable that the mechanism resulting in inward flow of current (depolarizing) is linked to one type of channel, and that resulting in outward current flow (hyperpolarizing) is linked to a different channel. Recent evidence (Ache et al. 1989) is consistent with the idea that in lobster olfactory neurons, at least, cAMP is linked to the hyperpolarization response, and IP_3 is linked to the depolarization response.

Why should a given cell have more than one transduction mechanism? One should not lose sight of the fact that outside of the laboratory the olfactory sensory apparatus in most animals rarely encounters pure odor compounds. Rather, the environment presents mixtures of odors, and the olfactory sensory apparatus analyzes the mixtures. Thus, a group of cells exposed to an odor mixture could respond as illustrated in Figure 3.6. Suppose odor A by itself elicited a depolarizing response in two cells, and odor B by itself had no effect on these two cells, but elicited a depolarizing response in another cell. If the two were mixed, the total response to the mixture would be additive. On the other hand, if odor B stimulated cell 6 to depolarize and elicited a hyperpolarizing response in one of the cells stimulated by odor A (Michel et al. 1991), the summated effect might result in a different pattern of responses reaching the central nervous system. Thus, the effect of one odor component of a mixture might be to suppress, at least partially, the contribution of another component. This phenomenon, well known to perfumers, could be explained by the presence of two sets of channels in sensory cells that are linked to different receptor/second messenger complexes.

The reaction of olfactory sensory cells to chemical stimuli that we know as odorants is not unique. It has been shown, for example, that certain odorants stimulate nonolfactory cells, such as neuroblastoma cells (Kashiwayanagi & Kurihara 1984, 1985) and melanocytes (Lerner et al. 1988), to increase their adenylate cyclase activity. The doses required for these changes are within the same range as those required for stimulation of olfactory cells, and the time course for stimulation of these other cells is similar. Lerner et al. (1988) suggest that there could

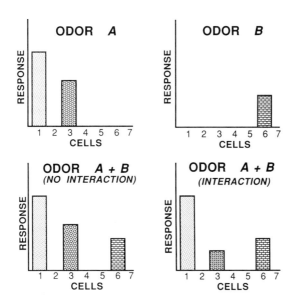

Figure 3.6. Hypothetical illustration of how a group of seven cells might respond to two odors in a mixture. In the two upper panels the response magnitudes for two odors, presented singly, by a group of cells are indicated by the heights of the bars. In the lower panels, two different patterns of responses are seen. On the left, if both odors elicit responses in separate cells, the responses are simply additive. On the right, if odor B depolarizes one cell (cell 6) but hyperpolarizes cell 3, the summated response intensity of cell 3 could be reduced, resulting in a different response pattern reaching the central nervous system. (Courtesy B. Ache)

be two mechanisms that are important for odorant detection, one based on specific receptors, the other nonspecific, but both working through cAMP (Anholt & Rivers 1990).

Returning to the earlier discussion (section 3.2.3.7) about the degree of participation of lipids and proteins as odorant receptors, we can incorporate the data on transduction mechanisms into our hypothesis. We suggested that immature cells, which presumably contain far fewer protein molecules within the membrane structure, respond to stimuli in a nondiscriminating way, primarily as a result of perturbation of lipids, and the selective or discriminating responses of mature cells, which contain greater numbers of intramembranous particles, are initiated by interactions between stimulus and protein receptor sites. This implies that at least some of the intramembranous particles added during maturation of the sensory cell are receptor proteins. It implies further that the cellular machinery for transduction is present in the cell *before* the protein receptor sites become dominant (i.e., in the immature cell). The earliest detection of G-protein in cells is at the time when action potentials are first detectable (Mania-Farnell & Farbman 1990), but electrical activity, in the form of EOGs, can be measured earlier (Gesteland et al. 1982). It should be noted that the immunohistochemical studies of Mania-Farnell and Farbman (1990) were based on labeling with an antibody against the α subunit of G_s that is found in olfactory sensory cells, but is not olfactory-specific. When an antibody to G_{olf} was used, it was not demonstrable in sensory cells until a day or two later in fetal life, whereas olfactory-specific adenylate cyclase was demonstrable at the same time as the α subunit of G_s (Dau et al. 1991). Given that the G-proteins involved with the adenylate cyclase cascade and the IP_3 system probably differ (Boekhoff et al. 1990b; cf. Kesbeke et al. 1990), it remains to be shown how the different G-proteins in olfactory epithelium are coupled to their effectors. Further, the roles of the inhibitory G-proteins remain to be defined.

3.4 Physiology of the olfactory epithelium

3.4.1 The EOG

When an appropriate stimulus is delivered to the olfactory sensory epithelium, several types of electrical responses can be measured. The first one to be studied extensively was the EOG, recorded from the epithelial surface. When a brief (1–2-sec) stimulus is presented to olfactory epithelium, it elicits an EOG response, which is described as a negative monophasic potential that shows a quick rise followed by a slow decline (Figure 3.7) (Ottoson 1956). Its amplitude is commonly 1–5 mV (rarely as much as 8–10 mV), and its duration usually is 2–4 sec or sometimes longer. Although the EOG typically shows a negative monophasic potential, it may be preceded by an initial short positive component. The latter may be an artifact related to the airstream bearing the stimulus (Ottoson 1956), an inhibitory potential generated by sensory cells in response to stimulation (Gesteland et al. 1965), or a secretory potential produced by supporting cell secretion (Takagi 1989). This issue has not yet been resolved (T. Getchell 1974b, 1986).

   A detailed discussion of the electrical characteristics of the EOG is beyond the intended scope of this book. The reader is referred to recent reviews by T. Getchell (1986) and Takagi (1989). In the following, some pertinent information about the cellular basis of the EOG is summarized.

   The EOG is a summed receptor cell-generated voltage change resulting from the activity of a *population* of sensory neurons. Adjacent respiratory mucosa does not respond (Ottoson 1956; T. Getchell 1974b). The preferred electrode for recording EOGs is a micropipette filled with Ringer-agar solution, and a tip diameter on the order of 0.1–0.2 mm (Mozell 1962). The cross-sectional area of the opening at the pipette tip

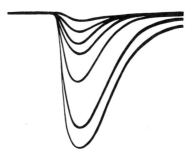

Figure 3.7. Schematic representation of typical EOGs. In this diagram, seven EOGs are superimposed, showing increased amplitude of response (in millivolts) with increased stimulus strength. EOGs may have amplitudes as much as 8–10 mV, although they are commonly 1–5 mV. (Redrawn after Ottoson 1956.)

is large enough to cover a field containing several hundred to perhaps a thousand dendritic knobs on the surface of frog olfactory epithelium (given an average knob density of about $5 \times 10^4/mm^2$) (Menco 1983). That the EOG is a true reflection of an *olfactory* response to odorant stimuli is indicated by its increased amplitude in response to increased concentration of stimulus (Figure 3.7) (Ottoson 1956).

The negative polarity of EOGs indicates that in response to the stimulus, the mucosal surface becomes electrically negative as a result of positive current flowing extracellularly from the depths of the mucosa to the surface; that is, the deeper region of the mucosa becomes more positive with respect to the surface (Gesteland 1971). This is confirmed experimentally by showing that the EOG becomes progressively less negative as the electrode is lowered into the epithelium (Figure 3.8) (Ottoson 1956).

Ion substitution experiments indicate that sodium and potassium ions are largely responsible for the EOG voltage change (Takagi et al. 1968, 1969b), but calcium ion is a necessary cofactor (Suzuki 1978). This implies the presence of monovalent cation channels in the ciliary membrane. The ion channels probably are nonselective (T. Getchell 1986). The EOG is not sensitive to tetrodotoxin, a neurotoxin that blocks voltage-sensitive sodium channels and is unaffected by local anesthetics in concentrations sufficient to block conduction in the olfactory nerve (Ottoson 1956; Takagi 1989). However, ion transport across frog olfactory mucosa in vitro is suppressed by amiloride, a sodium channel blocker (Persaud et al. 1988). The mechanism for this is unclear, because immunocytochemical electron microscopic localization of the amiloride-sensitive sodium channel shows that it is predominantly located on supporting cell microvilli (Menco & Benos 1989).

The presence of olfactory cilia is necessary for an EOG response, as

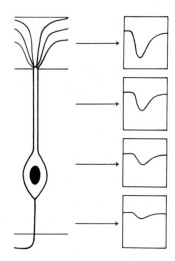

Figure 3.8. Diagram showing the changing shape of the EOG in response to a stimulus as the recording electrode tip changes in position. At the surface the negative-going EOG has its greatest amplitude. As the electrode tip is lowered into the epithelium, the EOG becomes progressively less negative. (Redrawn after Ottoson 1956.)

shown by both deciliation and degeneration–regeneration experiments. When the frog olfactory epithelium is deciliated by application of a detergent, the EOG potential disappears, and only reappears with ciliary regrowth (Bronshtein & Minor 1977; Adamek et al. 1984; Lidow et al. 1987). When olfactory nerves are sectioned, most of the sensory neurons degenerate and die, but supporting cells survive. The decrease in number of sensory neurons and the accompanying loss of cilia on the epithelial surface correspond with the decreased amplitude and ultimate disappearance of EOG responses (Takagi & Yajima 1964, 1965; Burns et al. 1981; Simmons & Getchell 1981a,b). Responses reappear with reconstitution of the epithelium and return of cilia. Similar results were seen after $ZnSO_4$ lavage of the epithelium, which destroys the sensory epithelium (Adamek et al. 1984; Lidow et al. 1987). These experimental data supported the view that the cellular origin of the negative EOG responses was in the cilia of sensory neurons (T. Getchell 1974b; T. Getchell et al. 1980).

EOGs to some stimuli can be positive. For example, Gesteland (1964) applied 62 different stimuli onto frog olfactory epithelium and found that 35 odors elicited negative EOGs, and the remaining 27 elicited strong positive potentials followed by negative potentials. He suggested that the positive potentials represented an inhibitory response involving hyperpolarization of the neurons. Takagi et al. (1969b) applied vapors of 122 odorants in three different concentrations onto frog olfactory epithelium and found that 106 elicited negative EOGs, and the remaining 16 elicited positive or positive–negative potentials. Takagi (1989) suggests that the origin of the positive EOG is not associated with the neurons, on the basis of the following: Electron microscopic evidence demonstrated that application of irritating stimuli (chloroform and others) to frog olfactory epithelia elicited vigorous secretion by supporting cells (Okano & Takagi 1974). Electrophysiological recordings showed that application of the same odorants elicited long-lasting positive EOG potentials. Takagi (1989, p. 179) explains the secretory mechanism as follows: Application of the stimulus results in entry of Cl^- through the surface membrane of the supporting cell. This is associated with the liberation of secretory granules and K^+ from the supporting cell into the mucus, resulting in a net gain in positive charge in the mucus. He suggests that the long-lasting positive EOG is associated with these events. Whether this is the case, or whether the positive potential represents hyperpolarization of the neurons, awaits further resolution.

3.5 Physiological properties of the olfactory sensory neuron

Among the earliest studies of the electrical properties of *individual* olfactory sensory neurons (as opposed to the ensemble response of many cells, measured by the EOG) were those of Robert Gesteland (1964; Gesteland et al. 1965). He and his collaborators used highly sensitive, platinum-coated metal-filled micropipette electrodes (Figure 3.9), with tip diameters of 5–10 µm, to make extracellular recordings of the signals from single sensory neurons within the olfactory epithelium (Mathews 1972; T. Getchell 1973, 1974a; T. Getchell & Shepherd 1978a,b). These metal electrodes proved to be superior to ordinary micropipettes in detecting the weak electrical signals generated by the small olfactory sensory cells.

Many of the response properties of the sensory cell were worked out with this technique for recording the activity of single cells. For example, it was shown that single olfactory sensory neurons responded to more than one type of stimulus, and the pattern of responsivity differed from cell to cell (Gesteland et al. 1965; Mathews 1972). As indicated in an earlier section, these observations helped to prove that the specificity of the olfactory response did not reside in sensory cells that were uniquely responsive to only one stimulus type.

Compared with other neurons, olfactory sensory neurons exhibit a relatively low spontaneous firing frequency, usually fewer than 20/sec, often as low as 2–3/sec, and some are altogether silent unless stimulated (e.g., O'Connell & Mozell 1969; T. Getchell 1973; Holley et al. 1974;

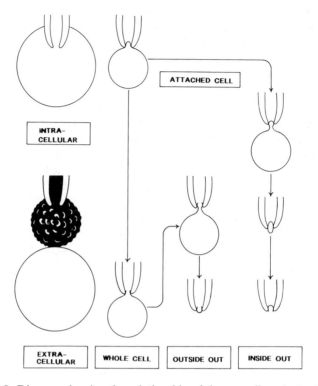

Figure 3.9. Diagram showing the relationship of the recording electrode to the membrane of the sensory cell in various methods that have been used to measure responses of olfactory cells to stimuli. For intracellular recording, the tip of the micropipette electrode penetrates the membrane of the cell, and a seal is formed on the lateral wall of the tip. For extracellular recording, the metal-filled glass pipette has a platinum-coated ball at the tip that makes contact with an adjacent cell membrane. For attached-cell patch recordings, a small piece of the membrane is drawn into the pipette by suction, and a seal is made between the smoothly polished pipette opening and the cell surface. In some whole-cell patches, the membrane drawn into the pipette is ruptured, thus establishing continuity between the electrolyte solution within the pipette and the cytoplasm; this method, essentially similar to intracellular recording, is referred to as whole-cell recording. In outside-out patches, the piece of membrane within the pipette is drawn away from the rest of the cell so that only a small piece of the membrane remains at the pipette tip. In this procedure, the original inner surface of the cell membrane faces the electrolyte solution within the pipette. In the inside-out patch, the original outside surface of the cell membrane faces the electrolyte solution in the pipette. (Parts of this figure redrawn after Hamill et al. 1981.)

T. Getchell & Shepherd 1978a; reviewed by T. Getchell 1986). A positive correlation was shown between responses of single neurons and odor concentration: When a cell was exposed to increasing concentrations of stimulus, it increased its rate of firing. In context, this early experiment was key in demonstrating that the recordings were actually measuring the responses of olfactory cells, responses that, as in any other biological system, ought to be proportional to the concentration of the stimulus, within a physiological range.

Another characteristic of the olfactory sensory cell shown with this method was that with increasing stimulus concentration, the onset latency of the response decreased (T. Getchell & Shepherd 1978a). It was also found that most sensory neurons adapted slowly; that is, the neurons typically responded to a stimulus of long duration with a prolonged discharge of impulses (T. Getchell & Shepherd 1978b). Further, the excitatory discharge was time-locked to the stimulus (i.e., it terminated abruptly within 1 sec of termination of the stimulus).

These pioneering experiments using extracellular electrodes to record single-cell responses were significant because they showed that the neurons responding to chemical stimuli differed from other neurons in several specific ways.

3.5.1 Properties of the sensory neuron membrane

Although the studies using extracellular electrodes to record from single units provided new information about sensory neuron responses, the physiological properties of the olfactory cell membrane itself could not be examined by this technique. Study of transmembrane events required development of a means to record changes from within the cell, using intracellular electrodes with very fine tips. Progress in devising a method for intracellular recording (Figure 3.9) had been hampered by the small size of the sensory neurons and by inability to obtain stable preparations. When the land-phase salamander (*Ambystoma tigrinum*) was identified as a useful animal for intracellular experiments because of the relatively large size of its sensory neurons and its relatively low mucus production (T. Getchell 1977), many investigators adopted it as a model for the study of membrane properties using intracellular electrodes (T. Getchell 1977; Masukawa, Kauer, & Shepherd 1983; Trotier & MacLeod 1983; Masukawa, Hedlund, & Shepherd 1985a,b; Hedlund, Masukawa, & Shepherd 1987). Later, when patch-clamp technology became available, new information was garnered about the types and properties of channels within the membrane.

Olfactory sensory neurons have a mean resting potential of approximately -30 to -60 mV (T. Getchell 1977; Masukawa et al. 1983; Trotier & MacLeod 1983; Anderson & Hamilton 1987; Trotier 1990). The cell interior is negatively charged relative to the exterior. When a stimulus is presented to the cell and current flows across the cell membrane, the membrane potential changes, usually by decreasing (depolarization). The depolarization induced in this way is the receptor potential, sometimes called the generator potential. This current flow can be stimulated experimentally by injecting the current (in the form of monovalent ions) by means of an intracellular microelectrode, and the cell is depolarized to a degree depending on the amount of current injected.

One of the most remarkable properties of olfactory sensory cells is that the amount of current required to elicit an action potential is relatively low. Using intracellular or patch-clamp micropipettes, researchers found that picoampere (pA; 1 pA $= 10^{-12}$ A) amounts of current could induce an action potential in olfactory sensory cells of the lamprey (Suzuki 1977), salamander (Masukawa et al. 1985a; Firestein & Werblin 1987; Hedlund et al. 1987), frog (Trotier & MacLeod 1983; Trotier 1986; Schild 1989), mouse (Maue & Dionne 1987), newt (Kurahashi 1989), rat (Lynch & Barry 1989; Trombley & Westbrook 1991), and lobster (Anderson & Ache 1985). For example, Firestein and Werblin (1987) showed that as little as 3 pA of injected current was required to elicit an action potential in larval salamander olfactory cells; Kurahashi (1989) found very similar results in the newt; and Lynch and Barry (1989) found that 1 pA was enough in rat. Depolarization in response to a stimulus of injected current, or in response to an odorant (in vivo or in vitro), was usually about 10 mV or less before action potentials were initiated (T. Getchell 1977, 1986; Masukawa et al. 1985a); comparable results were found in lobster olfactory cells by McClintock and Ache (1989a). In other words, if an odorant stimulus caused a + 10 mV depolarization from a resting potential of − 60 mV, this relatively small change in potential (the receptor potential) would spread passively along the cell membrane of the dendrite and soma and would be enough to trigger an action potential in the region of the axon initial segment.

If we apply Ohm's law, we can calculate a value, called the input resistance of the cell, as follows:

$$\text{input resistance } (R_i) = \frac{\text{receptor potential } (E_r)}{\text{injected current}}$$

If we substitute the values from salamander experiments (Firestein & Werblin 1987), we find

$$\frac{10 \times 10^{-3} \text{ volts}}{3 \times 10^{-12} \text{ amperes}} = 3.3 \times 10^9 \text{ ohms}$$

This sample calculation indicates that the olfactory sensory cell has a very high input resistance. When measured with intracellular micropipettes, the input resistance was usually 100–600 megohms (MΩ; 1 MΩ $= 10^6$ Ω) (reviewed by Takagi 1989), but because some cell damage is often associated with impaling a cell, a better estimate of input resistance has been gained with patch-clamp micropipettes using the whole-cell patch method (Figure 3.9), and it is about an order of magnitude higher, that is, 2–5 gigaohms ($G\Omega$; 1 $G\Omega$ = 10^9 Ω) (Trotier 1986, 1990; Maue & Dionne 1987; Schild 1989; Trombley & Westbrook 1991). The high input resistance is associated with the small size of the olfactory sensory neuron. Further, it indicates that the membrane of the sensory neuron is not very leaky to ions (i.e., few channels are open in the resting condition). Most neurons that have been studied by patch-clamp technology are larger than olfactory sensory neurons, and because of their size they have a lower input resistance (i.e., more current is required to elicit an action potential).

In terms of how the olfactory sensory cells function in situ, the significance of their unusual membrane properties is that a rather small conductance change in the ciliary membrane, elicited by an odor stimulus, can permit a small amount of current to flow across the membrane and can result in a depolarization of sufficient magnitude to reach the threshold required to fire an action potential (Hedlund et al. 1987; Schild 1989; Trotier 1990). Given that the membrane properties are such that a few picoamperes of current is enough to elicit an action potential in patch-clamped olfactory sensory cells, it should be apparent that under physiological conditions, a relatively small flux of ions into the cell in response to a stimulus would be enough to fire an action potential. Indeed, it has been argued that the stimulus-induced opening of only one or a very small number of ion channels can depolarize the olfactory sensory cell enough to trigger an action potential (Maue & Dionne 1987; Frings & Lindemann 1988; Lynch & Barry 1989; cf. McClintock & Ache 1989a).

3.5.2 Ion channels

Ion channels in cell membranes are constructed of protein molecules that span the cell membrane in such a way that they form a pore, through which specific ions can pass (Figures 3.4 and 3.5). But the channels are not open all the time; instead, they are "gated" (i.e., primed to open and close under certain conditions) (Figure 3.10). In neurons, the gating of some channels is responsive to the binding of chemicals. For example, at synapses, neurotransmitters bind to specific membrane receptors. The binding of transmitter to receptor elicits a cascade of events resulting in depolarization of the postsynaptic cell. Other channels are gated open by changes in the voltage across the cell membrane. When a membrane is depolarized, this type of channel permits ion transport across the membrane. An example of this is the sodium channel, found in abundance in the nodes of Ranvier of all myelinated nerve fibers and along unmyelinated axons, including those of olfactory sensory neurons.

The opening and closing of ion channels have important functional

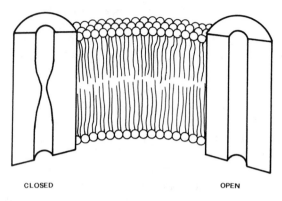

CLOSED   OPEN

Figure 3.10. Schematic representation of what a membrane channel might look like when gated closed (left) or open (right).

consequences in the olfactory sensory cell and other neurons. For example, in the olfactory sensory neuron, an odorant stimulus (ligand) interacts with the receptor, located on the cilia. This activates the signal transduction cascade, leading to the opening of ion channels. This initiates the voltage change, a depolarization, in the cell that ultimately results in generation of the action potential. The action potential is initiated at the axon initial segment, near the junction of the cell body and the axon. Generation of an action potential, the electrical impulse that carries the signal via the axon to the second-order neuron in the olfactory bulb, depends on the precisely timed opening and closing of voltage-sensitive channels for sodium and potassium.

The study of ionic flux across the membrane associated with electrical activity in olfactory sensory neurons has been aided by the utilization of the patch-clamp technique during the past decade. This technique is based on the use of glass micropipettes with very small (approximately 1–2 μm in diameter), fire-polished openings that permit tight, high-resistance seals to be formed between the electrode tip and very tiny pieces of cell membrane (Hamill et al. 1981). The pipette can be applied and sealed to the membrane of an intact olfactory cell that has been dissociated and isolated from its neighboring cells (Figure 3.9). This method allows control of the medium outside of the membrane by manipulation of the solution bathing the cell, while the intracellular contents of the cell remain intact. A variation of this method permits intracellular recording: After the cell is patched with a high-resistance seal, suction can be applied through the pipette, the membrane can be ruptured, and the contents of the pipette will come into contact with the cytoplasm. A variation of this method is the drawing of an individual cilium into the pipette and forming the tight seal at the base of the cilium, on the olfactory knob (Frings & Lindemann 1990, 1991).

Another variation of patch-clamp methods is the "inside-out" method. After a tight seal is established with the cell membrane, the patch of membrane can be torn from the cell, leaving the patch of membrane sealed to the pipette tip and exposing the cytoplasmic surface of the membrane to the bath medium (Figure 3.9). This permits the experimenter to control the solution bathing what had been the cytoplasmic surface of the membrane, as well as the pipette solution in contact with the external surface of the membrane.

What kinds of ion channels are present in olfactory sensory cells, and how are they involved in the biology of this cell? With the development of methods to dissociate olfactory tissue and isolate individual, viable, responsive cells (Kleene & Gesteland 1983; Dionne 1986; Kashiwayanagi et al. 1987; Maue & Dionne 1987; Nakamura & Gold 1987) it became possible to examine their specific ion channels by patch-clamp methods. The chief limitation of earlier patch-clamp studies was that the conditions under which cells were isolated may have altered the membrane properties to some degree, particularly if enzymatic methods were used for the isolation. Now several investigators have been successful in using gentle mechanical dissociation methods to isolate sensory cells, thus avoiding the use of enzymes (Firestein & Werblin 1989). As noted earlier, it is also possible to use a slice of olfactory mucosa and draw cilia into the micropipette (Frings & Lindemann 1990, 1991).

Nakamura and Gold (1987) reported some elegant studies involving recording from membrane patches of olfactory cilia of sensory cells isolated from the toad. They used the inside-out patch method. Essentially, their results showed that ciliary membranes have a monovalent cation-selective conductance pathway (or channel) that can be directly and reversibly activated by micromolar concentrations of cAMP or cGMP (Figure 3.11). This cationic channel did not select between sodium and potassium ions in the medium, but others have shown that the response to odorants can be blocked by amiloride, a sodium channel blocker (Frings & Lindemann 1988; cf. Persaud et al. 1987, 1988). These results suggest, then, that opening of this cationic channel occurs as a result of the direct action of a cyclic nucleotide, rather than by way of a phosphorylation requiring a protein kinase (Figure 3.4) (Nakamura & Gold 1987). The current entering the cell results in a depolarization (the receptor potential) spreading through the dendrite and soma; ultimately this depolarization excites the cell to generate an action potential at the initial segment of the axon (T. Getchell 1973).

Ion channels permit flow of ions both into and out of the cell. Concerning the inward flow, patch-clamp recordings from the soma and dendrite of olfactory sensory cells have shown a slow, sustained, inward

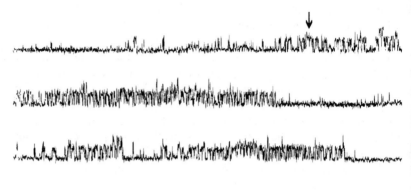

500 msec

Figure 3.11. Single-channel activity in an isolated frog olfactory cilium. The cilium was drawn into the patch pipette until a high-resistance seal was made near its base. The cilium, still inside the pipette, was broken away from the cell (inside-out patch) and voltage-clamped (cytoplasmic potential held at $+20$ mV). The bursts of single channels shown were activated when the cytoplasmic face of the ciliary membrane was exposed to 2mM ATP in the bath. Upward deflections (arrow) represent the channel opening. (Courtesy of Drs. S. Kleene and R. Gesteland.)

current carried by calcium ions, reflecting the activation of calcium channels; it was found that this current could be blocked by cobalt ions (Trotier 1986; Firestein & Werblin 1987; Maue & Dionne 1987; Schild 1989). A transient inward sodium current with a peak magnitude between 0.2 and 1.0 nA (1 nA $= 10^{-9}$ A) that was tetrodotoxin-resistant was demonstrated in studies on larval tiger salamanders (Firestein & Werblin 1987), but not in most of the other studies. In cells isolated from adult salamanders, the transient sodium current disappeared in the absence of external sodium (Trotier 1986; Firestein & Werblin 1987). The sodium current was activated at potentials beyond -40 mV and reached a peak in less than 3 msec. This transient sodium current was not seen in *Xenopus* cells (Schild 1989). In a few cells it was possible to demonstrate a rapid, inward, tetrodotoxin-sensitive sodium current (Trotier 1986; Schild 1989), but this probably was related to the action potential, because these cells had a short length of axon after they were isolated (Trotier 1986; McClintock & Ache 1989a; Trombley & Westbrook 1991).

The outward flow of current was carried by potassium ions and had three components: (1) a voltage-dependent potassium current that was activated rapidly at voltages more positive than -20 mV, (2) a calcium-dependent potassium current, and (3) a slowly activating and inactivating potassium current (Trotier 1986; Firestein & Werblin 1987; Maue & Dionne 1987; Schild 1989; Trombley & Westbrook 1991). The delayed potassium flux out of the cell is likely involved in repolarization after depolarization.

Another approach to the study of channels in olfactory tissue was pioneered by Vodyanoy and Murphy (1983). These investigators combined a crude homogenate, obtained from rat olfactory mucosa, with solvent-free planar lipid bilayers. Addition of the homogenate to the bilayer chamber resulted in the appearance of fluctuations in the current across the bilayer, indicating that ion channels had been inserted in the bilayer. Potassium ion was identified as the charge carrier in these experiments (reviewed by R. Murphy 1988). However, because the preparation was crude, there was no way to determine the precise origin of the channels (i.e., they could have come from membranes of any cells in the epithelium or lamina propria).

The development of procedures to obtain a highly enriched membrane preparation of olfactory cilia permitted reconstitution studies in preparations that had been subjected to biochemical scrutiny (Anholt, Aebi, & Snyder 1986; Pace & Lancet 1986; Sklar et al. 1986; Anholt et al. 1987). Labarca et al. (1988) fused ciliary membranes obtained from olfactory epithelium of frogs to a planar lipid bilayer and identified a cation-selective multiconductance channel that could be activated directly and reversibly by nanomolar concentrations of a stimulus. They used a derivative of pyrazine, the characteristic bell pepper compound. The channel did not discriminate between sodium and potassium ions and was activated in the absence of cyclic nucleotides. One of the difficulties associated with these reconstitution experiments is that even though the preparation has been enriched in ciliary membranes, there are many possibilities for contamination with membranes from other sources, such as respiratory cilia, sustentacular cell microvilli, and other cells in the olfactory mucosa homogenate. In other words, there was no check for homogeneity of the cilia preparation in these studies.

There are still gaps in our understanding of the ionic events occurring in membranes of olfactory sensory cells when they are excited by a stimulus. Labarca and Bacigalupo (1988) speculated on the possible sequence of events as follows: When a cell is stimulated by an odorant binding to its receptor, a transduction cascade is initiated, resulting in an increase in the concentration of cyclic nucleotides or possibly another second messenger (such as IP_3). In response to this, rapidly and transiently activated cation channels will open, permitting the inward passage of current, probably carried by sodium ions. This event results in a small depolarization of the membrane in the region of the dendrite and soma. This, in turn, is somehow associated with the initiation of an action potential in the axon initial segment. Calcium channel inactivation and opening of one or more of the slowly activating potassium channels will transiently repolarize the membrane. If the stimulus persists and the ciliary current continues to flow, this will repeatedly raise the membrane potential above the threshold, reopen the calcium channels, and lead to repetitive firing. This hypothesis is consistent with most of the known data, but the role of the calcium current remains unclear.

3.6 Electrical properties of the supporting cell

The studies on intracellular recording of sensory cells in olfactory epithelium were done on whole epithelium, either in the living animal (T. Getchell 1977; Trotier & MacLeod 1986) or in vitro (Masukawa et al. 1983). In this kind of preparation, as the microelectrode was passed into the epithelium, many penetrations were made of cells that had high resting membrane potentials but did not exhibit an action potential when current was injected. Cells with these properties were identified as supporting cells by injection of a marker dye (T. Getchell 1977). The membrane properties of supporting cells were thus examined, in most cases concomitantly with those of the sensory cells.

The resting potential of supporting cells usually was in the range of -50 to -100 mV (i.e., somewhat higher than that of sensory cells) (T. Getchell 1977; Masukawa et al. 1983; Trotier & MacLeod 1986). Responses to odorant stimulation elicited slow depolarizations (Trotier & MacLeod 1986), but in one study some cells responded by hyperpolarization (T. Getchell 1977). The input resistances in these cells were characteristically relatively low, usually 10–20 MΩ, as measured by passing a hyperpolarizing current and recording the voltage change produced (Masukawa et al 1985a). Input resistance often was difficult to measure because of an unfavorable signal-to-noise ratio (Trotier & MacLeod 1986).

The membrane properties of the supporting cell, with a relatively high resting potential and a relatively low input resistance, resemble properties of glial cells, and this may be related to their function (T. Getchell 1977; Masukawa et al. 1985a; Trotier & MacLeod 1986).

The most conspicuous characteristic of supporting cells is their high, voltage-independent potassium conductance (Trotier & MacLeod 1986). These cells are more than 20 times more permeable to potassium ions than are the neighboring sensory neurons. The data suggest that supporting cells play an important role in regulating extracellular potassium concentration within the epithelium and/or in the layer of mucus on the epithelial surface. This function, buffering of potassium ions, is also an important role of astroglial cells, which maintain the balance of potassium ions in the extracellular fluid of the brain (Kimelberg & Norenberg

1989). Trotier and MacLeod (1986) suggest that the buffering effect on extracellular potassium within the epithelium could provide a means of modulating the membrane potential of sensory neurons, thus regulating the low spontaneous firing frequency. Under these conditions, a weak odor stimulus might provide a small change in the rate of firing. Given the high convergence ratio of sensory neurons to secondary neurons in the olfactory glomerulus, a relatively large number of cells, with only a relatively small change in the rate of firing of each individual unit, could be summated, with a high signal-to-noise ratio, at the level of the mitral cell, thus activating the mitral cell.

3.7 Nonolfactory sensation in the nose: function of the trigeminal innervation

The trigeminal nerve is usually considered a somatosensory nerve, a mediator of the sensations of touch, temperature, pain, and proprioception. However, both nasal trigeminal branches, the ethmoid and the nasopalatine, carry chemosensory information by way of their smaller myelinated and unmyelinated fibers (Beidler & Tucker 1955; Moulton 1963; Tucker 1963, 1971; Silver 1987). For many years after Parker (1922) described the existence of a common chemical sense, it was assumed that the nasal trigeminal system responded only to irritating stimuli. However, certain nonirritating odorant stimuli, such as phenylethyl alcohol, can stimulate nasal trigeminal nerve endings at concentrations below those that elicit olfactory responses (Tucker 1971). For those compounds that stimulate both olfactory and trigeminal nerves, the thresholds for the latter are usually much higher (Tucker 1963, 1971; Silver 1987; Silver, Arzt, & Mason 1988). For example, the trigeminal threshold concentration for amyl acetate, a commonly used compound in olfactory experiments, is 4,000 times higher than the olfactory threshold (Tucker 1971).

It is interesting to note that lipid solubility may play a role in trigeminal response. In a study of responses to a homologous series of aliphatic alcohols it was shown that the response thresholds decreased with increasing carbon chain length (Silver et al. 1986). The threshold for methanol was 3,000 parts per million (ppm), whereas that for octanol was 3 ppm. Lipid solubility increases with increasing length of the carbon chain in these alcohols. One plausible explanation for the results is that the more lipid-soluble a substance, the more easily it can penetrate epithelial layers to reach the trigeminal nerve endings.

Is the trigeminal innervation of the nasal cavity capable of discrimination among several stimuli? In an electrophysiological study on salamanders, the trigeminal chemoreceptors exhibited complete cross-adaptation between two stimulus pairs, butanol versus *d*-limonene and amyl acetate versus cyclohexanone (Silver et al. 1988). These data indicated that trigeminal nerves were unable to discriminate between chemical stimuli (i.e., they were stimulated nonspecifically). Behavioral experiments confirmed that the salamanders were unable to discriminate between perceptually equivalent concentrations of paired stimuli (Silver et al. 1988).

The trigeminal responses to suprathreshold concentrations of vaporous stimuli often can be described as reflexive and protective. For example, responses can result in (1) secretion of mucus, (2) a change in the pattern of respiration, sometimes leading to short-term cessation of breathing, (3) autonomic reflexes (via sympathetic nerves) influencing circulation to the nasal mucosa (e.g., resulting in engorgement of intranasal erectile tissue, thereby reducing the airflow) (Tucker 1963), (4) changes in heart rate, (5) symptoms of cerebral arousal (Stone, Carregal, & Williams 1966; Stone, Williams, & Carregal 1968), (6) reflex changes in vascular permeability (reviewed by DeLong & Getchell 1987), and (7) sneezing.

In humans, trigeminal sensations have several qualities, described as pungent, tickling, warm, cool, burning, stinging, sharp, and so forth (Doty et al. 1978). Anosmic human subjects (i.e., individuals who have no olfactory sensation) can detect many odorants when they are presented in high enough concentrations (Doty 1975; Doty et al. 1978). Evidence from psychophysical studies on humans has led to the formulation of the notion that the trigeminal nerve contributes to the overall perceived intensity of an odorant (Cain 1974, 1976; Cain & Murphy 1980; Murphy 1987). For example, the pungency of carbon dioxide, which is odorless but stimulates the trigeminal nerve, can attenuate the response to odorants and is thought to act through the central nervous system (Cain & Murphy 1980). The degree of trigeminal contribution is concentration-dependent (i.e., at low concentrations it is relatively small, but it increases with increasing stimulus concentration). In electrophysiological experiments on rabbits, trigeminal stimulation both by odorants (Stone et al. 1966, 1968) and by direct electrical stimulation of the ethmoid nerve (Stone 1969) decreased the excitability of the olfactory bulb. Thus, the attenuating effect of trigeminal stimulation on olfaction can act at two levels: by autonomically driven reflexes at the level of the nasal cavity, and by an undefined mechanism at the level of the central nervous system (Stone et al. 1968).

It seems, then, that a major role of the trigeminal system in olfaction is to warn the organism of potential danger in the presence of high concentrations of irritating volatile stimuli. The effect is to elicit aversive behavior that would escape or avoid further exposure to the stimulus. However, in the awake animal, the trigeminal nerve probably affects bulbar activity even in the absence of irritating concentrations of stimuli (Stone et al. 1968; Cain & Murphy 1980).

3.8 Olfactory dysfunction in humans

Partial or complete loss of olfactory function (hyposmia or anosmia) is not an uncommon occurrence in the adult population, particularly in older age groups. Approximately 2 million adults in the United States suffer from disorders of smell and taste, and by far the majority of these are olfactory deficits. In a study of the smell identification abilities of nearly 2,000 individuals ranging in age from 5 to 99 years, Doty et al. (1984) found that the average ability to identify odors reaches a peak in the 20–40-year age group and begins to decline after age 50. There is a marked decrease in smell identification ability after the age of 60. They also found that, on the average, women outperformed men at all ages, and nonsmokers did better than smokers (Frye, Schwartz, & Doty 1990).

As with losses in other sensory systems, the anatomical and physiological bases for olfactory loss probably are very complex. They fall into two major categories: conductive loss and neural loss. Conductive changes involve a compromise in the patency of the nasal cavity, resulting in blockage and failure of stimuli to reach the sensory cells. In most cases conductive deficiencies are due to nasal polyps and can be easily remedied by surgical removal of the obstruction. Some are due to rhinitis, resulting from allergy or other cause, and can be treated with steroids to reduce inflammation. High doses of steroids can temporarily restore the sense of smell in these cases (Jafek et al. 1987).

Neural changes involve damage or loss of the sensory neurons or some region of the central nervous system associated with olfaction. There are several conditions that can lead to neural loss in the olfactory system.

The most common cause of olfactory deficits is upper respiratory infection (Schiffman 1983; Doty 1990). Upper respiratory infections and their aftermaths can account for cell death, and repeated infections, especially in people more than 50 years old, can lead to partial replacement of sensory epithelium by nonsensory respiratory tissue (Doty 1990; Jafek et al. 1990). In fact, examination of autopsied or biopsied human olfactory mucosa with the electron microscope reveals that it is really an irregular mosaic consisting of patches of respiratory and olfactory epithelium in the upper region of the nasal cavity (Morrison & Costanzo 1990). This is particularly true in patients with olfactory deficits following viral infections in the respiratory tract (Jafek et al. 1990).

Certain kinds of head injuries that involve rapid acceleration and/or deceleration of the brain are commonly associated with loss of the sense of smell. A sharp blow to the forehead or back of the head (e.g., impact resulting from a fall or automobile accident) can result in inertial displacement of the brain and shearing of the delicate olfactory nerves at the level where they pass through the cribriform plate. Posttraumatic anosmia occurs in 5–15% of serious head injuries (Sumner 1964; Douek 1974; Zusho 1982). This kind of trauma will result in degeneration of the olfactory epithelium, but replacement of the cells can occur as the epithelium reconstitutes itself. However, if, in the healing process, scar tissue blocks the tiny openings in the cribriform plate, the growing olfactory axons will be prevented from entering the cranial cavity, and connections to the bulb will not be reestablished (Zusho 1982; Moran, Jafek, & Rowley 1985a; Jafek et al. 1989). If this happens, loss of the sense of smell can be permanent.

Older patients with certain diseases exhibit some degrees of anosmia. For example, it is now known that most patients with Parkinson's disease have difficulty in detecting, recognizing, and identifying odorants (Ansari & Johnson 1975; Ward, Hess, & Calne 1983; Serby et al. 1985; Quinn, Rossor, & Marsden 1987; Doty, Deems, & Stellar 1988; Doty et al. 1989). The olfactory disorder of Parkinson's disease rarely results in total loss of the ability to smell (Doty 1990). Although the biological basis for this deficiency is not clear, it is established that the olfactory deficit is independent of any cognitive or motor manifestations of the disease (Doty et al. 1989). Significant deficiencies in olfactory ability also occur in patients with Huntington's chorea, an inherited movement disorder (Moberg et al. 1987), Korsakoff's psychosis, a condition associated with severe alcoholism (Jones et al. 1978; Mair et al. 1986), Down's syndrome (Warner, Peabody, & Berger 1988), and schizophrenia (Hurwitz et al. 1988). Alcoholism is associated with basic olfactory impairments that are only partially reversible with abstinence (Ditraglia et al. 1991).

In recent years a number of studies have shown olfactory dysfunction in patients with Alzheimer's disease. The ability to identify odors and the threshold of odor sensitivity are greatly impaired, even in early stages of this disease (Serby et al. 1985; Warner et al. 1986; Doty, Reyes, & Gregor 1987; Koss et al. 1987; Rezek 1987; Serby 1987; Murphy et al. 1990; Serby, Larson, & Kalkstein 1991). The olfactory deficits probably are related to structural changes in the olfactory pathway, either in the epithelium itself (Talamo et al. 1989) or in the more central structures, the bulb and cortex (Esiri & Wilcock 1984; Pearson et al. 1985; Reyes et al. 1987). Disproportionately high numbers of neurons in the olfactory regions of the brains of Alzheimer's patients contain the typical neurofibrillary tangles and neuritic plaques, compared with minimal abnormalities in visual and sensorimotor areas of the neocortex. Pearson et al. (1985) raised the possibility that the olfactory pathway is the site of initial involvement in Alzheimer's disease. It remains to be determined whether or not the olfactory regions of the brain are especially vulnerable to the etiological factors causing Alzheimer's disease for some intrinsic reason. It has also been suggested (Esiri 1982; Roberts 1986) that some dementia-related diseases can be related to the fact that the olfactory nerves can serve as a major conduit for viruses or other particles from the nasal cavity into the brain (see Chapter 2). It will be recalled that some particles can cross the synaptic space and be transported widely in the central nervous system (section 2.8.2).

Patients with Kallmann's syndrome (inherited hypogonadotropic hypogonadism) are usually anosmic (Males, Townsend, & Schneider 1973). These individuals never achieve sexual maturity (i.e., their gonads fail to mature). Autopsy study of a 19-week human fetus with Kallmann's syndrome showed an absence of the gonadotropic hormone, luteinizing hormone-releasing hormone (LHRH), in the hypothalamus. There were, however, groups of LHRH-positive cells and fibers in the nose (in the terminal nerve pathway; see section 2.12) and within the meninges beneath the forebrain (Schwanzel-Fukuda, Bick, & Pfaff 1989). The suggested reason why Kallmann's syndrome develops is failure of the migrating LHRH-positive cells, which originate in the olfactory placode, to reach the brain (see section 6.1.4). The LHRH cells must be in proximity to the portal circulation between hypothalamus and pituitary gland in order to reach cells of the pituitary gland in high enough concentration to be effective stimulating agents. The probable reason for the anosmia is the failure of the growing olfactory axons to reach the forebrain, where they would stimulate the growth and development of the olfactory bulb (section 6.4.2).

4

Vomeronasal Organ

The vomeronasal system plays a major role in the perception of stimuli related to social and/or reproductive behavior in many species of vertebrates. For example, stimuli of the vomeronasal system may alter the course of puberty, modulate estrus cycles of females, elicit attraction between males and females, elicit courtship and modulate reproductive, aggressive, or territorial behavior, and so forth. This book is not meant to review in depth all of the research on the several behaviors associated with the vomeronasal organ. The reader is referred to reviews by Estes (1972), Wysocki (1979, 1989), Meredith (1983), and Halpern (1987) for more detail on the role of the vomeronasal organ in vertebrate behavior. This chapter discusses some experiments in which certain behaviors were found to be correlated with the cell biology of the organ.

4.1 Distribution and location

The vomeronasal organ (VNO), also known as Jacobson's organ, is present in most terrestrial mammals, amphibians, and reptiles, but is absent from or vestigial in fishes, birds, higher primates, and some aquatic amphibians. In humans, vestigial VNOs have been described in fetuses (Kreutzer & Jafek 1980), and a VNO recess is present in many human adults (Johnson, Josephson, & Hawke 1985), but there is no evidence for sensory cells (Moran et al. 1985a). It is generally believed that a functional VNO is absent from adult humans (reviewed by Pearlman 1934; Wysocki 1979). For reviews on the evolution of VNOs in vertebrates, see Negus (1958) and Bertmar (1981).

The VNOs are paired, elongated, tube-shaped structures, each of which is enclosed in a bony or cartilaginous capsule within the anterior, ventral end of the nasal septum (see Figure 2.19). Posteriorly, the tube ends blindly; anteriorly, the VNO opens via a narrow duct either into the base of the anterior nasal cavity, as in rodents, lagomorphs, and some primates, or into the incisive (also called the nasopalatine) canal that connects the nasal and oral cavities, as in marsupials, monotremes, carnivores, ungulates (hoofed animals), insectivores, and some primates

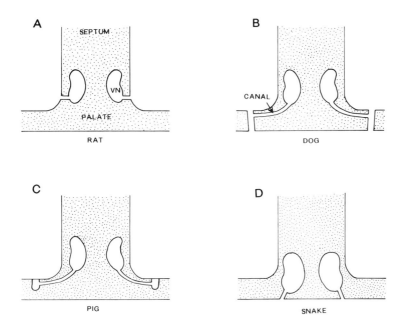

Figure 4.1. Diagram showing the vomeronasal organ (VN) opening in various vertebrates. A: The VNO opens directly into the nasal cavity, as in the rat. B: The VNO opens into the nasopalatine canal connecting the oral and nasal cavities, as in the dog. C: In this configuration, the nasopalatine canal does not open into the oral cavity, as in the pig; the VNO opens into the nasal cavity indirectly by way of the canal. D: In reptiles, the VNO opens directly into the oral cavity.

(Figure 4.1) (Wysocki 1979; Bertmar 1981; Wysocki & Meredith 1987; Meredith & O'Connell 1988). In some mammals the nasopalatine canal does not open into the oral cavity; consequently, the vomeronasal duct drains indirectly into the nasal cavity via the canal. In reptiles, the VNO is completely separated from the nasal cavity by the hard palate. The canals in these animals open directly into the oral cavity by a pair of palatal ducts located anterior to the internal nares (Estes 1972; Halpern & Kubie 1980). The VNOs of amphibians remain attached to the main olfactory chambers as ventromedial or ventrolateral diverticula.

4.2 Stimulus access

The sequestered position of the VNO renders it somewhat isolated from the airstream passing into the nasal cavity during respiration and/or sniffing. This is particularly true in animals in which the VNO does not open directly into the nasal cavity. Although volatile, airborne odor stimuli elicit responses in VNO sensory epithelium (Tucker 1963, 1971; Graziadei & Tucker 1970; Meredith & O'Connell 1979), many of the stimuli associated with social or sexual behavior appear to be nonvolatile. They are associated with urine or secretions from glands in the genital region. These are usually brought into the oral or nasal cavity by licking or nuzzling, as in rodents and ungulates, or tongue-flicking, as in snakes and lizards (Halpern & Kubie 1980).

The stimuli, whether airborne or nonvolatile, gain access to the sensory cells by way of the "vomeronasal pump" in some animals (e.g., rodents) and by the "flehmen" response (e.g., in ungulates). It is possible that a combination of the two mechanisms functions in some animals. The pump operates as follows: The lamina propria of the VNO in mice, rats, hamsters, and similar animals contains several large, thin-walled blood vessels (Figure 4.2) under autonomic nerve control. These cavernous vessels have been compared to those seen in erectile tissue. Meredith and O'Connell (1979) showed, in hamsters, that these vessels become constricted when their sympathetic innervation, entering via the nasopalatine nerve, is stimulated. Inasmuch as a relatively rigid bony capsule surrounds the vessels and the VNO, the vascular constriction allows expansion of the VNO lumen, thus creating a partial vacuum and permitting vapors and fluids to enter it from the nasal cavity or the nasopalatine canal. In this way, volatile and nonvolatile stimuli (dissolved in nasal or oral fluids) are drawn into the lumen and reach the sensory cells. Similar experiments on the cat supported the notion of the vomeronasal pump (Eccles 1982). This pump is functional in guinea pigs at birth (Mendoza & Kühnel 1989). Dilation of the vessels, under control of other autonomic nerves, results in reduction of the VNO size and expulsion of fluids via the duct connecting the VNO to the nasal vestibule or the nasopalatine duct. Thus, the volume of the vascular connective tissue is inversely related to the volume of the VNO.

In several orders of mammals, including ungulates (e.g., horses, deer, goats), "flehmen," a distinctive facial grimace involving a pronounced curl of the lips accompanied by closure of the external nares (and probably the internal nares as well), is thought to be associated with the mechanism by which stimuli gain access to the VNO (Melese-d'Hospital & Hart 1985). The VNO in these animals opens into a nasopalatine canal connecting the nasal and oral cavities, and the vessels associated with the VNO are relatively less prominent than those in rodents (Taniguchi & Mikami 1985); consequently, the vascular pump may be less important. It is clear, however, that these animals perform flehmen when stimulated to investigate urine from members of the same species, and it occurs repeatedly when the animals are sexually aroused (Estes 1972; Rasmussen et al. 1982). Flehmen behavior apparently is a means of determining whether or not the female is in estrus. Damage to the VNO

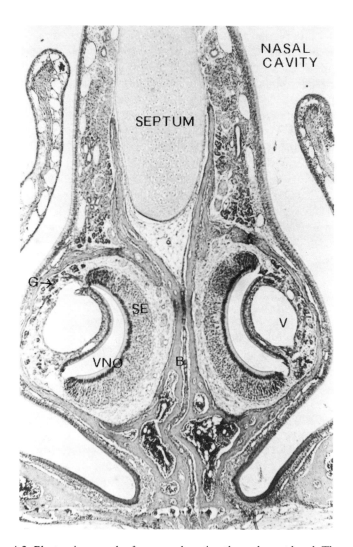

Figure 4.2. Photomicrograph of a coronal section through a rat head. The paired VNOs are located near the ventral end of the septum. Each is surrounded by a thin bony capsule (B). A large vein (V) and some other vessels are nearby. In this plane of section, the VNO is crescent-shaped. The thick sensory epithelium (SE) is located medially, and the nonsensory epithelium is lateral. Some glands (G) are seen lateral to the VNO.

duct or its nerve results in reduction of flehmen behavior. It was shown in rams that stimulation of the sympathetic innervation to the head activates a mechanism producing a negative intraluminal pressure in the VNO, thus drawing fluid in (Bland & Cottrell 1989). The mechanism was under α-adrenergic control. Further, stimulation of the maxillary trigeminal innervation to the nose increased the intraluminal pressure, thus expelling fluid. These experiments show that the intraluminal pressure in the VNO is under nervous control and suggest that it may at least partially explain how flehmen behavior draws stimuli into the VNO. It is not entirely clear whether or not a vascular pump, similar to that used by rodents, is an important component of the flehmen response (Meredith 1983).

In amphibians, the family Plethodontidae (land salamanders) have nasolabial grooves running from the upper lip to the lateral corner of each nostril. These may function as capillary tubes to convey chemicals to the VNO when the animals engage in a behavior described as nose-tapping (Dawley & Bass 1989). The VNOs in these animals lie anteriorly in the nasal cavity, near the end of the nasolabial grooves (Dawley & Bass 1988). Their position may ensure that nose-tapping behavior can be an effective means to promote stimulus access.

4.3 A dual olfactory system

The VNO develops relatively early in the embryonic life of most mammals as a medial fold or diverticulum from the main nasal cavity (Figure 4.3). Thus, even from early stages of development it becomes a more or less separate structure. The sensory cells of the VNO give rise to the vomeronasal nerve, which terminates in the accessory olfactory bulb (McCotter 1912); its projection does not overlap with the target of nerves from the main olfactory organ. The relative sizes of the main and accessory olfactory bulbs vary from species to species.

Much of the interest in the VNO during the past two decades was stimulated by the observation that this organ sends information via the accessory olfactory bulb to a region of the amygdaloid nucleus that in turn is connected with the medial preoptic and medial hypothalamic regions; these are involved in endocrine control mechanisms pertaining to sexual and reproductive behavior (Winans & Scalia 1970). The notion of a "dual olfactory system" has grown out of these and subsequent studies (Raisman 1972; Scalia & Winans 1975) showing that the secondary vomeronasal projections (i.e., the projections from the accessory olfactory bulb to higher centers, and the centrifugal fibers into the accessory bulb) are separate and distinct from those of the main olfactory system. The fact that the VNO is only one synapse removed from the amygdaloid nucleus has been well known for a long time (Herrick 1921). The significance of the studies during the early 1970s rested in the new evidence that the fibers projecting centrally from the accessory bulb were *not* intermixed with those of the main bulb; rather, they were segregated, and they terminated in different locations. Moreover, centrifugal fibers from the cerebral hemispheres that project back into the

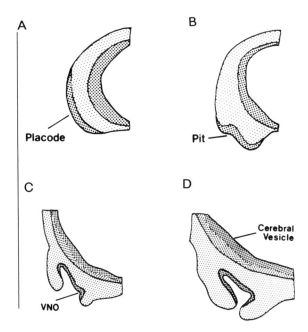

Figure 4.3. Diagrammatic representation showing four stages of development of the nasal cavity and the VNO. The placode and pit stages are depicted in A and B, respectively. In C, a vomeronasal diverticulum buds medially from the main nasal cavity and becomes somewhat more prominent in D. (From Farbman 1991, with permission.)

accessory olfactory bulb arise from the amygdala through the stria terminalis (Raisman 1972), a pathway that does not project to the main olfactory bulb. Thus, the separation of the olfactory sense into a sensory organ in the main nasal cavity and one in the VNO, and the separate central connections of their primary and secondary projections, permits processing of olfactory information by parallel, nonoverlapping circuits. Olfactory information from the main nasal cavity is processed primarily by cortical regions of the brain. On the other hand, input from the VNO is processed primarily by a subcortical circuit in the limbic system. To emphasize the distinction between olfactory and vomeronasal systems, the terms "vomerolfaction" and "vomodor" have been suggested to describe the specific sensation and stimuli, respectively, of the VNO (Cooper & Burghardt 1990), but these have not been generally adopted.

4.4 Cellular structure

In cross section, the VNO lumen usually is shaped like a crescent (Figure 4.2). The sensory epithelium lines the concave side of the lumen, and nonsensory epithelium lines the convex side. In many animals there are capillaries in very close proximity to the epithelium, sometimes within connective tissue ridges between basal epithelial projections (Wang & Halpern 1980a). The connective tissue surrounding the capillaries often is so thin that the vessels sometimes appear to be within the sensory epithelium itself (Breipohl et al. 1981). There are a few glands in the lamina propria of the VNO, mostly in the caudal region (Breipohl, Bhatnagar, & Mendoza 1979; Mendoza 1986). The glands are usually on the nonsensory side of the VNO and open near the junction between sensory and nonsensory epithelia. They are innervated by autonomic nerves (Mendoza & Kühnel 1987). Their secretion has a serous quality, but histochemical staining of the cells suggests that the secretory products differ from those of Bowman's glands in the main olfactory organ (Mendoza 1986).

The histology and fine structure of vomeronasal neuroepithelium have been described in many vertebrates, and they have been found to be remarkably similar (Graziadei & Tucker 1970; Kolnberger & Altner 1971; Kratzing 1971a,b, 1975, 1984; Ciges et al. 1977; Wang & Halpern 1980a,b; Miragall, Breipohl, & Mendoza 1981; Vaccarezza, Sepich, & Tramezzani: 1981; Bhatnagar, Matulionis, & Breipohl 1982; Miragall & Mendoza 1982; Naguro & Breipohl 1982; Taniguchi & Mochizuki 1982, 1983; Adams & Wiekamp 1984; Taniguchi & Mikami 1985). There are three cell types: sensory, supporting, and basal cells. The organization of the cells varies among animals, but two general schemes prevail. In most terrestrial mammals, the organization is similar to that of the olfactory epithelium in the main nasal cavity; that is, it is a pseudostratified epithelium, with columnar supporting cells and sensory cells reaching the lumen and stretching to the basal lamina. Basal cells are located along the basement membrane at the boundary between epithelium and connective tissue. In the rodent VNO, genesis of new neurons occurs primarily in the basal cell population at the junction between sensory and nonsensory epithelia (Barber & Raisman 1978; Wilson & Raisman 1980).

The other kind of organization is exemplified by the VNO of the garter snake (Wang & Halpern 1980a). The general shape of the lumen is similar, but the sensory epithelium has long extensions into the connective tissue, separated by ridges of connective tissue (Figure 4.4). Supporting cell nuclei are superficial, and the elongated sensory cells reach from the surface to the basement membrane in the long epithelial projections. Basal cells are found in the basal region of the projections. Cell division occurs here, giving rise to new sensory neurons (Wang & Halpern 1988).

The sensory cell is a bipolar neuron, with its dendrite reaching the epithelial surface. Although its general features are similar to those of

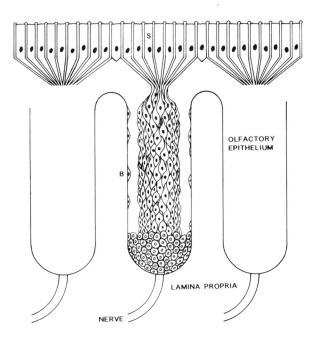

Figure 4.4. Sketch depicting the cellular organization of the VNO in the garter snake. A single row of supporting cells (S) lines the surface. The sensory cells are located within extensions penetrating into the lamina propria, and basal cells (B) are located along the sides and at the deepest parts of these extensions. (Redrawn after Wang & Halpern 1980a.)

sensory neurons in the main olfactory organ, it differs in that its apical surface appendages are mostly microvilli rather than cilia, although an occasional short cilium may be present (Graziadei & Tucker 1970; Kolnberger 1971; Kolnberger & Altner 1971; Loo & Kanagasuntheram 1972; Ciges et al. 1977; Miragall, Breipohl, & Bhatnagar 1979; Naguro & Breipohl 1982; Taniguchi & Mochizuki 1982, 1983; Adams & Wiekamp 1984; Taniguchi & Mikami 1985). Although there are few cilia or none, basal bodies are usually present near the apical surface (Figure 4.5). Basal bodies are, however, absent from sensory neurons in reptile VNO (Altner, Müller, & Brachner 1970). The microvilli, like those in other epithelial cells, have a cytoskeletal core of microfilaments (actin filaments) that pass down into the apical cytoplasm.

On the basal aspect of the cell body the sensory cell cytoplasm narrows into the axon. As in the main olfactory cavity, the individual axons come together to form small bundles in the connective tissue. These coalesce into one or two major bundles to ascend through the connective tissue of the nasal septum and pierce the cribriform plate at the roof of the

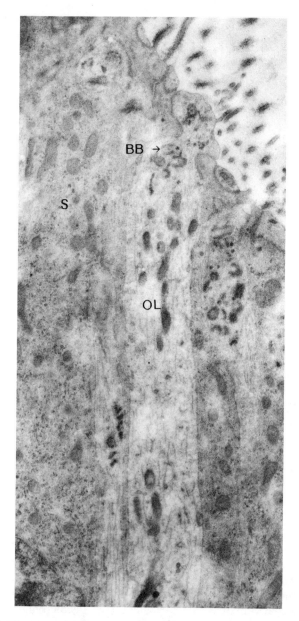

Figure 4.5. Electron micrograph of a section from a rat VNO. The dendrite of the sensory neuron (OL) ends in microvilli. Most sensory cells have basal bodies (arrow), but these rarely give rise to cilia, as in sensory neurons in the main nasal cavity; S, supporting cell.

nasal cavity. As the nerves pass caudally, they are very closely applied to the medial side of the main olfactory bulb. The vomeronasal nerve terminates in the glomerular region of the accessory olfactory bulb, which, in mammals, is on the dorsocaudal aspect of the main bulb (see Figure 5.11).

The receptor-free (convex) zone of the VNO is lined with a low columnar or sometimes pseudostratified epithelium, referred to by some authors as respiratory epithelium (Taniguchi & Mikami 1985), by others simply as nonreceptor epithelium (Breipohl et al. 1979). In mice, the cells are columnar, and their free surface is covered with 1-μm-long branched and unbranched microvilli. In addition, some cells have a few cilia about the same length as the microvilli. The apical membranes between the microvilli are highly folded, suggesting pinocytotic activity (Breipohl et al. 1979). It is believed that mice have only one basic type of cell and that the morphological variations merely represent different functional phases or ages of the cells, which are in a continuous process of turnover (Breipohl et al. 1979). However, the receptor-free zone in the VNO in horses and cattle (Taniguchi & Mikami 1985) and dogs (Adams & Wiekamp 1984) clearly has ciliated and nonciliated cells in a pseudostratified epithelium. In these animals, the receptor-free zone is truly similar to the respiratory epithelium of the main nasal cavity.

The growth and development of the VNO apparently are influenced by sex hormones in some species. In male rats, the total volume and the number of sensory neurons are greater than in females (Segovia & Guillamón 1982). Castration of neonatal male rats results in reduction in VNO size, whereas administration of male hormones to neonatal females leads to a size increase. As will be seen in Chapter 5, this is related to sexual dimorphism in the size of the accessory olfactory bulb.

4.5 Function

4.5.1 Mating behavior in mammals

Even the casual observer of domestic mammals, whether pets or barn-yard animals, is aware that olfaction is somehow related to sexual and/or reproductive behavior. It is now well established that the VNO plays a significant role in chemical communication in many terrestrial verte-brates. In the early 1970s, a series of important correlative morphological and behavioral studies from Winans's laboratory showed beyond doubt that the VNO played a key role in sexual behavior in at least some mammals. These workers were aware of earlier studies showing that bilateral ablation of the olfactory bulb abolishes or impairs copulatory activity in some mammals (e.g., Heimer & Larsson 1967). They rea-soned, however, that because the main olfactory bulb does not directly or indirectly project to the part of the hypothalamus concerned with sexual behavior, the explanation of the ablation studies lay elsewhere.

Ablation of the olfactory bulb results in destruction of several functions other than that associated with the first cranial nerve, which subserves the main olfactory system. Indeed, it results in removal of the vomeronasal system, the nervus terminalis, and the septal olfactory nerve. All of these structures are intimately related anatomically to the olfactory bulb and would be ablated in a routine bulbectomy. Winans and Scalia (1970) selectively lesioned the main olfactory pathway by flooding the nasal cavity with zinc sulfate. Selective lesioning of the vomeronasal pathway was accomplished by cutting the vomeronasal nerves as they pass along the medial aspect of the main olfactory bulbs; little damage to the main bulb was incurred. They discovered that the vomeronasal pathway projected from the accessory olfactory bulb specifically to the mediocortical complex of the amygdala. From earlier studies it was known that this region of the amygdala projects to the medial nuclei in the hypothalamus concerned with neuroendocrine regulation of sexual behavior.

In later studies Winans and her collaborators showed that the vomeronasal system plays a major role in controlling sexual behavior in male hamsters and that this system, acting in conjunction with the primary olfactory system, comprises an essential neural substrate for mating in these animals (Powers & Winans 1973, 1975; Winans & Powers 1977). Sexual behavior could not be entirely eliminated by ablation of only one of the two systems, but was eliminated by ablation of both. They proposed that both systems contribute to an arousal mechanism that when appropriately activated leads to copulatory behavior and that the vomeronasal system provides the greater part of the input to the arousal system.

Some notion of how the two systems may function together can be derived from studies on hamsters. Male hamsters become sexually aroused by the odors of vaginal fluids of females. The degree of arousal depends on the phase of the estrous cycle. A volatile component in these discharges, dimethyl disulfide, is attractive to males (Singer et al. 1976). This chemical signal or pheromone arouses the male hamster's interest in the female, resulting in licking and/or nuzzling behavior.

Another molecule, much larger than dimethyl disulfide, and nonvolatile, has recently been isolated from hamster vaginal secretions. This molecule, aphrodisin, is a protein of molecular mass 17,000 Da (Singer et al. 1986). Aphrodisin is a potent promoter of male copulatory behavior in hamsters (Singer et al. 1984), but fails to promote copulation in males with lesions of the VNO (Clancy et al. 1984). Thus, it seems reasonable to suggest that in hamsters, the complex behavior involving arousal and copulation is dependent on at least the two pheromones: dimethyl disulfide, a volatile component that attracts interest, probably by way of the main olfactory system, and aphrodisin, a nonvolatile component that acts on the VNO and leads to copulation (O'Connell & Meredith 1984). It will be recalled that in the goldfish, two stages, arousal and sperm release, have been observed, and they are also dependent on two distinct pheromones (see section 1.8.3).

Although arousal behavior has not been worked out in this kind of detail for other mammals, it is likely to be at least a two-stage process. In some animals (e.g., cattle and horses), males are attracted to females in urinating posture. Flehmen behavior ensues after the male samples the urine or the vaginal region of the female. Determination of whether or not the female is in estrus is accomplished by analysis of the urine by the vomeronasal system. Whether there is an effort to copulate or a loss of interest depends on the results of this analysis of the urinary components. This interpretation of the behavioral sequence fits the observations of many workers. However, we know much more about the details of the mechanism in hamsters than in other mammals.

It should be emphasized that although the VNO plays an important role in reproductive behavior in some mammals, it is also clear that it is not the only factor operating in this important function. Ablation of the olfactory bulb (with concomitant damage to the vomeronasal nerve) fails to eliminate normal copulatory behavior in many animals, including guinea pigs, rats, sheep, and pigs (Mykytowycz 1970). Obviously, in those mammals that have only a vestigial VNO or none at all (e.g., primates), other systems are involved in sexual arousal.

4.5.1.1 Hormonal consequences of vomeronasal stimulation. The vomeronasal system appears to be involved in the release of luteinizing hormone-releasing hormone (LHRH) in the brain and the subsequent release of luteinizing hormone (LH) (reviewed by Meredith 1991). LHRH release may be important in facilitating mating behavior in many animals, particularly those that may be in a suboptimal state for mating (e.g., males with low testosterone levels). There are many LHRH-containing neurons and fibers associated with the accessory bulb (and with the nervus terminalis and the main bulb) and with forebrain areas that have connections with both olfactory systems. Meredith (1991) suggests that LHRH release may be an important step in the pathway by which vomeronasal sensory stimulation facilitates mating behavior. He suggests further that this facilitation may not act by way of the pituitary gland. The important point to keep in mind when considering this hypothesis is that it applies to *facilitation of mating behavior,* not to other aspects of reproduction.

4.5.2 Pregnancy in mice

A pregnant mouse apparently can distinguish the difference in odor between its mating partner and a strange male. If the female remains with the stud male for 4–6 hr and then is placed in proximity to the strange male, it will first abort, and then mate with the new male (Bruce 1959). Susceptibility to this kind of pregnancy block varies among mouse strains (Hoppe 1975). The active agent in eliciting the pregnancy block is associated with the urine of the male (Parkes & Bruce 1962), and the response is mediated by the vomeronasal system. The proposed mechanism for this is discussed in more detail in section 6.6.3.3.

4.5.3 Mating behavior in reptiles

Bilateral denervation of the VNOs in male garter snakes abolishes sexual behavior (Kubie, Vagvolgyi, & Halpern 1978). This effect is not seen when the main olfactory nerves are cut. Garter snake males respond to a nonvolatile pheromone on the external skin of females (Halpern 1987; Mason et al. 1989). Courtship in the male is characterized by increased tongue-flick and chin-rubbing behavior in which the male moves up and down the female's back, repeatedly rubbing his chin along her dorsal skin (Mason et al. 1989). The sexual attractiveness of the female garter snake is mediated by a pheromone cocktail consisting of a series of nonvolatile saturated and monounsaturated long-chain (C_{28} to C_{37}) methyl ketones, extractable by hexane from homogenates of the dorsal skin of the female. It has been suggested that variation in the chain length and position, and the geometry of the double bonds of the methyl ketones, may impart species-specific information to courting male garter snakes.

Male garter snakes are quickly identified as unsuitable for mating by other males because of the absence of the pheromone or the presence of another specific chemical cue that is not present in females. The presence of squalene was detected in hexane extracts of male dorsal skin. Squalene is thought to be a component of the male sex recognition system (Mason et al. 1989). Unlike the situation in hamsters, no volatile attractant has been identified. Males must make direct contact with females before they display courtship behavior. Tongue-flicking brings the sampled substance into the mouth to the opening of the vomeronasal duct. Removal of the tongue also abolishes sexual behavior.

4.5.4 Feeding behavior in reptiles

It has been shown that water washings of earthworms contain chemoattractants to which garter snakes respond with tongue-flicking and attack. This behavior is mediated via the VNO (Halpern, Schulman, & Kirschenbaum 1986; Wang et al. 1988). Washings from other worms that are not part of the natural diet of garter snakes do not elicit this behavior. Further experiments have shown that a similar, but not identical, chemoattractant can be elicited by electric shock treatment of earthworms (Jiang et al. 1990). The latter chemoattractant has recently been isolated and characterized as a glycoprotein with an apparent molecular mass of about 20 kDa, as estimated from its migration on gel electrophoresis (Jiang et al. 1990). The protein binds specifically to membrane fractions of garter snake vomeronasal sensory epithelium in a saturable and reversible fashion with a relatively high affinity, but does not bind to membrane fractions of nonsensory epithelium of the VNO (Jiang et al. 1990). When the isolated protein is applied to the vomeronasal epithelium, it elicits an increase in firing rate of neurons in the accessory olfactory bulb of the garter snake (Jiang et al. 1990). These results provide clear evidence for a nonvolatile stimulus important to feeding in garter snakes and underline the importance of the vomeronasal system in reptiles living on the ground. These animals have a vomeronasal system that is better developed than their main olfactory system.

5

Olfactory Bulb

In this chapter we review the anatomy and cell biology of the main and accessory bulbs and examine their roles in the processing of olfactory information. More detailed reviews have been published by Macrides and Davis (1983), Mori (1987b), Scott and Harrison (1987), Takagi (1989), Halász (1990), Shipley and Reyes (1991), McLean and Shipley (1992), and Nickell and Shipley (1992).

5.1 Size of the bulb and its significance

The olfactory bulbs are rostral extensions of the cerebral hemispheres. They constitute the initial integrative network for the sensory information coming from the olfactory epithelium, and they send projections for further information processing to higher olfactory centers in the brain. In those animals that have a vomeronasal system, an accessory olfactory bulb is located on the dorsocaudal aspect of the main bulb. It varies in size, but usually is considerably smaller than the main olfactory bulb.

In aquatic vertebrates, fishes, and many amphibians, the olfactory bulb projects to a large part of the cerebrum (see Figure 1.6). In terrestrial animals, other parts of the cerebrum and diencephalon, concerned with vision, audition, and locomotion, have grown disproportionately and have become relatively large compared with the olfactory bulb (Negus 1958).

In mammals, the size of the bulb relative to the rest of the brain is variable. In general, the bulb is largest in monotremes and insectivores, small in most primates, and absent altogether from some marine mammals (e.g., toothed whales and porpoises) (Allison 1953). Among the primates, there is a trend toward reduction in relative size of the bulb from the prosimians to monkeys, apes, and humans (Baron et al. 1983).

There is a significant correlation between bulb size and the relative importance of olfaction in food-finding. For example, fruit-eating bats have a better-developed main olfactory bulb than bats that use echolocation to catch their prey in flight. When comparing terrestrial versus

132

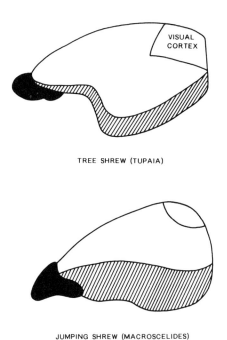

TREE SHREW (TUPAIA)

JUMPING SHREW (MACROSCELIDES)

Figure 5.1. Sketches of brain of a tree shrew and a terrestrial jumping shrew. The tree shrew has a smaller olfactory bulb (in black) and a smaller projection to the cerebrum (hatched). On the other hand, the part of the cerebrum devoted to vision is larger in this animal than in the ground-dwelling shrew. The species that lives on the ground depends more on olfaction for finding food, whereas the arboreal cousin depends more on vision. This is reflected in the relative differences in sizes of the olfactory- and vision-associated regions of the cerebrum. (Adapted from Negus 1958.)

arboreal animals or nocturnal versus diurnal animals within a taxonomic group, olfactory abilities are more likely to be better developed in those animals with the greatest limitations in vision (Harvey & Krebs 1990). For example, among insectivores, ground-dwelling species generally have better-developed olfactory bulbs and olfactory cortex than arboreal species; the jumping shrews (*Macroscelides*), which are terrestrial animals, have well-developed olfactory systems compared with tree shrews (*Tupaia*), which have relatively larger areas of the brain devoted to vision (Figure 5.1). Similarly, nocturnal bird families tend to have a larger olfactory bulb, relative to overall brain and body size, than diurnal families (Healy & Guilford 1990). The increased olfactory capacity of nocturnal birds is thought to be a response to the visual limitations of life under low light. On the other hand, in those animals that feed exclusively on animal prey and are well adapted to catch quickly moving animals, such as flying insects, vision and hearing are much more important than olfaction.

Social behavior mediated by olfaction is also significantly correlated with the relative size of the olfactory bulb and the extent of its connections in the cerebrum. For example, the ability to detect enemies at a distance, the ability to identify related individuals by their scent, and several aspects of sexual and reproductive behavior are important survival-related behaviors in many mammalian species. Thus, although the size of the bulb is largely linked to dietary adaptations, the importance of social behavior as a factor related to bulb size should not be overlooked.

5.2 Neural input to the olfactory bulb

The olfactory bulb is the first synaptic station in the olfactory system. Olfactory sensory cells in each half of the main nasal cavity give rise to bundles of axons that pierce the bony roof of the nasal cavity and terminate in the main olfactory bulb on the same side (ipsilateral). The principal output neurons of the bulb send their axons by way of the lateral olfactory tract to several regions of the ipsilateral olfactory cortex in the forebrain. The main olfactory bulb receives direct input also from several regions of the olfactory cortex and other regions of the central nervous system; most of these inputs are ipsilateral, but some come from the contralateral anterior olfactory nucleus via the anterior commissure (see section 5.6).

The accessory bulb receives input from vomeronasal nerve fibers originating from the sensory cells in the vomeronasal organ. The vomeronasal nerve ascends through the nasal septum, pierces the bony roof of the nasal cavity, and passes in a dorsocaudal direction along the medial surface of the main olfactory bulb. (In some animals it may pass along the lateral surface of the main bulb.) Its target, the accessory olfactory bulb, lies in the dorsocaudal region of the main bulb. The vomeronasal nerve enters the accessory bulb, and its fibers form the outermost layer of this structure. The output neurons from the accessory bulb project to different regions than those from the main bulb. As we have already noted in Chapter 4, the vomeronasal system constitutes a separate, parallel olfactory pathway. There is no overlap in the primary and secondary projection targets of the two olfactory systems.

5.3 Structure of the main olfactory bulb

Our knowledge of the detailed structure of the main olfactory bulb in mammals is derived mostly from laboratory animals, particularly rats, mice, and hamsters. The following description is based on work done on these small mammals, and many believe it can be applied to most mammals.

The main olfactory bulb is an elongated, paired structure lying on each side of the midline at the ventral, anterior end of the forebrain (see Figure 1.6). Anatomically the bulb is one of the most distinctively layered structures in the central nervous system (Figure 5.2), comparable in this respect to the cerebellum. Moreover, the synaptic contacts between nerve cells in the bulb are fairly precisely localized within particular layers. We shall first examine the layers and their cellular contents, from the outside in, as seen in most mammals. Some of the differences between bulbs in mammals compared with other vertebrates will be discussed. Later we shall discuss the organization of the synaptic contacts within the layers.

5.3.1 Layers of the main olfactory bulb

Olfactory nerve. This layer consists of the interweaving bundles of unmyelinated axons originating from the sensory cells in the main nasal cavity. The thickness of the nerve layer varies in different parts of the bulb. The ventral aspect of the bulb, which is closest to the roof of the nasal cavity, is thickest, particularly near the anterior pole of the bulb. During development (section 6.3.1), this layer grows onto the surface of the cerebral outpocketing that gives rise to most of the bulb. The only cellular elements in the olfactory nerve layer are glial cells, that is, astrocytes and ensheathing cells of the nonmyelinated olfactory axons (see section 2.8.4). The ensheathing cells, as their name implies, wrap the bundles of axons and are similar to the cells in the olfactory nerve projection from the sensory epithelium to the brain (Doucette 1990). The astrocytes in this layer of the bulb are found between the bundles of nerve (i.e., they do not wrap the nerve).

Glomerular layer. In the glomerular layer the synapse between axons of sensory epithelial cells and the second-order neuron occurs. The characteristic structures in this layer are the glomeruli (Figure 5.2), ovoid to spherical regions containing a dense network of nerve endings but no nerve or glial cell bodies. Glomeruli are about 150–250 μm in diameter in rabbits (Shepherd 1972), and about 50–120 μm in rats (Pinching & Powell 1971b). The glomeruli are enclosed by very thin glial sheaths that separate them from surrounding tissue. External to the glial sheaths, cell bodies of two types of neurons, periglomerular and external tufted cells, ring the glomeruli. The periglomerular cells, about 5–10 μm in diameter, send a spray of spiny dendrites into the glomeruli (Figure 5.3). Here the periglomerular cell dendrites receive synaptic input from sensory axons and form reciprocal (dendrodendritic) synapses with dendrites of mitral and tufted cells (section 5.7.2). Their

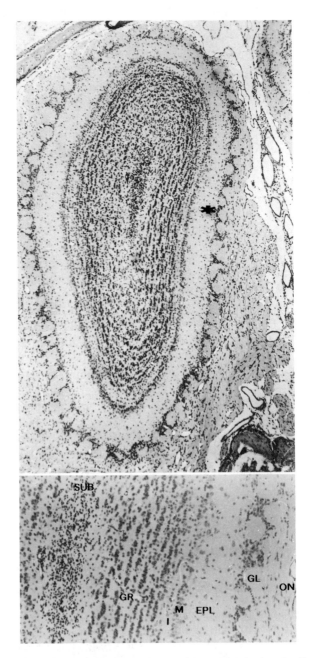

Figure 5.2. Top: Low-magnification photomicrograph of a longitudinal section through the main olfactory bulb of a rat. The well-defined layered structure of the bulb is apparent. The region around the asterisk is shown at higher magnification in the bottom photomicrograph. Bottom: The several layers from the outside in are the nerve layer (ON), the glomerular layer (GL), the external plexiform layer (EPL), the mitral cell layer (M), the internal plexiform layer (I), the granule cell layer (GR), and the subependymal layer (SUB).

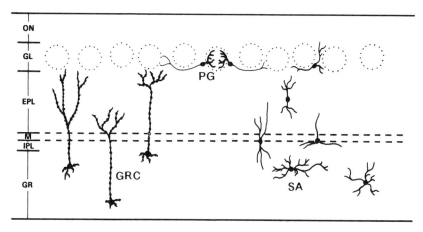

Figure 5.3. Diagram depicting the various locations of interneurons in the olfactory bulb. Granule cells (GRC) and their extensions into the external plexiform layer are depicted on the left side; periglomerular cells (PG), which send a branching dendritic tree into the glomeruli, are seen in the middle; and short axon cells (SA) at several levels of the bulb, from glomerular to granular, are seen on the right side. The levels of the bulb are indicated as follows: olfactory nerve layer (ON), glomerular layer (GL), external plexiform layer (EPL), mitral cell layer (M), internal plexiform layer (IPL), and granule cell layer (GR). The subependymal layer is omitted from this diagram.

axons terminate near neighboring glomeruli, thus providing a means for interglomerular communication (Shepherd 1972; Scott & Harrison 1987).

The external tufted cells, around and between glomeruli, are found more frequently in the deeper regions of the glomerular layer. These cells, 10–15 μm in diameter, send a primary dendrite into the glomerulus, where it branches and receives synaptic input from sensory axons and forms dendrodendritic synapses with periglomerular cells. External tufted cells differ from periglomerular cells in at least two ways: Their dendrites are not spiny, and their axons project to different regions. However, it is usually very difficult to distinguish clearly between these two populations of cells around the glomeruli simply by their morphology.

A third type of neuron in the glomerular layer is the superficial short-axon cell, about 8–12 μm in diameter (Pinching & Powell 1971a). These are found in small numbers between glomeruli. The small dendritic trees of these cells terminate in the periglomerular region, and their short axons terminate near neighboring glomeruli; some send their axons across the external plexiform layer into the granule cell layer (Scott, McDonald, & Pemberton 1987).

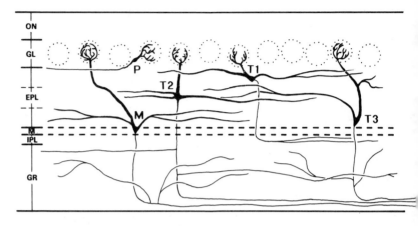

Figure 5.4. Diagram depicting the locations of tufted and mitral cells and their extensions in a mammalian olfactory bulb. Tufted cell bodies are found in the periglomerular region (P) and in the superficial (T1, external tufted cell), middle (T2, middle tufted cell), and deep layers of the EPL. Mitral cells (M) are found in the mitral cell layer. In this diagram, T3 represents an internal tufted cell or a displaced (Type II) mitral cell. The tufted and mitral cells each send an apical dendrite into a glomerulus, where it branches and engages in the synapses described in the text. The EPL can be divided into three layers (dashed lines at left) on the basis of where the lateral dendrites of tufted and mitral cells extend.

External plexiform layer (EPL). This layer is where much of the synaptic integration involved in olfactory processing occurs. The nerve cell bodies within this layer, as defined originally by Ramon y Cajal (1911), are the external, middle, and internal tufted cells, in the corresponding zones of the EPL (Figure 5.4). The internal tufted cells are sometimes referred to as displaced mitral cells (Mori, Kishi, & Ojima 1983), because in virtually every way they resemble the cells in the mitral cell body layer. The broad (3–5-μm diameter) apical (primary) dendrites of tufted and mitral cells pass through the EPL to terminate in the glomeruli. The lateral (secondary) dendrites of mitral and tufted cells are elaborated in this layer, where they enter into reciprocal synaptic relationship with dendrites of granule cells located deeper in the bulb. In addition, centrifugal axons from other parts of the central nervous system are found in the EPL.

    The EPL is often divided into three zones (Figure 5.4) on the basis of the branching patterns of the lateral dendrites of mitral and tufted cells (e.g., Mori et al. 1983; Orona, Rainer, & Scott 1984; Macrides et al. 1985; Scott & Harrison 1987) and cytochrome oxidase staining (Onoda & Imamura 1984; Mouradian & Scott 1988). The most superficial of the three zones, closest to the glomerular layer, contains the lateral (secondary) dendrites of external tufted cells. The intermediate layer contains lateral dendrites of middle and internal tufted (displaced mitral) cells, and of Type II mitral cells (Orona et al. 1984; Scott & Harrison 1987). The deep zone of the EPL contains the lateral dendrites of Type I mitral cells. (The mitral cells have been classified as Types I and II on the basis of where in the EPL their lateral dendrites branch.) Of the three zones, the middle one stains most intensely with the histochemical technique used to demonstrate the presence of cytochrome oxidase (Onoda & Imamura 1984; Mouradian & Scott 1988).

Other cells found in the EPL are short-axon cells (van Gehuchten cells). Dendrites from these cells remain in the EPL, and their axons terminate around mitral and tufted cells.

Mitral cell body layer. In the olfactory bulb in most mammals the mitral cell bodies are characteristically arranged as a single layer, usually interspersed with granule cells. These are the largest of the output neurons of the bulb (Figure 5.5), approximately 20–30 μm in diameter, and the first to form during development. Mitral cell size is not uniform throughout the bulb; those in the medial and lateral regions of the bulb are somewhat larger than those in dorsal and ventral regions (Panhuber et al. 1985). The size differences apparently are not related to the branching pattern of lateral dendrites, which forms the basis of the classification into Type I and Type II mitral cells mentioned earlier.

In other classes of vertebrates, such as the lamprey or teleost fish (see section 5.10 and Figure 5.12), the output neurons are found at all levels of the EPL (Allison 1953). Increasing differentiation into a single mitral cell body layer is seen in the bulbs of reptiles, amphibians, birds, and mammals (Andres 1970). Another important difference between mammals and other vertebrates is that mitral (and tufted) cells in mammals have a single apical dendrite, which projects to a single glomerulus. In other vertebrates these output neurons usually have more than one apical dendrite, and each may go to a different glomerulus (Figure 5.12) (Allison 1953; Andres 1970).

Internal plexiform layer. Deep to the mitral cell body layer is a very narrow layer with relatively few cell bodies, mostly short-axon or granule cells. This layer is similar to the EPL in that some processing may occur. Some of the centrifugal fibers ramify in this layer, as do the dendritic and axonal processes of some deep short-axon cells. This layer is the path traversed by axons of the association tufted cells connecting opposite sides of the bulb, as discussed later.

Granule cell layer. Granule cells are by far the most numerous cells in the bulb. All of their processes are dendritic; that is, they have no axons (anaxonic). In some animals the cell bodies are arranged in several layers or clusters. Granule cell aggregates have been shown to be coupled electrotonically by gap junctions, suggesting the possibility of synchronized neural discharges (Reyher et al. 1991). The granule cells have small cell bodies, 5–10 μm in diameter. Most have darkly staining nuclei, but some, especially in the deeper regions of the granular cell layer, have light-staining nuclei; this is the basis of one classification of granule cells into light and dark subtypes (Struble & Walters 1982). The granule cell layer continues to increase in size throughout life, at least in the rat (Forbes 1984; Kaplan, McNelly, & Hinds 1985), as a result of a net increase in the number of neurons (Kaplan & Hinds 1977; Kaplan et al. 1985).

Other cells found in the granule cell layer are the several categories of short-axon cells, which are about the same size as granule cells or a little larger. Short-axon cells usually have pale nuclei, but on this basis alone they cannot be distinguished unequivocally from granule cells. Usually, a Golgi impregnation showing a neuron with an axon and spineless dendrites is required for positive identification. Short-axon cells in the granule cell layer include the Blanes cells, the Golgi cells, the Cajal cells, and the horizontal cells (Schneider & Macrides 1978). Axons from these cells may terminate in the EPL, in the granule layer, or in some cases in the periglomerular layer (Figure 5.3) (Price & Powell 1970d; Scott et al. 1987). At least some of these axon terminals contain flattened vesicles and end in Gray Type II or symmetrical synapses, suggesting they are inhibitory (Price & Powell 1970d).

Nerve fibers from several sources pass through or terminate in the granule cell layer. These include centrifugal axons from other parts of the nervous system, mitral and tufted cell axons and their collaterals, and the axons of short-axon cells.

Subependymal layer. In the most central core of the bulb, small, densely packed cells, sometimes not fully differentiated even in adults, are often observed (Halász 1990). This region is much more prominent in younger animals, during the time when a portion of the lateral ventricle projects into the bulb. For example, in rats, the size of the subependymal layer reaches its peak at about 12 days after birth (Rosselli-Austin & Altman 1979), and about 8 days after birth in mice (Hinds & Hinds 1976a). This layer is probably the origin of new granule cell neurons formed throughout adult life (Kaplan et al. 1985).

5.4 Foci of activity in the olfactory bulb in response to stimuli (2-deoxyglucose studies)

Earlier, we noted the evidence for a spatial map in the olfactory mucosa that suggested there may be differential levels of sensitivity to stimuli, either "inherent" or "imposed" (section 2.6.1). We return now to the consideration of data from experiments on the bulb that suggest the existence of some degree of spatial patterning of neuronal activity in response to certain odorants. Most of these experiments are based on the use of radioactively labeled 2-deoxyglucose (2DG) as a marker for activity-related glucose uptake in the nervous system. This compound is taken into cells but not metabolized; thus it remains in the cells. When the metabolism of a neuronal cell increases, and it takes up more glucose, it is fooled into taking up 2DG as well. Because the 2DG is radioactive, it can be detected and localized by autoradiography. Rats (Sharp et al. 1975, 1977; Stewart et al. 1979), rabbits (Sharp et al. 1977), and tree shrews (Skeen 1977) have been the animals used in experiments to show that exposure to certain odors after injection with radioactively labeled 2DG leads to focal uptake of the labeled compound in certain glomerular regions of the bulb. The implication is that these foci represent odorant-specific regions of the bulb, because (1) 2DG is taken up in selective glomeruli or groups of neighboring glomeruli in response to different odor stimuli, and (2) the degree of 2DG uptake is directly related to the concentration of the odor stimulus. Supporting evidence for differences in glomerular metabolic activity in response to odorants comes from histochemical studies in which it was shown that cytochrome oxidase activity varied among glomeruli in response to stimulation (Shipley & Costanzo 1984).

The earlier 2DG studies were relatively crude in terms of their resolving power, and there is considerable overlapping of the spatial maps for different odors. However, the results do suggest at least some level of importance of spatial factors in the neural processing of odor quality and odor concentration at the level of the olfactory bulb. Later, higher-resolution studies showed that there was differential 2DG labeling in neighboring glomeruli, and differential labeling among mitral, periglomerular, and granular cells (Lancet et al. 1982; Benson et al. 1985). The differential labeling implies functional differences in activity (see section 5.5.5).

5.4.1 Spatial differences in acetylcholinesterase activity

Further morphological evidence for the existence of spatial differences in sensory information processing is seen in the report of an unusual type of olfactory axon terminal in rat bulbar glomeruli. The axons contain large, dense-cored vesicles and small, clear ones and are distributed more randomly within the glomeruli than is usually the case (Zheng & Jourdan 1988). These axons converge on specific glomeruli located in the medial dorsal region of the bulb (the region known as the modified glomerular complex) and in the ventrolateral region. These atypical glomeruli also receive an uncommonly high convergence of acetylcholinesterase-positive fibers from the central nervous system (Zheng et al. 1987). These data strengthen the suggestions for spatial specificity of information processing in the bulb (see section 6.5.2.1).

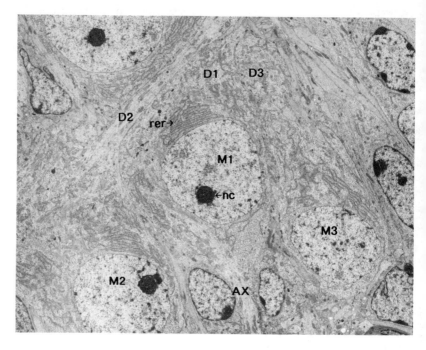

Figure 5.5. Electron micrograph showing parts of several mitral cell bodies (M1, M2, M3). The dendritic parts (D1, D2, D3) of the cells are labeled to correspond to the cell body from which they are derived. An axon (AX) is seen exiting from one of the mitral cells (M1). Mitral cell nuclei usually have at least one prominent dense nucleolus (nc). In the cytoplasm of the cell body are found many stacks of rough-surfaced endoplasmic reticulum (rer) that project into the dendrite but not into the axon (AX).

5.5 Principal neurons of the olfactory bulb: the tufted and mitral cells

5.5.1 Cell body

The cell body (soma) of the mitral cell comprises the nucleus and the surrounding cytoplasm and gives rise to the processes, namely, the dendrites and axon. The nucleus is spherical and fairly large and contains one or more prominent nucleoli (Figure 5.5). The chromatin is usually dispersed and granular. The cytoplasm around the nucleus (perikaryon) is characterized by large numbers of ribosomes, organelles that indicate the high rate of protein synthesis by the principal neurons. As visualized with the electron microscope, the ribosomes are arranged largely on stacked membranes of the endoplasmic reticulum. This so-called rough endoplasmic reticulum (because the membrane sacs are studded with ribosomes, as opposed to smooth endoplasmic reticulum, which is free of ribosomes), together with the free ribosomes, constitutes the Nissl substance characteristic of neurons. Distinct Nissl bodies are more characteristic of large neurons than of smaller ones such as granule cells. Prominent Golgi bodies are characteristically found in the perikaryon of a mitral cell. Other typical cytoplasmic elements include mitochondria, microtubules, microfilaments, and lysosomes.

In mitral cells, as in other large neurons, the boundary between the cell body and dendrite is difficult to define because the structural components of the perikaryon extend into the proximal portions of dendrites (Figure 5.5). This is not the case with axons, which usually arise from the soma at a conspicuous region, called the axon hillock. The most obvious feature of the axon hillock is the relative absence of free ribosomes and rough endoplasmic reticulum and the presence of microtubules and neurofilaments streaming from the perikaryon into the initial segment of the axon (Figure 5.5). The absence of ribosomes in axons indicates that protein synthesis is *not* carried out in this part of the neuron.

5.5.2 Dendrites

The apical dendrite of the mitral cell passes superficially, traverses the EPL, and enters into a glomerulus, where it branches profusely to form a dendritic tree of relatively thin, varicose processes (Figure 5.4). Apical dendrites from tufted cells follow a similar pattern, but the branching may be less profuse. The degree of branching of tufted cell apical dendrites within the glomerulus is related to the location of the cell body: the deeper within the bulb, the more elaborate the dendritic tree. The degree of branching of the lateral dendrites of tufted cells within the EPL follows a similar pattern: the deeper the cell body, the more extensive the dendritic field. Thus, different tufted cell subtypes can be defined by the position of the cell body, the extent of its apical and lateral dendrites, and its axonal projections.

Within the glomerulus the mitral and tufted dendritic terminals receive synaptic input from sensory nerve fibers and from periglomerular cell dendrites. In the rabbit bulb, about 50 million sensory cells converge on only 2,000 glomeruli (Allison & Warwick 1949), so there is enormous convergence of sensory input. Mori (1987b) estimates that about 24 different mitral cell dendrites project to a single glomerulus, and each mitral cell may receive input from 1,300 to 6,500 different sensory axons. In addition, about three times as many tufted cells project to each glomerulus. The convergence of sensory axons on mitral/tufted cells has an amplifying role on the responsiveness of the output cells and probably affects sensitivity (i.e., the ability to detect odors in low concentrations) (Bhatnagar 1977; Meisami 1989). In other words, the greater the number of sensory neurons carrying a signal to a single glomerulus, the greater the response amplitude of those mitral cells receiving input from that glomerulus (Duchamp-Viret, Duchamp, & Vigoroux 1989).

The lateral dendrites of mitral and (output) tufted cells extend tangentially within a more or less narrow sublayer of the EPL, as described in section 5.3.1. The projections pass in virtually all directions within a plane tangential to the mitral cell layer (Figure 5.5) and thus have a disklike projection field. A single mitral cell gives off two to nine lateral dendrites, which branch several times and may extend for long distances (up to 1 mm in rabbits, up to half the circumference of the bulb in turtles) (Mori 1987b). As noted earlier, the lateral dendrites make a number of dendrodendritic reciprocal synapses with granule cells (Price & Powell 1970a).

5.5.3 Axons and their projections

Until fairly recently the principal neurons of the olfactory bulb (i.e., the mitral and tufted cells) were thought to be rather similar to each other. They were believed to have an apical dendrite extending to one glomerulus (in mammals), and an axon projecting outside of the bulb. During the late 1970s and the 1980s it became increasingly clear that although mitral cells probably are similar to one another, tufted cells are rather diverse morphologically, neurochemically, and physiologically. For example, retrograde tracing studies showed that mitral and internal tufted cells have extensive projections to the olfactory cortex and, indeed, are the only bulbar output cells to project to the most caudal parts of the pyriform cortex, the amygdala, and entorhinal cortex; these output neurons also project axon collaterals to more rostral cortical regions (Figure 5.6). However, the central projections of the middle tufted cells are restricted to the more rostral cortical structures, the anterior part of the pyriform cortex, the olfactory tubercle, and the anterior olfactory nucleus (Schoenfeld & Macrides 1984; Macrides et al. 1985; Schoenfeld, Marchand, & Macrides 1985). Many of these anatomical tracing data confirmed the results of physiological studies (Scott, McBride, & Schneider 1980; Scott 1981; Schneider & Scott 1983).

Thus, some subtypes of tufted cells, especially the middle and internal ones, are indeed like mitral cells. In other words, they are true output cells in the sense that they project axons outside of the bulb. However, others, near the glomerular/EPL border, are association cells; that is, their projections are restricted to connections within the olfactory bulb (Macrides & Schneider 1982; Mori et al. 1983; Kishi, Mori, & Ojima 1984; Macrides et al. 1985; Schoenfeld et al. 1985). The most superfi-

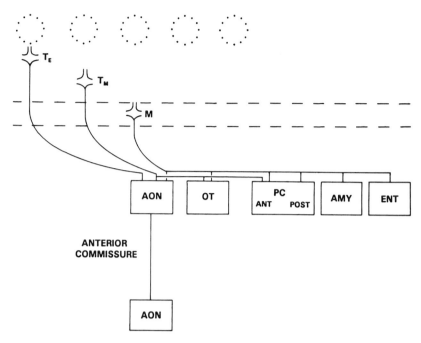

Figure 5.6. Diagram illustrating the projections of output mitral and tufted cells to various parts of the olfactory cortex. Mitral cells and deeply situated internal tufted cells project to the farthest reaches of the olfactory cortex: the posterior pyriform cortex (PC), the amygdaloid body (AMY), and the entorhinal cortex (ENT). Middle tufted cells (T_M) project to the more proximal regions: the anterior olfactory nucleus (AON), the olfactory tubercle (OT), and perhaps the anterior pyriform cortex. In both instances, the most distally projecting axons give off collaterals to the more proximal regions of the cortex. The most superficial tufted output cells (T_E) may project only to the AON. The latter is the only region of the olfactory system to project to the other side of the brain, by way of the anterior commissure. The dashed lines indicate the level of the mitral cell layer.

cially placed external tufted cells (i.e., those in the outer glomerular region) are more like interneurons. They lack lateral dendrites, and their axons project only short distances within the glomerular layer (Figure 5.4). Another group of external tufted cells have secondary dendrites and project axons via the inner plexiform layer to the opposite side of the same bulb; these constitute an intrabulbar associational system (Schoenfeld et al. 1985). Their fibers terminate at approximately the same rostro-caudal level within the bulb as their cells of origin. A fourth group of external tufted cells, responsible for communication with the contralateral bulb, send fibers a short distance outside of the main olfactory bulb to the retrobulbar region, where they synapse with cells in the anterior olfactory nucleus (pars externa). Pars externa neurons, in turn, give rise to axons that pass via the anterior commissure to the contralateral bulb (Figure 5.6). This is the only well-demonstrated pathway that crosses the midline to connect the bulbs with each other. As we shall see later (section 6.6.3.4), it may play a role in access to stored olfactory memories.

Axons of the output cells arise from the deep side of the cell body and run toward the deep region of the granular layer. The axons are myelinated. Within the granular layer they give off intrabulbar axon collaterals that pass laterally but remain within the granular or internal plexiform layers (Figure 5.4). These collaterals may branch several times and extend laterally for relatively long distances, some as far as the opposite side of the bulb (Kishi et al. 1984). Along the course of their projection these beaded axon collaterals make a number of *en passage* synapses with granule cell dendrites in the granule cell and internal plexiform layers, and they synapse on granule cells at their terminations. Output cell axon collaterals may also synapse on short-axon cells in the granular layer (Price & Powell 1970d; Mori 1987b). The mitral and tufted output cells, then, engage in synaptic relationships with granule cells at two levels, in reciprocal dendrodendritic synapses at their lateral dendrites in the EPL, and in excitatory (presumably one-way) synapses by way of their axon collaterals in the granule cell layer. The projection field of axon collaterals has a different spatial organization than the dendritic fields of their lateral dendrites.

It is estimated that each mitral cell has about 3.2×10^4 mitral-to-granule dendrodendritic synapses in the EPL and 3×10^3 output synapses by way of axon collaterals, a 10-fold difference (Mori 1987b). However, the synaptic influence of mitral cell axon collaterals extends more widely through the bulb than that of their lateral dendrites because they typically extend for a longer distance, up to 3,500 μm, compared with about 1,000 μm for dendrites (in rabbits) (Mori 1987b). Moreover, the axon collaterals tend to project in specific directions and sometimes make dense arborizations in particular regions of the granule cell layer. Consequently, their influence is more focally directed than that of the lateral dendrites, which project relatively evenly in all directions. The functional significance of this spatial organization in olfactory signal processing is unclear.

5.5.4 Chemical neuroanatomy

One of the major complications in understanding the function of bulbar output cells is the finding that tufted cells are a highly diverse population of neurons, based on their neurochemistry. There is, to date, no solid consensus on the neurotransmitter of these principal bulbar cells (Hori et al. 1981). Several lines of evidence indicate that the principal output neurons, both tufted and mitral, utilize glutamate and/or aspartate as an excitatory neurotransmitter (Harvey et al. 1975; Matsui & Yamamoto 1975; Bradford & Richards 1976; Yamamoto & Matsui 1976; Collins 1979; Godfrey et al. 1980; Collins & Probett 1981; Macrides et al. 1985; Fuller & Price 1988). The finding that N-methyl-D-aspartate (NMDA) receptors exist in the external plexiform layer and in many regions of olfactory cortex strengthens the argument that glutamate may be the output cell neurotransmitter (Monaghan & Cotman 1985). NMDA is an acidic amino acid that depolarizes neurons by selectively interacting with a distinct class of excitatory amino acid receptor, presumably that for glutamate.

However, the dipeptide N-acetyl-aspartyl-glutamate has also been proposed as a transmitter on the basis of findings that the excitatory response in the pyriform cortex to applications of this dipeptide is blocked by 2-amino–4-phosphonobutyric acid. Further, the concentrations of this dipeptide are significantly decreased in the pyriform cortex following ablation of the olfactory bulb (ffrench-Mullen et al. 1985). Immunoreactivity to N-acetyl-aspartyl-glutamate has also been localized in mitral cells of the olfactory bulb (Blakely et al. 1987).

More recently, the presence of another transmitter candidate, the peptide corticotropin-releasing factor (CRF), has been observed in several neuronal types of the olfactory bulb (Imaki et al. 1989; Bassett, Shipley, & Foote 1992). CRF has a widespread distribution, as it is expressed in a majority of mitral and tufted cells in the main and accessory bulbs, but it is also seen in subsets of granule and periglomerular cells. There is a dense network of fine CRF-positive fibers in the EPL of the bulb, where the lateral dendrites of the mitral and tufted cells are located, and in that part of the pyriform cortex (layer Ia) where axons of the bulbar output cells terminate. The broad distribution in most output cells suggests that CRF may serve an important function, perhaps as a modulator or co-transmitter (Imaki et al. 1989).

Many external tufted cells (i.e., those in the glomerular layer and EPL) are dopaminergic (Halász et al. 1977, 1981a; Halász, Ljungdahl, & Hökfelt 1978; Davis & Macrides 1983; Halász, Nowycky, & Shepherd 1983; Kream et al. 1984). Significant numbers of external tufted cells within the glomerular layer produce substance P (Burd, Davis, & Macrides 1982a; Davis, Burd, & Macrides 1982; Kream et al. 1984; Matsutani, Senba, & Tohyama 1988), and some others probably produce vasoactive intestinal peptide (Gall, Seroogy, & Brecha 1986; Sanides-Kohlrausch & Wahle 1990b). It has been suggested that some tufted cells may synthesize both dopamine and substance P, because of their similarity in morphology and distribution (Burd et al. 1982a; Davis et al. 1982), but this has been shown not to be the case, at least in the

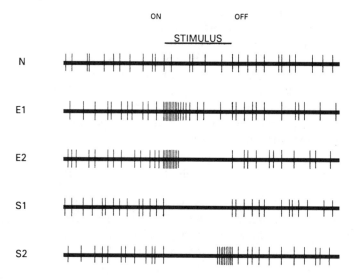

Figure 5.7. Schematic diagram to illustrate the variability of mitral/tufted cell responses to stimuli. The beginning and end of a stimulus are illustrated by a line segment. The uppermost tracing (N) illustrates only spontaneous neural activity, and no response to a stimulus. Two kinds of excitatory responses are indicated: One (E1) shows a rapid burst of activity, followed by a declining rate of firing in response to a stimulus; the second (E2) shows a period of suppression after an initial burst of excitation. Two kinds of suppressed responses are shown: One (S1) shows complete suppression throughout the stimulus duration; the other (S2) shows suppression, followed by an excitatory burst near the end of the stimulus. (Adapted from Kauer 1987.)

hamster (Baker 1986b). Some external tufted cells, not restricted to any specific level, contain cholecystokinin-like activity (Seroogy, Brecha, & Gall 1985).

For a detailed review of bulbar neurochemistry, see Halász (1990).

5.5.5 Physiology

In mammals, each mitral cell is related to a single glomerulus, unlike the situation in other vertebrates (compare Figure 5.4 with Figure 5.12). A glomerulus may be the target for several mitral cells, but the suggestion from 2DG studies (section 5.4) that individual glomeruli may be odor-specific has led to the notion that mitral and tufted cells connected to the same glomerulus may have similar response properties, at least in mammals. There is a degree of correlation between mitral cells and receptive field in the hamster olfactory epithelium. Costanzo and O'Connell (1980) showed, with intracellular recordings, that mitral cell inputs came from a fairly restricted region of the olfactory epithelium.

Recordings from single mitral cells have indicated that these cells exhibit various categories of responses to stimuli. Briefly, these can be classified into three basic groups: excitatory, inhibitory, and no response (Figure 5.7) (Kauer 1974; Kauer & Shepherd 1977; Meredith & Moulton 1978; Mair 1982; Meredith 1986). The response patterns are rather complex, because some cells respond with an excitatory burst at stimulus onset; then, if the stimulus persists, there may be a suppressive phase. Other cells exhibit other types of complex behavior. For example, at low concentrations of an odorant stimulus some mitral cells might exhibit an excitatory response, whereas the same cells might exhibit an inhibitory response at higher stimulus concentrations (Døving 1964; Kauer 1974; Mair 1982; Meredith 1986). The change from excitation to suppression has been suggested to reflect enhanced lateral inhibition induced by high stimulus intensities (Kauer & Shepherd 1977; Meredith 1986).

We have already noted the evidence from physiological (Adrian 1950; Leveteau & MacLeod 1966) and 2DG studies (Sharp et al. 1975, 1977; Jourdan et al. 1980; Greer et al. 1981; Lancet et al. 1982), as well as histochemical experiments (Shipley & Costanzo 1984), that glomeruli respond as functional units; that is, certain glomeruli respond preferentially to some stimuli and not to others. If there is a different spatial pattern of response to each stimulus, then one might expect the responses of the mitral and tufted cells associated with a given glomerulus to be similar when that glomerulus is excited by a stimulus. This hypothesis has been tested, and the results suggest that mitral cells associated with a single glomerulus indeed do have similar response properties (Meredith 1986; Wilson & Leon 1987; Buonviso & Chaput 1990).

Buonviso and Chaput (1990) made simultaneous recordings of the responses of pairs of mitral cells located close to one another (40 μm apart or less), as compared with pairs located at a greater distance (150–200 μm apart). The close cell pairs were thought more likely to be associated with a single glomerulus. Responses were classified simply as excitations, suppressions, or no response. The results revealed that the pairs of close cells had similar response patterns to odorants, whereas the pairs of distant cells showed opposite responsivities. The experimental data suggest that during odor stimulations, the mitral cell layer responds as foci of excited cells surrounded by zones of suppressed cells.

This type of central-excitation, surround-suppression behavior among mitral cells had been earlier demonstrated by others (Kauer & Shepherd 1977; Meredith 1986; Wilson & Leon 1987). The results of these studies, and of the high-resolution 2DG studies, suggest that lateral inhibition does occur in the bulb. Lateral inhibitory circuits are seen in several sensory systems. Depending on the sensory system, the inhibitory circuits are related to enhancement of the signal-to-noise ratio, gain control, contrast enhancement, and other factors.

Wilson and Leon (1987) suggested that lateral inhibition could take place at the glomerular level, by way of periglomerular cells inhibiting adjacent or nearby glomeruli, or at the level of the reciprocal synapses between granule and mitral cells in the EPL, or both. Thus, for example, in response to a given stimulus at high enough concentrations, activated glomeruli could suppress neighboring glomeruli, and/or the activated mitral cells could suppress their neighbors. This lateral inhibition could enhance the ability to identify the odor stimulus.

It seems clear that lateral inhibition does occur in the olfactory bulb, as well as self-inhibition. Activation of the mitral cell permits self-inhibition at the mitral/granule cell reciprocal synapse. The granule cell acts as a "feedforward" inhibitor of the mitral cell, and the mitral cell acts as a "feedback" inhibitor of itself (Nickell & Shipley 1992).

5.6 Interneurons of the olfactory bulb

5.6.1 Periglomerular cells

These small neurons distributed around the periphery of glomeruli are thought to be modulators of the mitral and tufted cells via the reciprocal dendrodendritic synapses. Many periglomerular cells contain glutamic acid decarboxylase (GAD) (Ribak et al. 1977) and are thought to use γ-aminobutyric acid (GABA) as a neurotransmitter (Ribak et al. 1977; Ribak, Vaughn, & Barber 1981; Halász, Ljungdahl, & Hökfelt 1979; Halász et al. 1981b; Mugnaini, Oertel, & Wouterlood 1984a; Mugnaini et al. 1984b; Gall et al. 1987). Some periglomerular cells are catecholaminergic and probably use dopamine as a neurotransmitter (Hökfelt et al. 1975; Halász et al. 1981a; Mugnaini et al. 1984a). A few periglomerular cells are thought to be opioidergic, on the basis of observations that they contain methionine-enkephalin immunoreactivity (Davis et al. 1982), and others contain another peptide, cholecystokinin (Seroogy et al. 1985; Matsutani et al. 1988).

In immunohistochemical studies to identify dopaminergic cells in the bulb, the antibody to tyrosine hydroxylase is usually used, because this enzyme is the rate-limiting enzyme in dopamine synthesis. Thus, a cell containing tyrosine hydroxylase immunoreactivity is thought to be a dopaminergic cell. Interestingly, virtually all of the tyrosine hydroxylase-positive periglomerular neurons in the rat olfactory bulb also contain GABA. However, it was found that about 27% of the periglomerular cells containing GABA did not contain tyrosine hydroxylase immunoreactivity (Gall et al. 1987). Hence, the two neurotransmitters, GABA and dopamine, may coexist in large numbers of periglomerular neurons (Kosaka et al. 1985). These findings underline the fact that our understanding of synaptic organization in the glomerulus is far from complete. The functional importance of the dopaminergic and GABAergic cells in receiving synaptic input from sensory neurons and providing synaptic output to mitral cells in the glomerulus is not at all clear (McLean & Shipley 1992). The presence of dopamine D2 receptors on axons of olfactory sensory neurons (Nickell et al. 1991) raises the possibility that sensory neurons may be responsive to dopamine. Moreover, there are species differences in immunoreactivity to various tested substances, such as tyrosine hydroxylase (Baker 1986a), thus complicating further our ability to make generalizations about synaptic organization in the glomerulus.

5.6.2 Granule cells

As already indicated, these small interneurons that constitute the largest number of bulbar neurons have been classified both by the staining density of their nuclei (Struble & Walters 1982) and by the branching pattern of their dendritic trees within the EPL (Mori et al. 1983; Orona, Scott, & Rainer 1983; Greer 1987). Most of the granule cells are GABAergic (Ribak et al. 1977, 1981; Halász et al. 1981b; Mugnaini et al. 1984a,b), and some contain methionine-enkephalin (Davis et al. 1982) or CRF (Imaki et al. 1989). So far, correlations between the morphological and neurochemical data have not been ascertained.

The distal dendritic processes of granule cells are characterized by spines (gemmules) that are engaged in reciprocal dendrodendritic synaptic relationships with mitral/tufted cell dendrites in the EPL, where they terminate (Price & Powell 1970b). The mitral and tufted cells, then, have two spatially segregated sets of reciprocal dendrodendritic synapses, one with spiny periglomerular cell dendrites in the glomeruli, the other with granule cell dendritic spines in the EPL.

The morphology of the distal dendritic field of granule cells in rabbit olfactory bulb has formed the basis of classification into three subtypes (Mori et al. 1983):

1. Type I extends its dendritic tree throughout the EPL and has spines throughout, though the branching pattern is such that more of the spines are in the deeper region of the EPL. Type I cell bodies are located at all levels in the granule cell layer (Figure 5.3).
2. Distal processes of Type II granule cells are limited to the inner half of the EPL, and their cell bodies are located throughout the granule cell layer.

3. Type III cells have distal processes that branch mostly in the superficial part of the EPL, and their cell bodies are located more superficially in the granule cell layer, in the inner plexiform and mitral cell layers.

Greer (1987) has seen a similar subdivision of granule cells in mice. However, Orona et al. (1983), in their study on rat olfactory bulb, showed that, in general, the branching patterns of granule cells with cell bodies deep in the granule cell layer tended to be deep in the EPL, whereas those with cell bodies located more superficially had most of their dendritic tree higher in the EPL. Thus, lamination of the granule cell dendritic tree also occurs in the rat, although there is good evidence for only two layers.

The significance of these detailed studies on the morphology of granule, mitral, and tufted cells is that they support the idea that parallel processing of olfactory information may occur in the main olfactory bulb (Macrides et al. 1985). Because of the distribution of their lateral dendrites in distinct zones of the EPL, the various subtypes of tufted and mitral cells come into synaptic relationship with specific subtypes of granule cells. External tufted cells, with lateral dendrites in the superficial zone of the EPL, engage in synapses with the Type I or III granule cells, not with Type II. The axonal projections of these tufted cells are mostly intrabulbar or to the parts of the olfactory cortex closest to the bulb. The internal tufted cells and mitral cells with lateral dendrites in middle and deep zones of the EPL are involved in synapses with Type II and Type I granule cells; their axonal projections are more widespread, to both rostral and caudal regions of the cortex. Similarly, at least in the hamster, the cortical input to the bulb is largely distributed in a laminar fashion, with that from more caudal parts terminating mostly in the deeper regions of the granule cell layer, and that from more rostral parts terminating mostly in more superficial regions (Davis & Macrides 1981). However, this kind of pattern was not seen in corticobulbar projections in the rat (Luskin & Price 1983a) or rabbit (Mori 1987b).

5.6.3 Short axon cells

This cell type has been identified in the glomerular layer (superficial short axon cells), the EPL, the internal plexiform layer, and the granular layer (deep short axon cells) (Figure 5.3). Although short axon cells are widely distributed, they are not found in large numbers anywhere in the bulb. The superficial short axon cells are thought to synapse on periglomerular cells, whereas the deep ones are thought to project to granule cells. These cells are thought to function as modulatory cells. For example, whereas GABAergic periglomerular or granule cells may act to inhibit mitral cells, activated GABAergic short axon cells may cause disinhibition (Halász 1990).

In the hamster, immunoreactivity to somatostatin (Davis et al. 1982) and enkephalin (Macrides & Davis 1983) is found in deep short axon cells, whereas in the rat these peptides are found in both deep and superficial groups (Bogan et al. 1982; Scott et al. 1987). Both superficial and deep short axon cells in rat bulb were immunoreactive when treated with antiserum to neuropeptide Y (Gall et al. 1986; Scott et al. 1987; Matsutani et al. 1988), although Gall et al. (1986) noted that neuropeptide Y-immunoreactive cells diminished in relative number from deep to superficial. Later studies reported the presence of neuropeptide Y-immunoreactive neuron cell bodies in the olfactory peduncle (the stalk connecting the bulb to the rest of the telencephalon; the anterior olfactory nucleus is in the peduncle) and many positively staining fibers in the deeper regions of the granule cell layer (Sanides-Kohlrausch & Wahle 1990a). In other words, not all of the neuropeptide Y-immunoreactive cells are thought to be short axon cells. For example, at early stages of postnatal development in the cat olfactory system, some neuropeptide Y-containing cells project axons contralaterally via the anterior commissure. These could originate from the anterior olfactory nucleus.

Halász (1990) has recently reviewed the literature on neuroactive substances reported to be in short axon cells. In addition to the peptides mentioned earlier, dopamine, GABA, vasoactive intestinal polypeptide (VIP), and cholecystokinin (CCK) are reported to be found in this rather heterogeneous cell type. Halász noted that because of their small numbers and dispersed locations within the bulb, it is difficult to recognize histochemical or immunocytochemical markers within them. In some cases, labeled short axon cells are identified by exclusion (i.e., if the labeled cell does not resemble any of the more common cell types, it is identified as a short axon cell).

5.7 Synaptology of the olfactory bulb

5.7.1 Sensory cell/principal cell synapses

Within the glomeruli, axons from the sensory cells form excitatory synapses on the dendrites of mitral and tufted cells and on dendrites of periglomerular cells (Figure 5.8) (Pinching & Powell 1971b). The sensory input to the bulb from the olfactory epithelium is known to have a spatial component (Land 1973; Land & Shepherd 1974; Costanzo & O'Connell 1978; Dubois-Dauphin, Tribollet, & Dreifuss 1981; Greer et al. 1981; Jastreboff et al. 1984; S. Fujita et al. 1985; Mori et al. 1985;

Astic & Saucier 1986, 1988; Saucier & Astic 1986). In general, sensory cells in the dorsal olfactory epithelium project to the dorsal olfactory bulb, those in the medial epithelium to the medial bulb, and so forth. Moreover, the projections are, as noted earlier, highly convergent (Allison & Warwick 1949; Mori 1987b; Meisami 1989). However, the topological projections are imprecise, compared, for example, with retinotectal or retinogeniculate projections in the visual system. Although many, or even most, olfactory sensory cells in one region of the nasal cavity may project to a specific region of the bulb, neighboring sensory cells may also project to different parts of the bulb (Astic & Saucier 1986; Saucier & Astic 1986; Kauer 1987). Thus, the projections from the olfactory epithelium in the nose are both convergent and divergent (Kauer 1987, 1991). Although a given glomerulus receives converging input from several sensory neurons that may be distributed quite far apart from one another in the olfactory epithelium, at the same time neighboring sensory cells may send divergent projections to widely separated glomeruli.

Identification of the neurotransmitter involved in this excitatory synapse has been a difficult problem. The most promising candidate is carnosine, a dipeptide (β-alanylhistidine) that has many characteristics of a neurotransmitter (section 2.2.6.3). It is found in high concentrations in olfactory epithelium and olfactory bulb (Margolis 1980a,b); it is synthesized in the epithelium and transported to the bulb (Margolis & Grillo 1977) and therefore is selectively present in the sensory neurons and their terminals (Burd et al. 1982b; Sakai et al. 1987). Ligand binding studies showed that binding of L-[^{3}H]carnosine was saturable, reversible, stereoselective, highly localized to the olfactory bulb, and inhibited by carnosine analogues (Hirsch, Grillo, & Margolis 1978). The number of stereoselective carnosine binding sites in olfactory bulbs declines significantly after ablation of the sensory epithelium (by zinc sulfate lavage of the nasal cavity) (Hirsch & Margolis 1979). Moreover, a specific binding site for carnosine was localized to the nerve and glomerular layer and was absent from other regions of the bulb (Nadi, Hirsch, & Margolis 1980). Thus, the characteristics of the carnosine binding site in the olfactory bulb fulfill most of the criteria considered relevant for a functional receptor. Moreover, carnosine in synaptosomes can be released by a calcium-dependent, potassium-induced depolarization (Rochel & Margolis 1982). Based on these data, the dipeptide carnosine was considered a strong candidate for the sensory neuron neurotransmitter (Margolis 1980a,b).

However, all of the foregoing data only indirectly support the suggestion that carnosine is a neurotransmitter in sensory neurons. Direct attempts to determine whether or not carnosine has neurophysiological effects have not yielded clear evidence in favor of its role as a neuro-

transmitter (Margolis 1988). If carnosine were the excitatory neurotransmitter of the sensory neurons, it would be expected to mimic the
excitatory effect of the sensory neuron on its postsynaptic target, the
mitral cell, when applied directly. However, iontophoretically applied
carnosine under some conditions blocks the mitral cell's excitatory response to olfactory stimuli (MacLeod & Straughan 1979; Tonosaki &
Shibuya 1979). On the other hand, it was noted that carnosine was a
particularly good inhibitor of periglomerular cells (MacLeod &
Straughan 1979). This suggests the possibility that carnosine exerts a
neuromodulatory effect on mitral cells by inhibiting periglomerular cells.
However, direct application of carnosine to the glomerular layer of the
reptilian olfactory bulb resulted in inhibition of the response of single
mitral cells, whereas application in the EPL resulted in enhancement
of the response (Shibuya, Aihara, & Tonosaki 1977). In turtle and
bullfrog olfactory bulbs, application of carnosine failed to elicit any
response in mitral cells (Nicoll, Alger, & Jahr 1980b). In rat olfactory
bulb, single mitral cells either were inhibited or were unresponsive to
application of carnosine (MacLeod 1978; MacLeod & Straughan 1979).
Studies of the effect of carnosine on bulbar neurons in dissociated cell
culture also showed no effect on input resistance, resting potential, or
synaptic activity of the neurons (Frosch & Dichter 1984), thus not supporting a role for carnosine as the excitatory neurotransmitter. However,
others showed that carnosine produced an increase in frequency of average evoked potentials resulting from stimulation of the lateral olfactory tract (González-Estrada & Freeman 1980).

  Taken together, the physiological results concerning the role of carnosine as an excitatory transmitter are inconclusive. The biochemical
evidence, however, suggests that it plays some role in the olfactory
system, possibly as a modulator or perhaps as a trophic molecule.

5.7.2 *Dendrodendritic and reciprocal synapses*

These types of synapses are found within glomeruli, between dendrites
of periglomerular and mitral/tufted cells, and within the EPL between
dendrites of granule and mitral/tufted cells (Rall et al. 1966; Shepherd
1972; Jackowski, Parnavelas, & Lieberman 1978). In both cases the
picture observed with the electron microscope reveals an asymmetrical
(Gray Type I) synapse, in which the mitral cell dendrite is presynaptic
and is found adjacent to a symmetrical (Gray Type II) synapse, in which
the granule cell is presynaptic (Figure 5.8). The mitral-to-granule synapse is excitatory, but, as indicated earlier (section 5.5.4), there is no
consensus on whether the transmitter is glutamate, aspartate, N-acetyl-
aspartyl-glutamate, CRF, or something else. The granule-to-mitral synapse is inhibitory and probably is mediated by GABA. However, be-

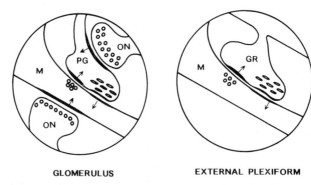

GLOMERULUS   EXTERNAL PLEXIFORM

Figure 5.8. Schematic diagrams of synapses within the glomeruli (left) and EPL (right). The direction of a synaptic impulse is indicated by an arrow. In glomeruli, olfactory nerve axons (ON) synapse on a dendrite of a mitral cell (M) and the spine of a periglomerular cell (PG). The spherical vesicles and the asymmetrical membranes in the synaptic region indicate that these are excitatory synapses. The mitral cell and the periglomerular cell are engaged in a reciprocal synapse, in which the mitral cell excites the periglomerular cell, and the latter inhibits the mitral cell (the flattened vesicles at the synapse and the symmetrical membranes indicate the presence of an inhibitory synapse). In the EPL, the reciprocal synapse between mitral and granule cells (GR) is shown. In terms of excitation/inhibition, these are similar to the mitral–periglomerular reciprocal synapses.

cause large numbers of periglomerular GABAergic cells are also dopaminergic, it is conceivable that dopamine plays a role in modulating the synaptic activity in the glomerulus (McLean & Shipley 1992). When a mitral cell is excited orthodromically or antidromically, it stimulates granule cells (and periglomerular cells) by release of its neurotransmitter. Granule cells, in turn, are stimulated to release GABA, which opens chloride channels in the mitral cell membrane and tends to hyperpolarize the mitral cell (Jahr & Nicoll 1980, 1982b). Thus, activation of the dendrodendritic reciprocal synaptic pathway results in an inhibitory postsynaptic potential (IPSP) produced in the mitral cells (Shepherd 1972). This tends to inhibit the generation of more action potentials in the mitral cell. How dopamine could modulate this response in the glomerulus is not known. As mentioned earlier (section 5.6.1), the presence of D2 receptors in sensory nerve axons (Nickell et al. 1991) has raised the possibility that transmitter release from the sensory nerve terminals could be modulated by dopamine release within the glomerulus.

Several experiments have shown that the IPSPs elicited by granule cells in mitral cells can be modulated. There is a considerable noradrenergic input to the olfactory bulb from the locus coeruleus in the brain stem (Shipley et al. 1985) that is thought to be involved in arousal. In turtle olfactory bulb slices, bathing in norepinephrine reduced the IPSP produced by the reciprocal synaptic pathway (Jahr & Nicoll 1982a), suggesting the norepinephrine acts directly on granule cells to reduce the amount of GABA released. On the other hand, iontophoretic application of norepinephrine onto mitral cells in an in vivo preparation of rat olfactory bulb resulted in the opposite effect, a suppression of mitral cell firing, implying an activation of GABA release by granule cells (McLennan 1971). McLean et al. (1989b) have discussed these contradictory findings in detail and concluded that the effect of norepinephrine on GABA release remains to be resolved by further physiological studies.

Another transmitter candidate for modulating the effect of GABA on mitral cells is enkephalin. This peptide, found in some periglomerular and granule cells (Davis et al. 1982), also decreases mitral cell IPSPs recorded in the slice preparation of turtle olfactory bulb (Nicoll et al. 1980a).

If we think of the bulbar granule cell as the primary modulator of mitral/tufted cell output to higher centers in the olfactory cortex, it is interesting to note that in mammals, at least, mitral cells are in place at birth, but most granule cells are produced after birth (section 6.3.2). Because of the absence of granule cells, the impulses from mitral cells are projected to higher centers with little modulation by interneurons (Mair & Gesteland 1982), unlike the mitral cell responses in adults (Mair 1982). Put simply, the input from sensory cells to mitral/tufted cells is minimally modulated because there are few modulating cells. One of the most important behavioral phenomena of newborn mammals, especially altricial mammals (those born in a "helpless" or less developed state, e.g., dogs, cats, rats, rabbits, mice), is that they use olfaction as their major sensory cue to find the mother's nipple. One wonders if there is a relationship between attenuation of olfactory responses associated with nipple-search behavior and the appearance of granule cells during the first few weeks of postnatal life.

5.7.3 Synaptic plasticity in mutant mice

A genetic mutant mouse, the Purkinje cell degeneration (PCD) mutant, has been described in which Purkinje cell loss occurs in the cerebellum about two weeks after birth. Beginning about 70 days after birth, a period of accelerated mitral cell loss has been observed, with nearly complete disappearance of mitral cells by 90–120 days of age (Greer & Shepherd 1982). Other olfactory bulb populations do not degenerate. This specific degenerative process is accompanied by a period of synaptic reorganization, during which granule cells, denervated because of mitral cell loss, establish new synaptic connections with tufted cells (Greer

1987; Greer & Halász 1987). Within the glomerulus, odor-induced patterns of 2DG uptake are not changed by mitral cell loss; this suggests that there is a reorganization of circuits within the glomerulus as well (Greer & Shepherd 1982).

5.8 Bulbar connections with the olfactory cortex

5.8.1 Efferent projections to the cortex

Projections from the main olfactory bulb pass to several regions of the olfactory cortex, which consists of the following: (1) the anterior olfactory nucleus; (2) the olfactory tubercle, which has been implicated in various functions of the limbic system (the tubercle also receives dopaminergic fibers from the midbrain); (3) the pyriform cortex, which is the main area for olfactory discrimination; (4) the cortical parts of the amygdala (the anterior and posterolateral divisions of the cortical nucleus and the nucleus of the lateral olfactory tract); (5) the entorhinal cortex, which receives olfactory input and projects to the hippocampus and may be related to memory (Macrides & Davis 1983). All of these projections from bulb to cortex are ipsilateral (i.e., to the same side of the brain). Most cortical areas have reciprocal projections to the ipsilateral bulb. One region of the anterior olfactory nucleus, the pars externa, projects fibers to the contralateral bulb (the bulb on the opposite side of the brain) by way of the anterior commissure.

The olfactory cortex is, in evolutionary terms, an older part of the forebrain. This is reflected in its relatively simple structure, compared with the neocortex (e.g., the frontal, parietal, and occipital regions of cerebral cortex). The olfactory cortex is histologically divided into three major layers, two of which are cellular, and a third that is relatively free of cells. Neocortex, on the other hand, has five cellular layers and a sixth peripheral layer that contains relatively few cells.

5.8.2 Afferent projections from the cortex

The centrifugal afferents to the main olfactory bulb originate from many parts of the central nervous system. Neurons projecting to the bulb have been found in all divisions of the ipsilateral anterior olfactory nucleus and the pars externa of the contralateral anterior olfactory nucleus. Tracing studies have revealed positive neurons in the taenia tecta (anterior hippocampal rudiment), the pyriform cortex, the lateral entorhinal cortex, the medial septum, the horizontal and vertical limbs of the diagonal band, the nucleus of the lateral olfactory tract, the posterolateral and medial cortical amygdaloid areas, the zona incerta, the locus coeruleus, and the dorsal and median raphé nuclei (Broadwell 1975b;

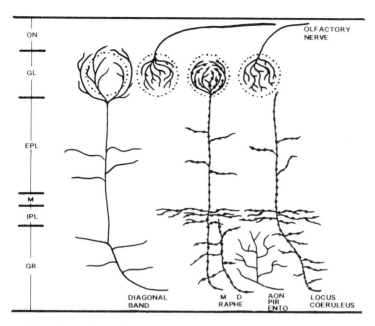

Figure 5.9. Central input into main olfactory bulb. The origins and the terminations of fibers originating in higher centers of the nervous system are shown diagrammatically: AON, anterior olfactory nucleus; M raphé, medial raphé; D raphé, dorsal raphé; PIR, pyriform cortex; ENTO; entorhinal cortex. (Redrawn after McLean & Shipley, 1992.)

Broadwell & Jacobowitz 1976; Davis & Macrides 1981; Luskin & Price 1983a; Macrides & Davis 1983; Schoenfeld & Macrides 1984; Shipley & Adamek 1984; Záborsky et al. 1986; McLean & Shipley 1987a, 1988, 1992; Mori 1987b; Senut, Menetrey, & Lamour 1989).

A summary of the inputs into the bulb from the central nervous system and the levels at which they terminate is provided in Figure 5.9 (Price & Powell 1970c; McLean & Shipley 1992). Cortical input includes cholinergic fibers from the horizontal limb of the diagonal band and medial septum in the basal forebrain that terminate in the glomerular layer, outside of the glomeruli, in the EPL, and in the granular layer (Carson & Burd 1980; Carson, 1984; Záborsky et al. 1986; Senut et al. 1989; McLean & Shipley 1992) on two types of identified cholinoceptive cells (cells responsive to acetylcholine) in these areas (Nickell & Shipley 1988). Cells from the same region of the basal forebrain also project GABAergic fibers to the olfactory bulb (Záborsky et al. 1986). It is difficult to determine where these terminate, because of the massive intrinsic GABAergic innervation of the bulb (McLean & Shipley 1992).

The medial septum/nucleus of the diagonal band region in the basal

PERIPHERY   OLFACTORY BULB   CNS

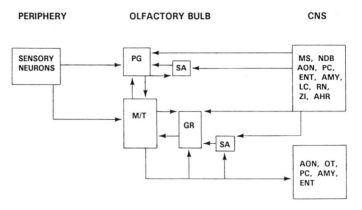

Figure 5.10. Box diagram of synaptic connections in the olfactory bulb: PG, periglomerular cells; SA, short axon cells; MS, medial septum; NDB, nucleus of the diagonal band; AON, anterior olfactory nucleus; PC, pyriform cortex; ENT, entorhinal cortex; AMY, amygdaloid cortex; LC, locus coeruleus; RN, raphé nucleus; ZI, zona incerta; AHR, anterior hippocampal rudiment; M/T, mitral/tufted cells; GR, granule cells; OT, olfactory tubercle. (Redrawn after Halász 1990.)

forebrain projects fibers containing several other neuroactive substances to the olfactory bulb. For example, immunohistochemical studies have shown that fibers immunoreactive for galanin, calcitonin gene-related protein, LHRH, and dynorphin B also project to the olfactory bulb (Senut et al. 1989). The roles of these neuroactive substances in the bulb are not known.

Neurons from the ipsilateral anterior olfactory nucleus terminate primarily in the granule cell layer, although a few ascend to the periglomerular region (reviewed by Mori 1987b), whereas those from the contralateral anterior olfactory nucleus terminate in the granule cell layer. The anterior olfactory nucleus, then, is the component of the olfactory system where bilateral olfactory representation of olfactory information occurs, because it receives and projects information from both sides of the nasal cavity and from the higher cortical centers on both sides. Input from other cortical areas, including the pyriform cortex, the entorhinal cortex, the taenia tecta, and the posterolateral and medial cortical amygdaloid areas, also project to the bulb (Shipley & Adamek 1984; McLean & Shipley 1992) and terminate mostly in the granule cell layer (Luskin & Price 1983a). The transmitters of the cortical projections into the bulb are not known, although glutamate is suspected (McLean & Shipley 1992).

From the brain stem, serotonergic fibers from the raphé terminate preferentially in the glomeruli and less heavily in other layers of the bulb, particularly the internal plexiform layer and the granule layer (McLean & Shipley 1987a). Most of the ingrowth is within the first few postnatal weeks in the rat (McLean & Shipley 1987b). Noradrenergic neurons from the locus coeruleus in the brain stem project abundantly to the external and internal plexiform layers and the granule layer (Shipley et al. 1985; McLean et al. 1989b).

A box diagram summarizes the synaptic connections within the olfactory bulb (Figure 5.10).

5.9 Accessory olfactory bulb

The accessory olfactory bulb (AOB) in mammals is located on the dorsocaudal surface of the main olfactory bulb (Figure 5.11). It is absent from or vestigial in higher primates, including humans. The laminar pattern is very similar to that in the main olfactory bulb, although it is less distinct. The outermost layer consists of fiber bundles of the vomeronasal nerve. These fibers terminate in the next deeper layer, the glomerular layer. Glomeruli are more tightly packed and somewhat smaller than those in the main olfactory bulb, and usually there are few or no periglomerular cells, so that the glomeruli are not distinctly outlined as in the main bulb.

The EPL is somewhat less thick than that seen in the main olfactory bulb and is mostly neuropil, with scattered principal neurons. The principal output cells may be distributed in the EPL, somewhat like the distribution of principal neurons in the main olfactory bulb in lower vertebrates (see section 5.10), rather than in a single distinct layer of mitral cells as seen in the mammalian main olfactory bulb. Tufted cells are not identified as a distinct cell group; principal cells in the AOB are often referred to as mitral/tufted (Mori 1987b) or simply as mitral cells. Deep to the EPL/mitral cell layer is a granule cell layer.

The principal cells of the AOB project to the bed nucleus of the accessory olfactory tract, the medial and posteromedial cortical nuclei of the amygdala, and the bed nucleus of the stria terminalis (Price 1973; Broadwell 1975a; Scalia & Winans 1975; Davis et al. 1978; de Olmos, Hardy, & Heimer 1978). The projection targets of AOB mitral cells, in turn, project centrifugal afferent fibers back to the AOB. There is some evidence that the output neurons of the AOB may use aspartate as a transmitter (Fuller & Price 1988). This is based on the observation that relatively more AOB mitral cells take up extrinsically applied [^{3}H]-D-aspartate than do those seen in the main bulb.

The afferent input to the AOB from higher centers comes primarily from the medial and posteromedial cortical amygdaloid nuclei, the bed nucleus of the stria terminalis, the locus coeruleus, the diagonal band nucleus, the raphé, and the nucleus of the accessory olfactory tract (de

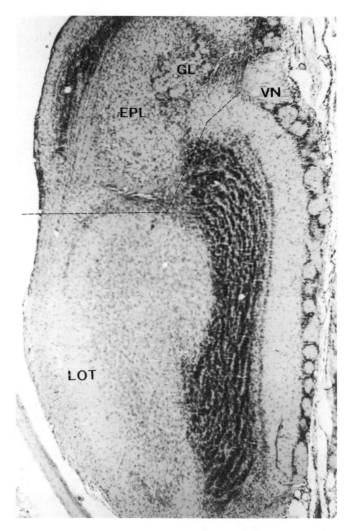

Figure 5.11. Photomicrograph of a section through the olfactory bulb of a rat. The AOB is on the dorsal aspect; the dashed line roughly separates the AOB from the main bulb. The vomeronasal nerve (VN) enters from the medial aspect. Glomeruli (GL) are grouped together near the medial surface, and the external plexiform (EPL) and granular layers are seen laterally. On the medial aspect of the main bulb (right side of photograph) the layered structure is well defined (compare with Figure 5.2). On the lateral aspect, this has disappeared as the lateral olfactory tract (LOT) has formed. The lateral olfactory tract is the principal pathway for nerve fibers leaving and entering the bulb, projecting to and from the central nervous system.

Olmos et al. 1978; Shipley & Adamek 1984; Shipley et al. 1985; McLean & Shipley 1992). The cholinergic input from the diagonal band and the serotonergic input from the raphé do not reach the glomeruli, in contrast to the situation in the main olfactory bulb. The noradrenergic input from the locus coeruleus terminates in the EPL and granular layer and is thought to play a role in memory formation in preventing pregnancy block in mice (see section 6.6.3.3). The transmitters from the other centrifugal inputs to the AOB have not been identified. Projections from the anterior olfactory nucleus to the AOB have not been identified. Unlike the main bulb, then, there is no evidence for bilaterality of olfactory processing associated with the AOB at this level.

When comparing the main olfactory bulb and the AOB, one sees similarities in the laminar pattern, but there are differences as well. For example, the principal output cells of the AOB are somewhat smaller and have fewer and smaller secondary dendrites, but each of their apical dendrites may send branches to three to seven glomeruli (Mori 1987b). The AOB in mammals sometimes exhibits sexual dimorphism. In the rat, the male AOB is somewhat larger than that of the female; this corresponds with a slightly larger vomeronasal organ in the male (Segovia et al. 1984; Roos et al. 1988). However, in bats and shrews there is no evidence for sex-related size variation in the AOB (Frahm 1981; Frahm, Stephan, & Baron 1984). Moreover, there is no evidence for size differences related to sex in the main olfactory bulb.

It is important to emphasize that the main and accessory olfactory systems are essentially independent of one another. The afferent connections into the two systems and the secondary projections process olfactory information in parallel with each other, and there is no anatomical or physiological overlap, at least until the amygdala. In the amygdala, the posterolateral cortical nucleus (PLCN) and the anterior cortical nucleus (ACN) both receive input from the main olfactory pathway. Both project to the amygdala posteromedial cortical nucleus (PMCN) and medial nucleus, which receive input from the vomeronasal system by way of the AOB. In hamsters, single units have been recorded in the PMCN that have responded to electrical stimulation of either the main olfactory bulb or the vomeronasal system (Licht & Meredith 1987).

5.10 Olfactory bulb in fish

Because the fish olfactory system differs in some key respects from that of the mammal, which has been the model for the foregoing discussion, and because the fish has been used as a model for studies of olfaction, this section is devoted to a brief discussion of some of these differences. For a comprehensive review of the fish olfactory bulb, see Satou (1990).

One obvious difference is that olfactory stimuli for fish are not volatile. Another was mentioned in Chapter 1, namely, that in fish the olfactory and respiratory systems are more often than not separate, unlike the situation in terrestrial vertebrates. Further, fish and other water-dwelling vertebrates do not possess a separate vomeronasal organ.

In most teleost families the olfactory bulb is closely apposed to the telencephalon ("sessile" bulb), but in some (Gadidae, Mormyridae, Siluridae, Cyprinidae) it is connected to the telencephalon by a long stalk or peduncle ("pedunculated" bulb) (see Figure 1.6). Because the latter are easier to manipulate, they have most often been used in anatomical, physiological, and behavioral experiments (Satou 1990). Thus, catfish, goldfish, carp, and codfish are the most commonly used teleosts in olfactory experiments.

It is interesting to note that the olfactory system in teleosts, as in mammals, is divisible into two topographically separate and parallel systems. The olfactory nerve can be divided into a lateral and medial olfactory bundle, each of which goes to its respective part of the olfactory bulb. The projections from the bulb and the centrifugal projections into the bulb pass via the olfactory tract. Fibers in the olfactory tract comprise two distinct fiber bundles, a lateral and a medial, which project to different higher centers, suggesting that they may have different functions.

Indeed, in the carp it has been shown that the olfactory bulb is subdivided functionally into two parts (Thommesen 1978; Satou et al. 1983). Either the medial or the lateral part of the bulb can be activated, leaving the other part quiet (Satou et al. 1983). Besides the topographical relationship between the input and output pathways, the anatomical basis for such a functional segregation may be the limited extension of the mitral cell lateral dendrites, which extend less than 400 μm (Fujita, Satou, & Ueda 1988), compared with 1,000 μm or more in mammals (Macrides & Schneider 1982; Mori et al. 1983). Consequently, the dendritic field of individual mitral cells in the medial or lateral part of the teleost olfactory bulb is confined to that part and does not overlap into the other (Fujita et al. 1988; Satou 1990).

Behavioral studies have also suggested a functional differentiation between the medial and lateral parts of the teleost olfactory bulb: the medial part being involved in sexual behavior, and the lateral part being involved in feeding behavior (Døving & Selset 1980; Stacey & Kyle 1983). This is consistent with the finding, in salmonids, that the lateral part of the olfactory bulb is responsive to amino acids (associated with feeding), whereas the medial part is responsive to bile salts (Thommesen 1978).

Thus, although bony fishes have no separate vomeronasal system, as seen in other classes of vertebrates, the spatial organization of the olfactory system into topographically segregated functional domains presages the evolutionary development of anatomically separate olfactory systems, as seen in amphibians, reptiles, and mammals.

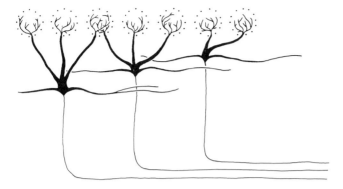

Figure 5.12. Diagram to show that mitral cells of fish olfactory bulb are located at different levels, rather than in a single layer as in mammals. Further, the apical dendrites of fish mitral cells typically are distributed to more than one glomerulus.

The teleost bulb exhibits a concentric laminated structure similar to that in mammals, except that the laminar pattern is less distinct. Four layers are distinguishable: (1) the olfactory nerve layer, where incoming olfactory nerve fibers are distributed, (2) the glomerular layer, (3) the external plexiform/mitral cell layer, in which the mitral cells are scattered, and (4) the internal or granule cell layer. The major difference is that mitral cells do not constitute a separate and distinct layer that divides an EPL and an internal plexiform layer, but are scattered throughout the equivalent of the plexiform layers, much as tufted cells are scattered in mammalian olfactory bulb. In one study, two types of mitral cells were identified in the teleostean olfactory bulb (Alonso et al. 1988): Type I had widely divergent dendritic branches and was found in greater numbers in the medial part of the bulb. Type II, found more frequently in the lateral part of the bulb, had dendrites that terminated in tufts relatively closer together. It is possible that the morphological differences are correlated with the different functions in the two regions of the teleost bulb, as discussed earlier. However, other studies, although reporting variations in mitral cell structure, have not discerned distinct morphological types (Kosaka & Hama 1982).

Convergence of sensory neurons onto secondary neurons in the bulb in teleosts is roughly equal to that in mammals (i.e., about 1,000 : 1 if we use the ratio between the number of sensory neurons and the number of myelinated fibers in the olfactory tract) (Gemne & Døving 1969). However, because of certain differences between teleosts and mammals, there are caveats in accepting this estimate of the convergence ratio. Teleost mitral cells generally have two or more primary or apical dendrites that terminate in different glomeruli (Figure 5.12) (Andres 1970).

Figure 5.13. Diagram of a Golgi preparation of a ruffed cell, showing several cytoplasmic protrusions in the area of the initial segment, beneath the cell body. (Adapted from Alonso et al. 1987.)

Further, some fibers projecting from the bulb to higher centers may not be myelinated, and some of the centrifugal fibers in the olfactory tract may be myelinated.

The teleost bulb contains an unusual cell, referred to as the ruffed cell, in the glomerular and plexiform layers. This cell is relatively large, about the same size as mitral cells. Its most distinctive feature is the region near the initial segment of the axon, where there are several closely intermingled protrusions (Figure 5.13) Together these protrusions form a field with a round or ovoid contour (Kosaka & Hama 1979, 1982; Alonso et al. 1987). This is called the ruff, and it has a diameter of 15–40 μm. Ruffed cells seem to be peculiar to teleosts; they are not found in bulbs of other species.

6
Development and Plasticity

6.1 Early development of the sensory epithelium

6.1.1 Origin of the olfactory placode

The earliest sign of the olfactory organ in the developing vertebrate embryo is the appearance of bilateral, oval-shaped epithelial patches in the anterolateral region of the head (Figure 6.1A). These patches, called olfactory placodes, are delineated by a ridge of mesenchyme, which grows rapidly, resulting in the formation of depressions, the olfactory pits (Figure 6.1B). The floor of each pit lies over the roof of the developing oral cavity and is separated from it by a thin structure, the oronasal membrane (= buccopharyngeal membrane). Lining each pit is a pseudostratified columnar epithelium, much like that lining the subjacent cerebral vesicle. The ridges outlining the pits are known as the lateral and medial nasal processes (Figure 6.1C). The medial nasal processes are separated from each other by a midline structure, the frontonasal process, which forms the nasal septum. The growth of the frontal, medial, and lateral nasal processes defines the nose or snout, the olfactory pits, and the later stages of nasal cavity formation (Figure 6.1A–D).

At the placodal and pit stages, the presumptive olfactory epithelium exhibits a slightly reduced rate of cell division, compared with adjacent epidermis (Smuts 1977), and acquires a surface coating containing carbohydrate-rich substances (Smuts 1977; Burk, Sadler, & Langman 1979). In mammals, the future olfactory and supporting cells can be distinguished from one another by electron microscopy, even at this early stage (Figure 6.2) (Cuschieri & Bannister 1975a,b; Farbman 1977).

What induces formation of the olfactory placodes? Several investigators in the first half of the twentieth century attempted to answer this question. Their experiments were based on the assumption that placodal origin was linked to the inductive action of a stimulus originating from a nearby tissue. There were two favored candidates for the origin of the stimulus: the adjacent mesenchyme, and the cerebral vesicle (the most

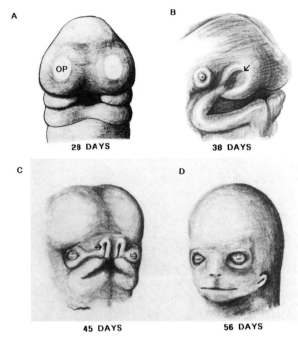

Figure 6.1. A: Frontal view of the head of a human embryo, approximately 28 days old. The paired olfactory placodes (OP) are oval areas on the surface of the head, above the mouth opening. The eyes are not visible in this view because they are on the side of the head. B: View of human embryo head, approximately 38 days old. A horseshoe rim of tissue surrounds the olfactory pit (arrow). The lateral part of the rim (closest to the eye) forms the edge of the lateral nasal process; the medial part of the rim belongs to the medial nasal process. (A and B redrawn after Johnston & Sulik 1980; from Farbman 1991, with permission.) C: View of human embryo head, approximately 45 days old. The horseshoe-shaped rim defining the opening to the nasal cavity is still evident, but is relatively much smaller compared with the size of the face at this stage. The face has broadened, and the eyes are beginning to look toward the front of the face. D: At approximately eight weeks, the embryo face is much larger, the nasal openings are relatively much smaller, the eyes are facing front, and the opening of each external ear is still located in the neck region.

anterior end of the neural tube), as reviewed in detail by Farbman (1988).

Zwilling (1934) tested the effect of neural tissue on induction of olfactory placodes by transplanting fragments of the anterior neural plate (the precursor of the neural tube and cerebral vesicle) from *Rana pipiens* embryos to various parts of recipient embryos of the same species and showed that the transplants could induce formation of nasal sacs from

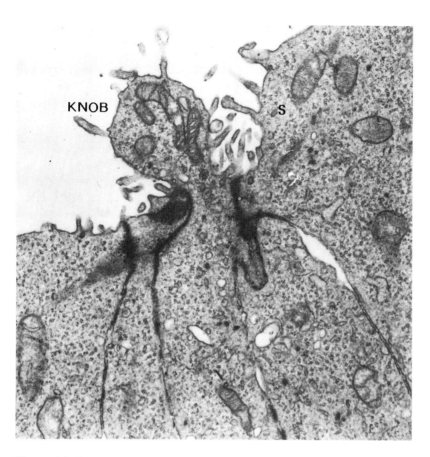

Figure 6.2. Electron micrograph of olfactory pit epithelium from 11-day rat embryo, showing presumptive olfactory dendritic knob and supporting cell (S). (From Farbman 1991, with permission.)

ectoderm covering parts of the body other than the head. In later studies, however, he repeated the experiments using heterotypic transplants, in which cells from the donor and recipient embryos could be distinguished by different patterns of pigmentation. He found that the donor neural plate tissue often developed into olfactory epithelium. These results forced him to reinterpret the results of his earlier studies, and he concluded that the transplants in his earlier experiments must have contained fragments of ectoderm that had already been "determined" to become olfactory epithelium (Zwilling 1940). However, when he took greater care to remove what he thought was presumptive olfactory epithelium from the transplants, the neural plate tissue did not induce nasal sacs in ectopic sites.

Further experiments on *Rana pipiens* embryos suggested that the ectodermal tissue destined to become olfactory placode had already begun to differentiate in the mid-gastrula stage (i.e., before the neural plate was obvious) (Zwilling 1940; Emerson 1944). Moreover, when fragments of presumptive neural plate ectoderm from late-gastrula embryos were grown in tissue culture, nasal sacs did develop (Emerson 1945). These studies suggested that (1) the neural plate was not necessary for the early determination of the olfactory organ, and (2) the neural plate and the olfactory placodes probably developed simultaneously under the influence of the roof of the primitive gut (archenteron) (Zwilling 1940; Emerson 1945; Farbman 1988).

Other experiments on a different species of amphibian, *Ambystoma punctatum* (Haggis 1956), and on chick embryos (Street 1937) led to different conclusions. In these animals the ectodermal tissue destined to become the olfactory placode does not appear to be programmed until well after the neural plate stage of development. When the nasal precursor region of the *Ambystoma* was dissected and grown in vitro, nasal organs did not develop unless the tissue came from late embryos, after the neural plate stage (Haggis 1956). These studies implicated the neural tissue as the inducer of olfactory placode development.

However, because the developmental timing of the olfactory system varies from one species to another, the results of these experiments do not permit generalizations to be made concerning the origin of the inductive stimulus. Some authors have suggested the possibility that the inductive stimulus is present both in the primitive forebrain and in the mesenchymal tissue associated either with the roof of the archenteron or the head and that some or all of these tissues participate in the induction of the olfactory organ (Yntema 1955; Jacobson 1963).

In fact, the possibility that an inductive stimulus originating from either the neural plate or the mesenchyme actually exists has been questioned by the results of a detailed investigation of placode development in mouse embryos. That study concluded that the placodes originated from the anterolateral end of the neural plate, not from ectoderm modified by an extrinsic inductive stimulus. Patches of neural plate cells from this region presumably become isolated as a result of differential growth of adjacent nonneural tissues before closure of the cranial end of the neural tube (Verwoerd & Van Oostrum 1979) (Figure 6.3). Experimental support for this view comes from experiments in which fragments of the anterolateral neural ridge were grafted from Japanese quail embryo donors to chick embryo hosts. Japanese quail cells have a prominent nuclear marker not present in chicks; this permits their detection in the host embryo. The results indicated that the olfactory placode and

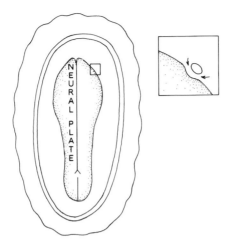

Figure 6.3. Diagrammatic representation of a vertebrate embryo in the neural plate stage. The neural plate is defined by a ridge. The opposite sides of the ridge will ultimately grow toward one another and fuse in the midline, thus changing the neural plate into a tube. One theory holds that the olfactory placode comes from a piece of neural plate (in the region of the square) that becomes detached before fusion of the neural tube. The arrows indicate how the patch of neural ectoderm may become isolated by growth of nonneural tissue around it.

the sheath cells of the olfactory nerve arose from the neural ridge tissue (Couly & Le Douarin 1985).

6.1.2 Formation of the nasal cavity and vomeronasal organ

The medial and lateral nasal processes continue to grow, resulting in deepening of the olfactory pit. On the medial aspect of the pit a small outpouching into the septum is formed, the future vomeronasal organ (see Figure 4.3). The region where this ingrowth originates is marked in the adult by the location of the ductlike opening into the vomeronasal organ. In air-breathing vertebrates, the oronasal membrane at the bottom of the olfactory pit ruptures, creating two funnel-shaped internal nasal openings, the primary choanae, which establish a direct communication with the oral cavity.

The continued growth of the head results in an increase in its breadth, and, in effect, that shifts the anterior nasal openings relatively closer to the midline (Figure 6.1C,D). Accompanying this is a relative narrowing of the nasal septum. In mammals, a palate is established that separates the mouth from the nasal passages, thus making it possible for the young to suck (or chew) and breathe at the same time (Arey 1974). The palate is organized by growth of the lateral palatine processes out of the maxillary processes of the upper jaw. The lateral palatine processes first grow downward and are separated from each other by the tongue. Later the tongue is withdrawn, and the processes are rotated medially so that they grow toward each other across the dorsal aspect of the tongue and fuse with each other in the midline. Soon afterward, they fuse with the nasal septum (Figure 6.4). Formation of the palate effectively separates the nasal and oral cavities, except posteriorly, where the nasal cavities open into the pharynx via the secondary choanae (also called posterior nares).

If we look more closely at the fusion of the palatal processes with the nasal septum it becomes easier to explain the species differences in location of the vomeronasal opening. As noted in Chapter 4, in rodents this opening is directly into the nasal cavity, whereas in ungulates it is into the nasopalatine canal. If the fusion of the palatine processes with the nasal septum is below this opening, the vomeronasal opening will be into the nasal cavity. If, on the other hand, the palate-septum fusion covers the vomeronasal opening, the latter will be within the fusion line, marked in adults by the nasopalatine canal, as in ungulates.

In mammals, the surface area of the nasal cavities is increased by the formation of folds, known as turbinates (or conchae), on the lateral wall of the cavity. The turbinates arise in the nasal cavity of the developing embryo as a series of elevated folds on the lateral wall. The folds acquire a skeletal support of cartilage in the early embryo; later, the cartilage is replaced by bone. In humans, there are three turbinates, a superior, middle, and inferior, but in macrosmatic animals, such as the rabbit and pig, there may be several (see Figure 1.10); often they are elaborately branched and scrolled, thus amplifying their surface area enormously. Because most of the turbinate surface is lined with olfactory epithelium rather than respiratory epithelium, this specialization permits an augmentation in size of the olfactory sense organ. In postnatal life, as the snout or nose grows and the turbinates become more elaborate, the area of the nasal cavity occupied by olfactory epithelium increases significantly (Meisami 1989).

6.1.3 Cell division in embryonic olfactory epithelium

Unlike the situation in mature epithelium, most of the cell division in the embryonic epithelium occurs near the epithelial surface (Smart 1971; Cuschieri & Bannister 1975a). This pattern resembles that seen in the developing neural tube (Sauer 1935). In the neural tube, which has been studied extensively, after a cell divides at the epithelial surface, the two daughter cells elongate and become thinner, and the nucleus moves

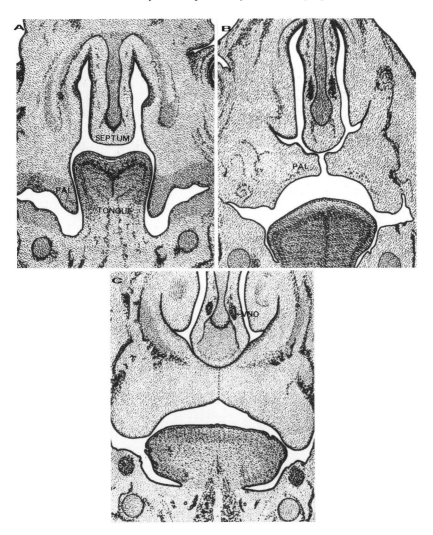

Figure 6.4. Diagrams of sections in the frontal (coronal) plane representing three stages in the formation of the palate in the human embryo. A: The lateral palatine processes (PAL) are oriented vertically on both sides of the tongue. The tongue is virtually in contact with the ventral end of the nasal septum. B: The tongue has withdrawn, and the lateral palatine processes approach each other in the midline. They are also approaching contact with the nasal septum. C: Fusion has occurred between left and right palatine processes, and the two processes have fused with the nasal septum. This effectively separates the oral cavity from the left and right nasal cavities. The remnant of the vomeronasal organ (VNO) in the developing embryo is seen in part C. (Adapted from Johnston & Sulik 1980.)

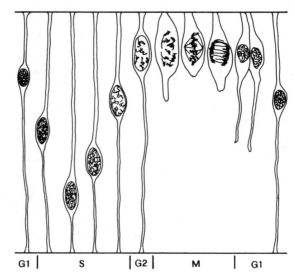

Figure 6.5. Diagram showing the path an olfactory cell nucleus follows through the stages of the cell cycle. In the S phase, during synthesis of DNA, the cell nucleus descends within the epithelium and begins to rise again toward the free surface. When the S phase is complete, the cell enters the G2 stage as its nucleus is near the surface. The cell enters mitosis (M phase) and divides into two cells. Mitotic figures in olfactory (and neuro-) epithelium during the embryonic stage are almost always found near the surface of the epithelium. The nucleus then enters the G1 stage, and the nucleus begins its descent into the epithelium, where it will begin the cycle anew.

deeper into the cytoplasm (i.e., away from the surface of the epithelium). The cells remain elongated as they pass through the G1, synthetic, and G2 stages of the cell cycle. During the latter stage, cells begin to round up, and the cytoplasmic mass and the nucleus move toward the free surface of the epithelium, where they divide (Figure 6.5). The mode of cell division changes as the animal matures and a basal cell population is established (Smart 1971).

It should be emphasized again that the differentiation of olfactory sensory neurons is not a synchronous process (i.e., not all receptor cells develop simultaneously). Therefore, at any given time during embryonic life, cells in several stages of development may be seen in a single olfactory organ. Moreover, because neurogenesis continues throughout the life of the animal, even after growth ceases, the population of olfactory receptor cells at any given moment consists of cells of many ages. This important fact must be dealt with fully in consideration of the observations that there are subpopulations of receptor cells (Allen

& Akeson 1985b; S. Fujita et al. 1985; Mori et al. 1985, 1987; Schwob & Gottlieb 1986; Mori 1987a; Akeson 1988; Plendl & Schmahl 1988; Shinoda, Shiotani, & Osawa 1989). In other words, could the so-called subpopulations represent cells at different stages of development? We have already mentioned that any interpretation of data from single-unit electrophysiological experiments must take into account the relative state of maturity of the sensory cells (Gesteland et al. 1982).

6.1.4 Growth of olfactory axons

An early event in sensory cell development is the genesis of the axon, a narrow process growing out of the basal pole and penetrating the basal lamina of the epithelium (Mendoza, Breipohl, & Miragall 1982; Farbman & Squinto 1985). At the same time, or perhaps somewhat earlier, entire cells leave the epithelium and begin a migration toward the rostral end of the cerebral vesicle (Figure 6.6) (Disse 1897; Van Campenhout 1937; Romanoff 1960; O'Rahilly 1967; Bossy 1980; Mendoza et al. 1982; Taniguchi, Taniguchi, & Mochizuki 1982; Farbman & Squinto 1985). In the chick embryo, it is clear that the migration of epithelial cells precedes axonal growth (Mendoza et al. 1982), and they traverse the short distance to the telencephalon before axons begin their growth. The sequence of events in the chick embryo suggests that these cells form a substrate on which the axons can grow to find their way to their appropriate target.

In early mouse embryos, some of the migrating cells, from the medial end of the placode, contain luteinizing hormone-releasing hormone (LHRH) and (1) migrate ultimately to the hypothalamus, where they take up residence as the LHRH neurons of this part of the diencephalon (Schwanzel-Fukuda & Pfaff 1989; Wray et al. 1989); (2) some of these LHRH-immunoreactive neurons also become ganglion cells of the terminal nerve (Schwanzel-Fukuda & Silverman 1980; Witkin & Silverman 1983); (3) others become the ensheathing (Schwann) cells of the nerve (Chuah & Au 1991); these include the cells that ensheath the olfactory nerve axons within the outer nerve layer of the bulb (Doucette 1989). The progenitor cells in the original olfactory placode, then, give rise to a surprising variety of cell lineages (Table 6.1).

6.1.5 Dendrite development

At the other end of the differentiating sensory cell, the dendrite extends from the cell body to the epithelial surface. Even in early stages, many of the presumptive sensory cells display a bulbous surface process that will ultimately become the ciliated dendritic terminal, whereas the supporting cells have a broad apical process (Figure 6.2) (Waterman &

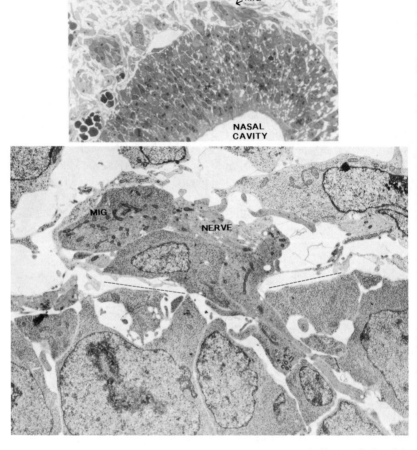

Figure 6.6. Light micrograph (top) and electron micrograph (bottom) showing groups of cells caught in the process of migrating (MIG) out of the fetal rat olfactory epithelium. In the bottom figure, a dashed line indicates the boundary between the epithelium (below) and the lamina propria. Olfactory axons (NERVE) are seen accompanying the migrating cells.

Meller 1973; Cuschieri & Bannister 1975a; Farbman 1977; Menco & Farbman 1985a). A primary cilium may extend from the surface of placodal epithelial cells. Primary cilia are expressed transiently on many cell types during development, usually during the G1 phase of the cell cycle (Menco & Farbman 1985a). In differentiating olfactory cells, they are short (1–2 μm) extensions of one or, rarely, both centrioles. It is not known if primary cilia can become true olfactory cilia.

Table 6.1. *Cells derived from the olfactory placode*

Neuronal
   Olfactory sensory cells, main nasal cavity
   Olfactory sensory cells, vomeronasal organ
   LHRH-secreting neurons, hypothalamus
   Ganglion cells, terminal nerve
   (At least some basal cells of olfactory epithelium are progenitors of neurons.)

Nonneuronal
   Olfactory supporting cells, main nasal cavity
   Olfactory supporting cells, vomeronasal organ
   Ciliated cells in respiratory epithelium in the main nasal cavity and paranasal
     sinuses
   Nonsensory region of vomeronasal organ
   Glandular cells of respiratory mucosa in main nasal cavity, vomeronasal organ,
     and paranasal sinuses
   Bowman's glands in olfactory mucosa of main nasal cavity
   Ensheathing (Schwann) cells of olfactory and vomeronasal nerves and nerve
     layer of olfactory bulb
   Submucosal glands, nonsensory region of main nasal cavity
   Brush cells and other microvillous cells in both olfactory and respiratory
     epithelium

Source: From Farbman (1991).

The differentiation of the dendritic knob has been studied in mice (Noda & Harada 1981) and humans (Pyatkina 1982), but the most detailed study has been in the rat fetus (Menco & Farbman 1985a,b; Menco 1988a,b). The characteristic pattern of organization and distribution of the cells in the epithelium begins to become apparent in the rat on the 14th embryonic day (E14); that is, the supporting cell nuclei are organized as a single layer, superficial to the level of olfactory nuclei, and tight junctions develop among adjacent cells (Kerjaschki & Hörandner 1976; Menco 1980b, 1988c). At an equivalent stage of development in the mouse, around E12 (Cuschieri & Bannister 1975b; Noda & Harada 1981), the number of mitotic figures decreases markedly (Smart 1971). This probably denotes a time in development when some cells leave the cell cycle and begin to differentiate (Menco & Farbman 1985a). At this stage the dendritic knobs may contain several centrioles that migrate from the perikaryon, where they are formed (Farbman & Squinto 1985; Menco & Farbman 1985a; cf. Mulvaney & Heist 1971a). Another event occurring at the same stage of development is the expression of a cell surface glycoprotein that is immunoreactive with a monoclonal antibody referred to as 2B8 (Allen & Akeson 1985a). This antibody recognizes many vomeronasal sensory cells as well. The relationship between the expression of this glycoprotein and early differentiation is not known.

In rat fetuses, dendritic knobs first develop multiple cilia on E15–E16 (Menco & Farbman 1985a). In mice this occurs on day E13 (Noda & Harada 1981), and in humans, in the 9th week of gestation (Pyatkina 1982). In mice and in rats, development of cilia occurs at approximately the time when axons first make contact with their bulbar targets (Cuschieri & Bannister 1975b; Farbman 1986). Although no cause-and-effect relationship between these two events has been established definitively, the timing suggests that differentiation of dendritic terminals may be related to contact between axons and the bulb (Cuschieri & Bannister 1975b), possibly brought about by a factor retrogradely transported along the olfactory axons from the bulb to the sensory cell bodies. Notwithstanding the experimental evidence for a bulbar influence on ciliogenesis in sensory cells (Chuah et al. 1985), it is also clear that ciliogenesis can occur to a limited extent in the absence of bulb (Farbman 1977; Chuah et al. 1985).

In the rat fetus, ciliogenesis becomes more evident after E16 as an increased number of dendrites sprout several cilia (Menco & Farbman 1985a), and the ciliary membranes differentiate, as illustrated by insertion of increasing numbers of intramembranous particles (Menco 1988a). The average number of cilia per dendrite increases linearly from a value of less than 1 (because some knobs have no cilia) at E16 to about 7 at E22, a rate of about 1 per day (Menco & Farbman 1985b). At E18–E19, increasing numbers of the growing cilia begin to taper and resemble adult cilia. At the same embryonic age, the supporting cell microvilli also undergo an increase in length (Menco & Farbman 1985b).

G-protein (Mania-Farnell & Farbman 1990) and adenylate cyclase (Dau et al. 1991) are first demonstrable in cilia of rat olfactory cells shortly after ciliogenesis begins, thus supporting the evidence that there is a role for the G-protein-mediated cyclic AMP cascade in transduction of the olfactory stimulus.

6.1.6 Biochemical aspects

Several other events have been shown to occur in developing olfactory epithelium of prenatal rats. Adenosine deaminase immunoreactivity appears in placodal cells at E12 (Senba, Daddona, & Nagy 1987). Two blood-group antigens are demonstrable with immunohistochemical techniques in olfactory receptors: The H antigen first appears on E14 in the entire cell, from dendritic terminal to axon terminal, and the B antigen appears on E16 in some but not all of the cells expressing the H antigen (Mollicone et al. 1985; Astic et al. 1989). In one study, olfactory marker protein (OMP) was detected immunohistochemically on E14 (Baker &

Farbman 1989a); in another, its expression was linked to ciliated olfactory knobs, because it was apparently absent before ciliogenesis (Menco 1989). Although OMP is expressed before axons reach the bulb, and is not dependent on contact with bulbar elements; the number of cells expressing this protein is doubled when olfactory mucosa is co-cultured with the bulb (Chuah & Farbman 1983).

Carnosine has been demonstrated in rat olfactory tissue at E16 (Margolis et al. 1985). Its presence in receptor cells at E16 indicates that the enzyme carnosine synthetase has been expressed. The expression of this enzyme, incidentally, does not appear to be dependent on any influence of the olfactory bulb; the enzyme is expressed in organ cultures of the olfactory mucosa from which the bulb is absent (Margolis et al. 1985).

Immunohistochemical studies using monoclonal antibodies have shown that several membrane-related antigens are expressed early in development of rodent olfactory epithelium. In mice, neural cell adhesion molecule (N-CAM) is seen on olfactory cell bodies as early as embryonic day 9, and another cell adhesion molecule, L1, has been seen on E10.5 (Miragall et al. 1989); in rats, N-CAM is seen on olfactory axons at E13 and prominently on sensory perikarya at E14 (Carr et al. 1989). In both species, the immunoreactivity is demonstrable just before the outgrowth of axons begins. Two other monoclonal antibodies, Neu–4 and Neu–9, reveal the expression of their respective antigens on both sensory and supporting cells of rats on E13–E14. Another monoclonal antibody, 1A–6, binds specifically to the apical microvilli of a special microvillous cell type in adults, but binds to the apical surface of the entire olfactory epithelium at E13 (Carr et al. 1991).

During the last week of gestation of the rat embryo, several anatomical and physiological events have been documented: (1) the first synapses between sensory cell axons and second-order neurons in the bulb (Farbman 1986); (2) the electro-olfactogram can be recorded (Gesteland et al. 1982); (3) receptor cells become selectively responsive to odorant stimuli (Gesteland et al. 1982); (4) in supporting cells, the characteristic dumbbell-shaped particles are inserted into the membrane at E16–E17 (Menco 1988a). Figure 6.7 is a time line, correlating the expression of several molecules in rat olfactory epithelium with the ages at which several important developmental events occur.

6.2 Postnatal growth of epithelium

Several lines of evidence from behavioral (e.g., Tobach, Rouger, & Schneirla 1967; Singh & Tobach 1975; Rudy & Cheatle 1977; Alberts 1981; Pedersen & Blass 1981; Hudson & Distel 1983, 1986), electrophysiological (Gesteland et al. 1982; Mair & Gesteland 1982), and morphological (Cuschieri & Bannister 1975a,b; Hinds & Hinds 1976a,b;

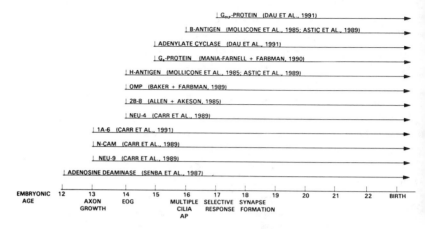

Figure 6.7. A time line showing the date on which several molecules are expressed in rat olfactory epithelium during fetal development. On the horizontal axis, some important anatomical and physiological developmental events are marked.

Menco & Farbman 1985a,b; Menco 1988a,b, 1989) studies support the notion that mammals have functional olfactory systems at birth or shortly before birth. At birth, mammals use their sense of smell for nipple search and localization, and somewhat later for other interactions with the mother and the nest (e.g., Hofer et al. 1976; Alberts 1981; Pedersen & Blass 1981, 1982; Hudson & Distel 1983, 1984, 1986).

One of the bases for increased olfactory sensitivity with age is thought to be the increase in size of the olfactory sheet and the greater convergence ratio between the numbers of sensory cells and mitral cells in the olfactory bulb (Meisami 1989). Within the first 25 postnatal days, the olfactory epithelial sheet in the rat undergoes an 8-fold increase in surface area and a 12-fold increase in sensory cell number, indicating an increase in epithelial thickness as well as area (Meisami 1989). Increases of the same order occur in rabbit between birth and weaning (Meisami et al. 1990). Given that the number of mitral cells does not increase after birth, the number of sensory cells converging on each mitral cell increases by a factor of more than 10 during the first 25 days in the life of the rat. The expansion of the olfactory sheet occurs with an increase in ventral and rostral growth of the septum and greater elaboration of the turbinates. The surface area of rat septal olfactory epithelium alone expands about threefold between the ages of 2 and 29 months, after which it declines (Hinds & McNelly 1981). This increase in septal olfactory area is accompanied by a threefold increase in the number of sensory cells, as estimated by counting the dendritic knobs (Hinds & McNelly 1981). In general, the most caudal regions of the epithelium differentiate first (i.e., mature, ciliated sensory cells are observed here before they are seen more rostrally) (Breipohl & Fernandez 1977; Breipohl & Ohyama 1981; Breipohl et al. 1986; Farbman & Menco 1986).

In a study of amphibians (*Xenopus laevis*), the number of olfactory neurons was shown to increase 16-fold between late larval stages (stage 58) and adulthood (Byrd & Burd 1991). There was only a twofold increase in the number of mitral/tufted cells during the same developmental period. The convergence ratio for olfactory axons onto mitral/tufted cells increased from 5 : 1 in the late larval stage to 34 : 1 in adults. In these animals as well, the increased convergence was linked with increased olfactory sensitivity.

6.3 Development of the olfactory bulb

6.3.1 Anatomy

The main and accessory olfactory bulbs develop as an outpocketing of the rostral end of the cerebral vesicles (Figure 6.8). In this phase, each cerebral vesicle is made up essentially of a pseudostratified columnar epithelium surrounding the (lateral) ventricle. In mice, the outpocketing begins on the 12th day of gestation (Hinds 1968b), and in rats on E14–E15; in both animals it occurs at about the same time as, or slightly after, the growing olfactory nerves reach this region. At this early stage there are only two layers in the bulb: the highly cellular ventricular layer, where neuronal genesis occurs in the earliest stages of development, and the marginal (acellular) layer (Hinds 1968b). The part of the projection destined to become the accessory bulb is relatively large and is developmentally more advanced than the primordial main bulb. The region destined to become the olfactory nerve layer of the bulb is outside of this telencephalic outpocketing, but apposed to it, thus forming its outermost layer.

On E15, in mice, recognizable mitral cells appear in a layer in the primordial bulb (Hinds 1968b). A subventricular layer consisting of darkly stained cells lies between the mitral cells and the ventricular layer (Hinds 1968b). This subventricular layer is the site for genesis of later-forming neurons, particularly the granule and periglomerular cells. With continued growth of the bulb prenatally and postnatally, the relative position of the ventricle moves caudally, so that by the end of the first postnatal month the most rostral tip of the lateral ventricle is caudal to

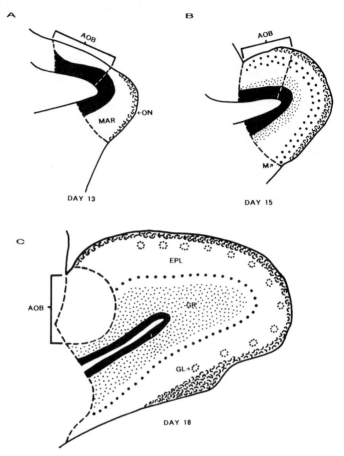

Figure 6.8. Diagrams summarizing the early development of the mouse olfactory bulb. A: On the 13th day of fetal life, the ventricular layer (shown in black), the marginal layer (MAR), and the cap of the olfactory nerve fibers (ON) at the tip of the bulb are seen. The accessory bulb (AOB) occupies a major portion of the telencephalic outpocketing, as indicated by the dashed lines and the bracket. B: On the 15th day of fetal life, a subventricular layer (small dots) and mitral cell layer (M, large dots) make their appearance. C: On the 18th day, some glomeruli (GL) are seen, the granular layer (GR) and external plexiform layer (EPL) are recognizable, the ventricular layer is mostly gone, and the ventricle has receded into the posterior part of the bulb. (Redrawn after Hinds 1968b and Farbman 1991.)

the bulb. A subventricular layer of cells does, however, remain in the bulb to mark the former position of the ventricle (see Figure 5.2).

In human embryos, the accessory bulb begins to develop relatively early, but undergoes regressive changes in the second and third trimesters of gestation (Humphrey 1940). The accessory bulb is the target of the nerve from the vomeronasal organ, which is also vestigial in humans, although it, too, is present in early development (Nakashima et al. 1985).

6.3.2 Proliferation and migration of neurons

The genesis of bulbar neurons follows a stereotypic pattern. As in other parts of the central nervous system, the large output neurons are formed first, followed by intermediate-sized and smaller neurons. Autoradiographic data from rats and mice injected with tritiated thymidine have shown that of the larger neurons, the output neurons of the accessory bulb are born (i.e., have undergone their final cell division) slightly earlier than the mitral cells of the main bulb (Hinds 1968a; Bayer 1983). In fact, significant proportions of the output neurons in both the accessory and main bulbs are born before the actual outpocketing of the primordial bulb from the cerebral vesicle (Hinds 1968b; Bayer 1983).

After the mitral and the larger internal tufted cells are formed, the smaller tufted cells in the more superficial parts of the external plexiform layer make their appearance. The mitral and tufted cells are progeny of cell divisions in the ventricular layer of the bulb. Their cell bodies migrate from the ventricular layer to the site where they will take up their permanent positions. It should be noted that the later-forming tufted cells must migrate past the mitral cell layer as they travel from the ventricular layer to the external plexiform layer, in a manner similar to the inside-out order of formation of neurons in the layers of the cerebral neocortex (Rakic 1974).

The last cells to form are the small periglomerular and granule cells, most of which, in rodents and probably in most other mammals, form after birth. Neurogenesis reaches a peak during the second postnatal week and continues until the end of the third week (Hinds 1968a; Rosselli-Austin & Altman 1979; Bayer 1983; Frazier & Brunjes 1988). These cells are generated in the subependymal layer, mostly near the anterior horn of the lateral ventricle; in the postnatal rodent, the relative position of the ventricle has receded, so that it is somewhat caudal to the bulb proper. Thus, the postnatal increase in number of granule cells arises from cells in the subependymal layer of the lateral ventricle. These must migrate rostrally into the bulb; they move in a "rostral migratory stream" (Altman 1969; Frazier-Cierpial & Brunjes, 1989a,b; Kishi et al. 1990) from a region near the anterior horn of the lateral ventricle into the granular layer of the bulb. As noted earlier (section 5.3.1), a low level of mitosis has been observed in the granule cell layer in adult rats (Kaplan & Hinds 1977; Forbes, 1984; Kaplan et al. 1985).

Bulbar glial cells are formed throughout the fetal and postnatal periods. Scattered proliferative cells are seen in the olfactory nerve layer; these do not leave the nerve layer and are thought to be glial cells (Hinds 1968a; Bayer 1983; Frazier-Cierpial & Brunjes 1989a,b), most likely astrocytes.

6.3.3 Formation and growth of laminae

A few individual glomeruli are first distinguishable in the rodent olfactory bulb a few days before birth (Figure 6.8), but the glomerular layer at birth is relatively indistinct (Meisami 1979). However, on the caudal, mediodorsal aspect of the main olfactory bulb, near the boundary with the accessory bulb, there is a group of glomeruli, constituting the so-called modified glomerular complex, that are well defined at birth. The precocious development of these glomeruli has been associated with odor cues important in suckling behavior in rats (Greer et al. 1981, 1982). In rats, the number of glomeruli increases about sevenfold during the first 25 days after birth (Meisami 1979; cf. LaMantia & Purves 1989), and they form a well-defined glomerular layer.

The next deeper layer, the external plexiform, is visible but narrow at birth, and it grows in width significantly, about 20-fold, during the first three weeks after birth (Hinds & Hinds 1976a; cf. Meisami 1979; Mair et al. 1982; Brunjes & Frazier 1986). The increase in volume of the internal granular layer during the first three weeks of postnatal life is of the same order (Hinds & Hinds 1976a).

The data from studies on several species, including mice (Hinds & Hinds 1976b; Benson, Ryugo, & Hinds 1984), rats (Rosselli-Austin & Altman 1979; Leon et al. 1984; Brunjes 1985; Brunjes & Frazier 1986), lizards (Garcia-Verdugo et al. 1989), and Mongolian gerbils (Leon et al. 1984) indicate that rapid postnatal growth of the bulb may be a common vertebrate feature, even in precocial animals (Brunjes 1983; Leon et al. 1984). In all of these animals, the main olfactory bulb undergoes considerable increase in size and maturation, particularly during the first two weeks after birth. Thereafter, growth continues well into maturity, but at a slower rate (Hinds & McNelly 1977, 1981). What is true of the main olfactory bulb is also true of the accessory olfactory bulb, which, in rats, undergoes a sixfold increase in volume from birth to the 18th postnatal day, and then shrinks somewhat. In two-month-old animals, it is still four times as large as it was at birth (Rosselli-Austin, Hamilton, & Williams 1987).

The postnatal growth in the bulb can be attributed to several factors: First, there is an increase in the volume of the olfactory nerve layer (Hinds & McNelly 1977; Rosselli-Austin & Altman 1979), because of a 10-fold expansion in the number of olfactory sensory neurons (Hinds

& McNelly 1981; Meisami 1989) and an augmented number of glial cells (Hinds 1968a; Bayer 1983).

Second, as already noted, for the first three weeks or so after birth there are increases in total numbers of interneurons, both granular and periglomerular (Altman 1966, 1969; Hinds 1968a; Bayer 1983). Accompanying this is an increase in number of glomeruli (Meisami 1979; LaMantia & Purves 1989). The elaboration of periglomerular and granule cell dendritic processes contributes to the growth in thickness of the three neuropil layers: glomerular (Hinds & McNelly 1977, 1981; Ros-selli-Austin & Altman 1979), external plexiform, and internal granular (Hinds & McNelly 1977, 1981; Meisami 1979; Rosselli-Austin & Altman 1979; Brunjes, Schwark, & Greenough 1982; Brunjes 1985).

Third, there is an increase in volume of individual mitral cells, as well as increased elaboration of their organelles, particularly rough-surfaced endoplasmic reticulum, Golgi bodies, and mitochondria (Hinds & McNelly 1977; Singh & Nathaniel 1977). The increase in size of mitral cells is reflected not only in the mitral cell layer, where the cell bodies reside, but also in the external plexiform layer, where the expanded dendritic trees of mitral cells contribute to the volume increase (Hinds & McNelly 1977).

Finally, part of the increase in postnatal growth, at least during the first three weeks in rats, is due to the growth of centrifugal fibers from higher centers into the bulb (Schwob & Price 1984). It is interesting that although many of these are already present at birth, there is a delay in their maturation. This is evident because the levels of expression of several parameters are still relatively low. These include (1) choline acetyltransferase and muscarinic acetylcholine receptors (Large et al. 1986), (2) serotonin (McLean & Shipley 1987b), and (3) acetylcholinesterase (Van Ooteghem, Schumacher, & Shipley 1984).

6.3.4 Synaptogenesis

The most detailed study of synaptogenesis in the olfactory bulb has been done in the mouse (Hinds & Hinds 1976a,b). Essentially all synaptic types seen in adults (albeit in much smaller numbers) have been found on the day of birth, which in the mouse occurs after a gestation period of 19 days. This is consistent with the idea that olfaction is an important sensory modality for newborn mice, and, likely, for most mammals.

The first synapses in mouse olfactory bulb are seen on the 14th embryonic day, between sensory axons and mitral cell dendrites in the presumptive glomerular layer. Olfactory axons reach the bulb at E12 and remain in contact with bulbar cells for two days before actual synaptic specializations can be seen with the electron microscope (Hinds

& Hinds 1976b). At E15, considerably greater numbers of axodendritic synapses are seen.

Mitral cell axons grow out and reach the olfactory cortical areas before their dendrites become involved in dendrodendritic synapse formation. The first dendrodendritic synapses involving mitral cells in mouse olfactory bulb are seen at E15 in the presumptive glomerular layer. In this layer, only periglomerular cells engage in dendrodendritic synapses with mitral cells (Pinching & Powell 1971b). This suggests that the first periglomerular cells must migrate into position and differentiate rapidly (Hinds & Hinds 1976a).

The overall pattern for the development of synaptic density in the three major neuropil regions of the olfactory bulb (glomerular, external plexiform and internal granular layers) is from the outside in (i.e., synaptogenesis in the glomeruli is the most precocious, followed by the external plexiform layer and then the internal granular layer). The first dendrodendritic synapses are seen in the external plexiform layer at E18 in the mouse, but reciprocal synapses are not seen until the day of birth (Hinds & Hinds 1976a). Development of the two kinds of dendrodendritic synapses [mitral-to-granule (M/G) cell and granule-to-mitral cell G/M] appears tightly linked, although the number of M/G cell synapses is generally higher in early stages. In fact, the evidence suggests that the G/M part of the reciprocal synapse in the external plexiform layer forms in the presence of a preexisting M/G synapse on the same gemmule of the granule cell dendrite (Hinds & Hinds 1976a).

6.4 Cellular interactions between the sensory epithelium and bulb during development

Several references have already suggested the possibility that the sensory neurons and the bulb may interact in such a way that the development of one may be regulated by the other. In this section we shall examine the evidence for this in more detail as we focus on two major questions:

Do the sensory neurons in the epithelium influence formation of the bulb? More specifically, what is the evidence for participation of afferent neurons, glia, and secondary neurons in formation of bulb glomeruli?

How does the bulb influence growth and maturation of the epithelium?

6.4.1 General considerations of cell–cell interactions in development of the nervous system

There is consensus among developmental biologists that cellular or tissue interactions are extremely important in the development of the organism and its many parts, including the nervous system. For example, it is well known that target organs have major effects on the survival, differentiation, or maturation of neuronal cells. Embryonic motor neurons are unable to survive if they fail to make connections with skeletal muscle targets (Hamburger 1934, 1958; Barron 1943, 1946, 1948; Cowan & Wenger 1967; Oppenheim, Chuwang, & Maderut 1978). In sensory systems, the dorsal root ganglion neurons of chick embryos do not survive if their target tissue in the limbs is removed (Hamburger & Levi-Montalcini 1949). Similarly, the parasympathetic neurons of the ciliary ganglion (Landmesser & Pilar 1974; Pilar & Landmesser 1976) and the sensory neurons in the cochleovestibular ganglion (Ard, Morest, & Hauger 1985) are dependent on their target tissues for survival.

Not all instances of neuron/target interactions are a matter of survival, however. An example in which targets have a modulating influence on development is the relation between salivary glands and sympathetic neurons in the superior cervical ganglion, which provides some of their innervation. When superior cervical ganglion cells are grown with salivary glands in vitro, they show greater elaboration and directionality of nerve fiber outgrowth than do control explants (Coughlin et al. 1978). Similarly, it is clear that in the olfactory system there are developmental interactions between the sensory neurons in the nasal cavity and their target, the olfactory bulb. Each of the two participant tissues in this interaction affects the development and differentiation of the other, so that it is a true interaction.

Cellular interactions in the nervous system are somewhat easier to analyze when neuronal populations have a single target organ, as is the case for olfactory sensory neurons. The axons of all olfactory sensory neurons terminate and form synapses within the glomerular layer of the bulb, primarily on dendrites of mitral or tufted cells, but also on the periglomerular cells. In light of the many studies from other neural systems, it is pertinent to ask if the olfactory bulb influences the survival or differentiation of sensory cells, and vice versa. Some of these issues have been examined in detail in recent reviews (Farbman 1988, 1991).

6.4.2 Participation of sensory afferent axons in development of the bulb

In normal development, the growing olfactory axons enter the bulb primordium, and glomeruli are formed where the axons synapse with their target neurons. The questions whether or not and how sensory neurons affect bulb development have been analyzed largely by ablation and transplantation studies, both in developing animals and in adult animals, and by experimental teratology.

In amphibian embryos, ablation of the olfactory placode early in development results in reduction in size of the anterior region of the telencephalon, where the bulb is located (Burr 1916; Piatt 1951; Stout & Graziadei 1980; Venneman, Van Nie, & Tibboel 1982). On the other hand, when additional placodes are transplanted to the head region of the amphibian embryo, the axons growing from these transplants to the brain can promote cell proliferation in the diencephalon and telencephalon (Burr 1924; Stout & Graziadei 1980). One can conclude from these and other studies (Giroud, Martinet, & Deluchat 1965; Byrd & Burd 1991) that formation of the bulb in early development is dependent on influences exerted by growing axons of sensory neurons. Indeed, olfactory sensory neurons in adults retain their ability to elicit plastic changes in several parts of the central nervous system, as discussed later.

6.4.2.1 Sensory neuron influence on glomerulus formation Several studies, mostly from the laboratory of Graziadei and Monti Graziadei, have shown that olfactory axons can organize glomeruli in other parts of the brain, including the telencephalon, diencephalon, mesencephalon, and myelencephalon, and even from synaptic junctions (Burr 1930; Graziadei et al. 1978, 1979; Graziadei & Kaplan 1980; Graziadei & Samanen 1980; Magrassi & Graziadei 1985). In some of these experiments, cerebral or cerebellar cortex was transplanted into the region of the bulb after the bulb had been partially or completely removed. Olfactory axons invaded and formed glomeruli in whatever neural tissue they encountered during their growth. When regenerating axons grew into the remnant bulb in a partially bulbectomized animal, the glomeruli were often found in atypical layers, such as the external plexiform layer or granule cell layer. Glomeruli were also formed in the transplanted cerebral or cerebellar cortex, or even in the host's frontal cortex. When there was a fragment of olfactory bulb remaining in the host, this did not preferentially attract the sensory axons (Graziadei & Kaplan 1980; Graziadei & Monti Graziadei 1980b; Monti Graziadei & Graziadei 1984; see also the in vitro studies of Gonzales, Farbman, and Gesteland 1985). This indicated that the appropriate target of the sensory epithelium, namely, the olfactory bulb, is not necessary for the formation of glomeruli, because the olfactory axons seem to be able to organize these structures independently. However, when tissue other than the bulb became the target, the animals did not recover their sense of smell (Butler et al. 1984).

Although reconstituting olfactory epithelium left in situ can form glomeruli with many brain targets, sensory epithelium transplants do not. When excised mucosal fragments containing sensory epithelium were transplanted into several regions of the brain, axons invaded the surrounding tissue, but did not form glomeruli (Morrison & Graziadei 1983; Monti Graziadei & Graziadei 1984; Graziadei & Monti Graziadei 1986).

Further, when olfactory mucosa fragments were co-transplanted with pieces of the brain into the anterior chamber of the eye in rats, no glomeruli were found (Heckroth, Monti Graziadei, & Graziadei 1983). In each of these experiments, only fragments of the olfactory mucosa were included in the transplants. However, when the entire olfactory pit in the embryo rat was transplanted into the cerebrum or other parts of the brain, glomeruli did form (Graziadei & Monti Graziadei 1986).

The extensive analysis of glomerular formation in transplantation studies led to the formulation of a hypothesis relating to morphogenesis of the glomerulus, namely, that the "glomerulus is determined by the reciprocal recognition of subsets of complementary axons originating from different areas of the sensory sheet and converging in the fiber plexus" (Graziadei & Monti Graziadei 1986). Formation of a glomerulus by olfactory axons clearly was not target-dependent. The glomerulus was "redefined as a tangle of branching olfactory axons derived from the entire olfactory sheet and forming a discrete, self-limiting globose structure" (Graziadei & Monti Graziadei 1986). In other words, the totality of evidence from the several transplantation studies done in the laboratory of Graziadei and Monti Graziadei suggests that the glomerulus can form, irrespective of its target, but only when all or most of the olfactory sheet is present, because each glomerulus receives axonal contributions from many parts of the olfactory sheet. This interesting hypothesis awaits further testing.

6.4.2.2 Role of glia in glomerulus formation Recent experiments on the antennal lobe of the insect brain have indicated that glial cells play a pivotal role in the formation of glomerular structures. The antennal lobe in insects receives the input of olfactory sensory cells in the antenna. The sensory cells form synapses on dendrites of second-order neurons within glomeruli, similar to the situation in vertebrates. In the moth *Manduca sexta,* the growth of axons from the antennal sensory neurons into the antennal lobes of the brain initiated a stereotyped sequence of changes in shape and position of glial cells in the neuropil (Oland & Tolbert 1987). These changes preceded the formation of antennal glomeruli, which were separated from each other by glial borders. When sensory axons were prevented from entering the lobe, the glial cells remained in their immature configuration, and glomeruli did not form (Oland & Tolbert 1987). This suggested the possibility that glial cells played an important role in glomerular formation and were acting under the influence of incoming sensory axons. In a subsequent experiment, glial cells were prevented from dividing by irradiation of the animal at the critical time before glomeruli would have been formed. In the lobe rendered deficient in glia the neuropil did not reorganize into glomeruli, despite the presence of afferent axons in numbers sufficient to induce glomeruli in a control animal (Oland, Tolbert, & Massman 1988; Tolbert & Oland 1989). Similar results were obtained by using hydroxyurea to prevent glial division (Tolbert & Oland 1989).

How might the olfactory sensory axons from the moth antenna cause the glia to change their shape and position in the developing antennal lobe? Preliminary results from tissue culture studies suggest that a soluble factor or factors from dissociated sensory neurons elicit changes in glial cells (Tolbert, Oland, & Or 1989).

There is some evidence that glia in the olfactory bulb in mammals may participate in the formation of glomeruli. Bailey, Paston, & Shipley (1989) showed that the radial glia in rat olfactory bulb branched profusely and formed a plexus at the level of the forming glomeruli. The glial plexus interacted with the developing neurons to form glomeruli.

6.4.2.3 Neuron–neuron interactions affecting bulbar cell differentiation It
is clear from the foregoing that incoming sensory axons play an important role in the organization of glomeruli in the olfactory bulb. At a more cellular level, there are indications that the sensory neurons play a key role in regulating the neurotransmitter phenotype of bulbar neurons. Sensory neurons apparently regulate gene expression in at least three types of bulbar cells. In deafferented olfactory bulbs, both substance P in external tufted cells (Kream et al. 1984) and dopamine in periglomerular, external, and middle tufted cells were significantly reduced (Nadi et al. 1981; Kawano & Margolis 1982; Kream et al. 1984). Moreover, the number of juxtaglomerular neurons expressing tyrosine hydroxylase (TH), the rate-limiting enzyme in catecholamine synthesis, was severely reduced (Baker et al. 1983). This was correlated with a reduction in TH messenger RNA in the deafferented bulbs (Ehrlich et al. 1990). The changes in dopaminergic function do not occur as a result of death of dopaminergic neurons, because the number of neurons expressing dopa decarboxylase, the second enzyme in the catecholamine synthesis pathway, remains unchanged after deafferentation. This indicates that the neurons do survive (Baker et al. 1984). The return of dopamine levels and TH-immunoreactive cells after reinnervation of the bulb supports this conclusion (Nadi et al. 1981; Kawano & Margolis 1982; Baker et al. 1983). That afferent nerve regulation of transmitter expression was specific to juxtaglomerular cells of certain phenotypes was shown by the lack of a deafferentation effect on GABAergic neurons (Baker, Towle, & Margolis 1988).

During development of the bulb, the dopaminergic neurons begin to express TH only after they have reached the juxtaglomerular region, not during their migration from the subependymal layer, where they are born (McLean & Shipley 1988). The expression of TH in the neonatal olfactory bulb seems to be dependent on the olfactory nerve – in other words, the olfactory nerve appears to be necessary for both the initiation and maintenance of TH expression in bulbar neurons. This suggestion, that regulation of the dopaminergic neuron is mediated by the olfactory nerve, was confirmed by organ culture experiments in which bulb was explanted alone and in combination with mucosa. TH was expressed in many more cells in the co-cultures (Baker & Farbman 1989b). It seems that the regulation is contact-mediated.

In bulbectomized newborn rats the olfactory nerve grows to the forebrain, where it induces novel expression of TH in local neurons (Guthrie & Leon 1989). These forebrain neurons do not ordinarily express TH after birth. However, these TH-immunoreactive cells do not express other enzymes in the catecholamine synthesis pathway, suggesting they are not catecholaminergic. Thus, the afferent regulation of TH by olfactory nerves is not limited to bulbar cells. The basis for this specific induction of TH gene expression, even in cells that do not ordinarily express it, is not understood.

6.4.3 How does the bulb influence growth and maturation of the epithelium?

When the bulb is removed surgically (Ferriero & Margolis 1975; Graziadei et al. 1978; Booth et al. 1981; Costanzo & Graziadei 1983; Butler et al. 1984), or when the axons of the sensory neurons are cut (Harding et al. 1977; Harding & Wright 1979; Monti Graziadei & Graziadei 1979; Graziadei & Monti Graziadei 1980a; Graziadei et al. 1980; Monti Graziadei et al. 1980a; Booth et al. 1981; Kawano & Margolis 1982; Doucette et al. 1983a,b; Baker et al. 1984; Càmara & Harding 1984; Costanzo 1984; Gozzo & Fülöp 1984; Samanen & Forbes 1984), olfactory sensory cells degenerate and die, but after a period of time they are replaced by new sensory cells. Reinnervation of denervated glomeruli occurs when the new axons reach the olfactory bulb (Graziadei & Monti Graziadei 1980a; Monti Graziadei et al. 1980a; Doucette et al. 1983b). Thus, the olfactory system is not like other peripheral nerves, which, after injury by axotomy, simply regenerate axons that grow to the target (see section 2.8.3). Olfactory neurons suffer death after injury, and the sensory epithelium is reconstituted with new olfactory sensory neurons (Doucette et al. 1983b).

Reconstitution of the epithelium after injury does not usually lead to complete recovery in mammals. In some experiments, recovery of chemical markers (Harding et al. 1977) and cell numbers (Samanen & Forbes 1984) approached control values, but in others recovery was less impressive, reaching only 60–70% of control values (Costanzo & Graziadei 1983). Other studies have shown that biochemical (Harding & Margolis 1976) and immunohistochemical markers (Monti Graziadei 1983; Hempstead & Morgan 1985) also do not return to control values. The reason for the variation from experiment to experiment probably lies in the several technical problems associated with axotomy and bulbectomy in mammals (Costanzo 1984; Farbman 1988): The olfactory axons are bundled into large numbers of small fascicles that pass through the bony cribriform plate to reach the bulb, which is closely applied to the plate. This anatomical arrangement of large numbers of very short nerve fascicles makes it particularly difficult to sever *all* axons, particularly those entering the most ventral aspect of the bulb. The other difficulty arises from the fact that the axotomy procedure usually results in some scarring on the cribriform plate, thus preventing the regrowth of axons into the bulb.

There are fewer technical difficulties associated with axotomy in fish, amphibians, and birds, because the axons are bundled into a single, easily accessible olfactory nerve that is considerably longer than nerves in mammals. Nevertheless, the results from studies to determine whether or not postaxotomy recovery has been complete have been inconclusive (Oley et al. 1975; Bedini, Fiaschi, & LanFranchi 1976; Kiyohara & Tucker 1978; Graziadei & Okano 1979; Simmons & Getchell 1981b; Simmons, Rafols, & Getchell 1981). What has come out of these experiments, however, is the idea that *sensory neurons pass through two phases of functional maturity. The first is independent of bulbar contact, but the second is dependent on presumed synaptic contact with bulbar neurons* (Simmons & Getchell 1981b; cf. Doucette et al. 1983a). In effect, then, sensory neurons can differentiate up to a point without bulbar influence, but the final stages of differentiation apparently are dependent on that influence.

Support for this idea comes from organ culture experiments. When olfactory mucosa is explanted alone, the sensory neurons undergo differentiation, but full maturation of all receptors does not occur. Axons grow out, a few cilia are expressed at the dendrite, OMP is synthesized by some cells, and electro-olfactograms can be recorded (Farbman & Gesteland 1975; Farbman 1977; Chuah & Farbman 1983, 1986; Chuah et al. 1985; Gonzales et al. 1985). Similar findings were reported in experiments in which olfactory mucosa was transplanted to the anterior chamber of the eye (Barber, Jensen, & Zimmer 1982; Morrison & Graziadei 1983; Novoselov et al. 1984).

The basis for these interactions is, at present, not understood. The question whether or not the bulbar factor is a diffusible molecule has been investigated in organ culture experiments. The results suggest that the bulbar enhancing effects on ciliogenesis and on OMP synthesis require contact between olfactory axons and bulbar tissue. When contact is prevented by insertion of a very thin (10- or 25-μm-thick) porous filter between the mucosal fragment and the bulb in combined explants, the influence of the bulb is prevented (Chuah & Farbman 1983; Chuah et al. 1985). Furthermore, other brain tissue cannot substitute for the bulb in eliciting the maturation of sensory cells.

The experiments cited have shown that sensory cell maturation, including ciliogenesis, is dependent on bulbar contact. In light of this, an interesting possibility to consider is that the bulb may actually specify the types of receptors on the cilia. The argument in favor of this might be stated as follows: The bulbar glomeruli have some degree of response specificity, as discussed in section 5.4. If direct contact with a postsynaptic target cell is required for ciliogenesis and complete maturation of the sensory cell, perhaps the response code for odor quality resides in the postsynaptic cell, and a signal from the postsynaptic cell travels retrogradely along the olfactory axon to the sensory cell body and specifies the types of odorant receptors it should synthesize and place in the ciliary membrane. Alternatively, or possibly in addition, this kind of retrograde signal could also specify the type of second messenger system used by the cell. This, of course, is entirely speculative, but not inconsistent with existing experimental evidence.

Recently it has been shown that the bulb may exert other influences on the epithelium. In bulbectomized rats, the rate of cell division in the epithelium of the operated side increases to twice that of the unoperated side and remains at that level for at least seven weeks (Carr & Farbman 1992). It is thought that the bulb participates in some way in the regulation of the rate of mitosis in the dividing cell population in olfactory epithelium.

In another study, the enzymes for synthesis and degradation of an excitotoxin, quinolinic acid, were shown to be differentially expressed by astrocytes in the olfactory bulb (Poston et al. 1991). Quinolinic acid is a product of the metabolism of the amino acid tryptophan and is endogenous to the brain. It is known to have a retarding effect on growth cone motility. The cellular localization of its synthetic enzyme was strongest in astrocytes at the junction between the glomerular and external plexiform layers, whereas the localization of the degradative enzyme was strongest in astrocytes at the boundary between the glomerular and olfactory nerve layers. If the localization of these two enzymes reflects a gradient of quinolinic acid concentration, low where the nerves enter the glomeruli and high at the glomerular–external plexiform boundary, it is possible that quinolinic acid may have an effect on regulating the ingrowth of new olfactory axons in the adult olfactory bulb and limiting them to glomeruli.

Taken together, these experiments strongly suggest that cellular interactions between sensory neurons and bulbar cells are indeed reciprocal and may play a critical role in the differentiation of both elements. Sensory neurons can influence the bulb at all stages, from early development to recovery after deafferentation. On the other hand, the sensory cell does not reach a fully mature structural and biochemical state unless it makes synaptic contact with the appropriate cells in its target organ, the olfactory bulb. Moreover, neuron–glial interactions play an important role in development of the olfactory nervous system. Thus, the olfactory system is no exception to the general rule that cellular interactions are extremely important in the development of neuronal systems.

The effect of the bulb on development and maturation of neurons connected to it is not limited to the sensory neuron population. As indicated in Chapter 5, regions of the basal forebrain, including the medial septum, project cholinergic fibers to the bulb. It was shown recently that soluble proteins from rat olfactory bulb promote survival and differentiation of cultured basal forebrain neurons (Lambert et al. 1988). The activity of the bulb extract was resistant to nerve growth factor antiserum. The identity of the active agent(s) has yet to be characterized.

6.5 Functional maturation

6.5.1 Sensory epithelium

It is clear that olfaction in mammals is functional at birth or earlier. Electrophysiological experiments to determine the time of onset of odorant-induced electrical activity in sensory neurons have shown that electro-olfactogram (EOG) readings are first detectable in rat fetuses on E14, and they mature relatively early, at about E19, well before most of the developing axons have reached the bulb (Gesteland et al. 1982). The first action potential in single sensory cells was recorded from E16 fetuses. However, at that time, the cells from which recordings were made were responsive to all odor stimuli administered (i.e., the responses were nonselective). A few days later, at E18–E19, increasing percentages of cells became responsive to only some stimuli in the stimulus set; in other words, at that later stage, cells were able to *discriminate* among odorants (Gesteland et al. 1982). This supports behavioral experiments suggesting that rodents have the ability to detect odors in utero (Stickrod, Kimble, & Smotherman 1982).

6.5.2 Olfactory bulb

Most newborn mammals have a limited repertoire of olfactory responses. Correlations between olfactory bulb organization and the ability to respond to certain odor stimuli have been shown in a study of behavioral differences between precocial and altricial rodents (Leon et al. 1984). For example, the attraction to maternal odor is demonstrable in two altricial species: in Norway rats at two postnatal weeks, and in gerbils at three weeks. In spiny mice, which are precocial, this behavior is demonstrable at one to two days after birth. The three animals are roughly similar in terms of morphological organization of the bulb (i.e., almost mature) at the time when this behavior is demonstrable (Leon et al. 1984). The results of this and other studies (Alberts 1981) argue that the ability to process olfactory information by the central nervous system undergoes changes during development, probably because of the postnatal changes in the anatomical substrate. For example, the output neurons (mitral cells) in the medial quadrant of the bulb are the first to send their axonal projections, via the lateral olfactory tract, to the olfactory cortex (Grafe & Leonard 1982). The relatively late appearance of granule cells and the reciprocal synapses between granule and mitral cells must also contribute to functional changes in behavior. So, too, must the postnatal growth of centrifugal fibers into the bulb (Schwob & Price 1984).

6.5.2.1 Glomeruli Behavioral studies have shown that neonatal mammals respond to olfactory cues to locate their mothers' nipples, although their olfactory acuity is low (e.g., Teicher & Blass 1976, 1977; Rudy & Cheatle 1977; Pedersen & Blass 1981; Hudson & Distel 1983, 1986). Experiments using the 2-deoxyglucose (2DG) technique showed that foci of increased metabolic activity could be elicited in the olfactory bulb in neonatal rats by the odors of pure chemicals, such as amyl acetate and ethyl acetoacetate, or by nest odor (Astic & Saucier 1982; Greer et al. 1982), or by stimulation of the olfactory nerves (Greer et al. 1981). The foci were localized over the glomerular layer, which, at this age, is poorly differentiated. The modified glomerular complex, a group of well-defined glomeruli located in the caudal, dorsomedial main olfactory bulb near the boundary with the accessory bulb, also exhibited focal 2DG uptake on the first postnatal day. The observations that glomeruli in the modified glomerular complex are well differentiated earlier than other glomeruli in the main olfactory bulb, and that they are metabolically active, have led to the postulate that this region of the bulb may be important in processing odor cues related to suckling behavior (Teicher et al. 1980; Greer et al. 1982) (see section 5.4.1).

6.5.2.2 Mitral cells Mitral cells exhibit both spontaneous activity and stimulus-evoked activity in newborn animals (Math & Davrainville 1980; Mair & Gesteland 1982). The synapses between sensory neurons and mitral cells are, therefore, apparently functional at birth, and the activity of this afferent pathway is linked to respiratory (rather than sniffing) behavior (Mair & Gesteland 1982).

However, the mitral cell response properties in newborn animals are qualitatively different from those seen in adults. In adults, single units recorded from cells in the mitral cell body layer exhibit three types of on-responses during the stimulus event. Type I on-responses are characterized by increased activity within 1 sec of the onset of a weak stimulus; stronger stimuli evoke two bursts of action potentials separated by a relatively quiet period. Type II on-responses are characterized by a diminished firing rate in response to weak stimuli and by a period of decreased firing followed by a sudden burst of action potentials after stronger stimuli. Type III on-responses are characterized by diminished activity for all concentrations of odorants that affect the rate of neural activity (see Figure 5.7) (Mair 1982). None of these response patterns are seen in newborn animals. On the contrary, response patterns of mitral cells in neonates are qualitatively similar to those described for sensory neurons. The paucity of granule cells and the lack of significant numbers of granule-to-mitral synapses in newborn animals may explain the observation that mitral cells transmit the temporal patterns of activity exhibited by receptor neurons with little modification (Mair & Gesteland 1982).

6.6 Plasticity in the olfactory system

In section 6.4.2.1 we discussed some aspects of plasticity, namely, the ability of sensory neurons to organize glomeruli in other parts of the nervous system and even to form synapse-like contacts with "incorrect" neurons in the central nervous system. Also discussed (section 6.4.2.3) was the role of the sensory neuron in inducing the expression of neurotransmitters in some bulbar neurons. In this section we turn our attention to other aspects of olfactory system plasticity that can be modified by experience.

6.6.1 Rearing in a single-odor environment

For several years it has been known that normal functional development of the olfactory system depends on the animal receiving a varied sensory experience. This has come out of studies in which animals were either deprived of olfactory input or continuously exposed to a limited odorous environment. When neonatal or two-week-old rats are exposed to a single predominant odor for 2–11 months, changes in size and morphology are observed in specific mitral cells in the olfactory bulb (Døving & Pinching 1973; Pinching & Døving 1974; Laing & Panhuber 1978). It is particularly significant that different subsets of mitral cells are affected, depending on the odor used during exposure (Pinching & Døving 1974; Panhuber & Laing 1987). The density of spines on granule cell dendrites is also reduced under these experimental conditions (Apfel-bach & Weiler 1985; Rehn et al. 1988). Other anatomical data on rat pups exposed to a single-odor environment showed morphometric mod-ifications in the volumes of bulbar layers. In rats that were exposed for one month after birth to a single odorant, ethyl acetoacetate, the vol-umes of all layers in the anterior and middle regions of the bulb were significantly reduced more than those in the posterior part of the bulb. This suggests that neuronal pathways activated by ethyl acetoacetate stimulation are mainly located in the posterior region of the bulb (Royet, Jourdan, & Ploye 1989a; Royet et al. 1989b).

These anatomical experiments, coupled with behavioral data (Laing & Panhuber 1978, 1980), indicate that mitral cells, glomeruli, and in-terneuronal cells associated with active mitral cells that remain un-changed during prolonged exposure to an odor are those stimulated by that odor, whereas the other cells become altered because of lack of olfactory input (Laing 1984; Laing et al. 1985; Royet et al. 1989a,b). The inclusion, in some of these experiments, of a group of animals receiving deodorized air made it possible to determine that mitral cells are smaller in animals reared under these conditions. Further, behavioral studies showed that the animals had a lower acuity for test odors (Laing & Panhuber 1980; Panhuber & Laing 1987).

Other studies, using 2DG injections in rats exposed to a single-odor environment, revealed bulbar glomeruli that had high glucose uptake, corresponding to increased neural activity (Sharp et al. 1975). Indeed, the use of 2DG has shown that the regions of odor-induced increases in metabolic activity in the glomerular layer can be as small as single glomeruli, the positions of which appear to be odor-specific and con-sistent among the animals in the group (Stewart et al. 1979; Jourdan et al. 1980; Greer et al. 1981; Royet et al. 1987).

These data, taken together, support the notion that glomeruli (and the mitral cells associated with them) are arranged, more or less, in an odor-specific pattern in the olfactory bulb. The ability to modify mitral cells during postnatal development indicates that the full maturation of the system is, at least in part, dependent on experience.

6.6.2 Sensory deprivation studies

Several workers have studied how restricting the sensory input affects the development of the olfactory system in neonatal and young rodents. When one naris in a neonatal animal is experimentally occluded, and the animal is allowed to survive for 30 days, the postnatal increases in size and weight of the ipsilateral olfactory bulb are significantly less than those for the bulb on the unoperated side (Meisami 1976; Meisami & Safari 1981; Meisami & Mousavi 1982; Brunjes & Borror 1983; Benson,

Ryugo, & Hinds 1984; Brunjes 1985; Brunjes, Smith-Crafts, & McCarty 1985; Skeen, Due, & Douglas 1985, 1986; Brunjes & Frazier 1986; Meisami & Noushinfar 1986). The volume change is seen in most layers of the bulb; the mitral cells are reduced in size (Benson et al. 1984), and there are dramatic decreases in the density and number of granule cells and glia in the granule cell layer (Frazier & Brunjes 1988). This is due to an increase in cell death in the granule cell population (Frazier-Cierpial & Brunjes 1989a). In addition, the bulb experiences reductions in total protein, DNA, RNA, acetylcholinesterase, sodium-potassium ATPase (Meisami & Mousavi 1982), total dopamine content (Brunjes et al. 1985), and succinic dehydrogenase and cytochrome oxidase activities (Cullinan & Brunjes 1987). On the other hand, there is an increase in nerve growth factor receptors (Gomez-Pinilla et al. 1989). Even when naris occlusion was done in older animals, four to nine weeks old, and continued for 10–14 days, a drastic reduction in the number of TH-immunoreactive cells in the glomerular layer of the bulb was shown (Kosaka et al. 1987). The morphological and biochemical changes in the bulb after unilateral naris occlusion were attributed to a reduction in activity of the olfactory afferent nerves to the bulb.

Some recent studies have documented functional changes in the odor-deprived bulb. Rat noses were unilaterally occluded on postnatal day 2 and reopened 20 days later. Recordings from single units in the deprived bulb showed that the mitral/tufted cells exhibited increased responsiveness when compared with those in the control bulb (Guthrie, Wilson, & Leon 1990). Furthermore, there was greater uptake of 2DG on the deprived side in response to odorant stimuli.

Paired pulse stimulation of the lateral olfactory tract ordinarily produces inhibition of mitral cell firing in untreated rats. In unilaterally odor-deprived rats the inhibition was significantly enhanced (Wilson, Guthrie, & Leon 1990). This was surprising in light of the data showing enhanced granule cell death resulting from unilateral odor deprivation (Frazier-Cierpial & Brunjes 1989a). A possible reason for this result may be that in addition to a reduction in number of granule cells, there is a reduction in the extent of lateral dendrites of mitral cells; there could be an actual increase in relative numbers of granule/mitral synapses, because the surviving granule cells do not undergo reductions in their dendritic trees (Wilson et al. 1990). These data demonstrate that unilateral deprivation modifies several aspects of olfactory function.

Unilateral naris occlusion also results in changes in the cell dynamics in the olfactory sensory epithelium. These are expressed as a reduction in epithelial thickness (Farbman et al. 1988; Stahl, Distel, & Hudson 1990) and a reduction in the number of not fully mature sensory cells, although no change is demonstrable in the number of supporting cells or mature sensory cells (Farbman et al. 1988). The diminished number of cells is attributable to a reduction in the rate of neurogenesis in the epithelium on the occluded side (Farbman et al. 1988). However, because the rate of mitosis in the respiratory epithelium on the occluded side is diminished to the same extent as that of the sensory epithelium, the changes in cell dynamics probably result from a protective effect of naris occlusion (i.e., chemical, physical, and infectious agents are prevented from entering the nasal cavity), rather than from a change in functional activity of sensory cells (Farbman et al. 1988; cf. Hinds et al. 1984). In other words, the protective effect of blocking the nostril reduces the stress on both the sensory and nonsensory epithelium lining the nasal cavity, perhaps resulting in longer-lived cells, and a lessened requirement for replacement cells. However, closure of the entrance to the nasal cavity also results in blockage of stimulus access; this almost certainly results in diminished functional activity, which is likely responsible for the bulbar changes.

6.6.3 Olfactory "learning"

An active area in neuroscience research involves analysis of the anatomical and physiological substrates of learning. It is generally believed that the memory of an experience can be retrieved or the learning of a behavior can occur only if there are stable changes in synaptic effectiveness. In this section we shall discuss two systems in which progress is being made in describing the mechanisms involved in increasing the efficacy of certain identified synapses. In both systems, a stable behavioral change is induced under the influence of norepinephrine selectively released from centrifugal fibers originating from the locus coeruleus and terminating in either the main or accessory olfactory bulb. There is further evidence in a third system, the sheep olfactory bulb, that a learned behavior is induced as a result of norepinephrine release at a critical time during the formation of bonding between the ewe and the newborn lamb.

6.6.3.1 Mother–infant bonding in rats   Within the olfactory system it is well known that young rat pups exhibit a strong behavioral preference for the specific maternal odor they experience early in life (Leon 1974, 1983). Several experiments have shown that rat pups will develop an attraction to artificial odors paired with a tactile, reinforcing stimulus to simulate the kind of stimulation they receive from maternal care, such as grooming (licking) or intraoral infusions of milk or sucrose (Leon, Galef, & Behse 1977; Brunjes & Alberts 1979; Johanson & Hall 1982; Caza & Spear 1984). Rat pups exposed to brief daily pairings of an (unfamiliar) odor and a reinforcing tactile stimulus exhibited enhanced behavioral attraction to the odor upon its subsequent presentation alone (Sullivan & Hall 1988).

Are these behavioral preferences in rat pups associated with changes in the nervous system that can be measured by anatomical or physiological methods? Some very interesting correlations between learning and measurable changes in the olfactory system have been reported. The pairing of a tactile stimulus and an unfamiliar odor, peppermint, results in enhanced bulbar uptake of 2DG in certain glomeruli (Sullivan & Leon 1986). The enhanced uptake of 2DG was associated with a 21% enlargement of the cross-sectional areas of the individual glomeruli affected (Woo, Coopersmith, & Leon 1987). The development of these anatomical changes occurs during a sensitive period, the first week after birth (Woo & Leon 1987). None of these changes are apparent in bulbs of control groups exposed to the peppermint odor alone or to the tactile stimulus alone.

Analysis of this phenomenon by single-unit recording from mitral and tufted cells in the bulb indicated that the response patterns of these secondary neurons to the conditioning peppermint odor stimulus were specifically reduced in frequency, while they remained unchanged to other odor stimuli (Wilson, Sullivan, & Leon 1985, 1987; Wilson & Leon 1988a). The mitral cells exhibiting changed response patterns were associated with glomeruli in those regions of the bulb where increased uptake of 2DG had been seen. The anatomical and physiological changes are thought to be the basis of the learned behavioral preference for peppermint odor.

In subsequent experiments, norepinephrine was shown to be important in acquisition of the learned behavioral and bulbar responses. Injection of propranolol, a norepinephrine β-receptor antagonist, prevented acquisition of the several changes associated with pairing of the tactile stimulus and the peppermint odorant. Mitral cell response changes did not occur after propranolol injection, nor did enhanced uptake of 2DG or the learned behavior. On the other hand, injection of isoproterenol, a β-receptor agonist, had no effect (Wilson & Leon 1988b; Sullivan, Wilson, & Leon 1989).

The significance of these experiments is that the norepinephrine input to the bulb may affect the reciprocal synapses between mitral and granule cells. The mechanism by which these synapses are affected is not entirely clear. There is evidence that a change in the granule cell does indeed occur, because when the animal is presented with the conditioning stimulus, the firing rate of the mitral cells is specifically reduced by way of the granule/mitral synapse. This mechanism has been proposed to explain the olfactory learning. It is possible that the norepinephrine acts to stimulate transmitter release from granule cells, thus reducing the firing rate of the mitral cell (see section 5.7.2) (McLennan, 1971; cf. McLean et al. 1989b).

Whatever the mechanism, the consequence of the norepinephrine input is that mitral cell firing is reduced after learning. The most interesting aspect of this phenomenon is that a series of stimuli during a critical period of development induces long-lasting modifications of synaptic activity. The cellular mechanism for the changes is not known, though it has been suggested that the norepinephrine activation of β-adrenergic receptors may induce an increase in cyclic AMP formation (Gervais, Holley, & Keverne 1988). This probably comes about by an up-regulation in the number of β-adrenergic receptors and the transduction apparatus (including G-proteins and the enzyme adenylate cyclase) involved in generation of cyclic AMP.

Given that the norepinephrine input to the olfactory bulb comes from the locus coeruleus (Shipley et al. 1985), one might wonder if activation of specific locus coeruleus cells, at the level of their cell bodies, could somehow be responsible for the norepinephrine effect on the bulb in olfactory learning. This is not likely, because (1) it would be difficult to explain how specific mitral cells could selectively signal the locus coeruleus, and (2) the locus coeruleus is a relatively small region in the hindbrain with relatively few cells that project widely throughout the entire brain. Consequently, if a cell body fired, there would be demonstrable effects elsewhere in the brain.

The data might be explainable by local modulation of norepinephrine release within the bulb, and there is evidence supporting this possibility. In anesthetized or awake immobilized rats, presentation of odor stimuli elicited a 3–10-fold rise in the efflux of [³H]norepinephrine above the baseline level, measured in the bulb with a push–pull cannula method. This did not result from a general arousal reaction (Chase & Kopin 1968). In the rabbit, antidromic electrical stimulation of mitral cell axons produced a similar effect on [³H]norepinephrine release in the olfactory bulb (Brennells 1974). The results of these two studies suggest that odor-induced activation of olfactory mitral cells could lead to a modulation of norepinephrine release in the bulb. Further support for the hypothesis that norepinephrine release may be modulated locally is derived from results of experiments on superfused slices of rat bulb. These studies have shown that GABA significantly enhances evoked release of [³H]norepinephrine without affecting its spontaneous release (Gervais 1987). This implies that GABA receptors on locus coeruleus nerve terminals could establish a functional link between mitral cell activity and local release of norepinephrine. Although this has not been demonstrated experimentally, it would provide a logical explanation for all of the data. Essentially this would mean that GABA-induced release of norepinephrine locally would be restricted to those noradrenergic terminals located in areas responding to the odor stimulus (Gervais et al. 1988). Hence, in an animal in which there was sustained activation of certain mitral cells responsive to a biologically meaningful stimulus, there could be local activation of norepinephrine release by a GABAergic mechanism that would not require a signal from the noradrenergic cell body in the locus coeruleus (Gervais et al. 1988). This could provide the norepinephrine necessary for establishing the olfactory memory.

6.6.3.2 Ewe/lamb recognition in sheep During the first few hours after parturition, the ewe establishes a selective bond with her own lamb, based exclusively on olfactory recognition (Poindron 1976). When the bond is established, the ewe will reject other newborn lambs. Depletion of norepinephrine in the ewe's main olfactory bulb during the bonding period prevents bonding, and she then will accept newborn lambs other than her own for suckling (Pissonnier et al. 1985). As was the case in the paired stimulus experiments described earlier in the rat model, activation of β-adrenergic receptors in the bulb evidently is required to establish olfactory recognition of the lamb (Lévy et al. 1990). The treated sheep still were able to discriminate control odors; that is neither the depletion of bulbar norepinephrine nor the blockage of β-adrenergic receptors resulted in a general impairment of olfactory functioning, only impairment of the formation of the particular olfactory memory.

In both sheep (Keverne et al. 1983) and rats (Yeo & Keverne 1986), vaginocervical stimulation is thought to be important for inducing maternal behavior at the time of birth. The peptide oxytocin, from the posterior lobe of the pituitary gland, stimulates maternal behavior in these animals. Studies of changes in olfactory bulb chemistry were carried out in nonpregnant sheep that had been primed with estrogen and stimulated in the vaginal and cervical regions with a plastic probe, and the levels of oxytocin, aspartate, GABA, and glutamate were all found to be elevated in these animals (Kendrick et al. 1988). These studies indicated that vaginocervical stimulation in sheep produces increases in olfactory bulb oxytocin to levels similar to those seen during parturition. Further, they indicate that at the critical time for olfactory recognition of the lamb, the transmitter systems at the bulbar dendrodendritic synapses are active. The release of these substances may reflect activation of the noradrenergic input to the bulb.

6.6.3.3 Pregnancy block in mice During a 4–6-hr critical period after mating, female mice form an olfactory imprint to the odor of the stud male. Once this memory is formed, exposure to an odor from a strange male, or even from the urine of a strange male, prevents implantation of the embryo (Bruce 1959, 1960; Parkes & Bruce 1962). This olfactory recognition is able to distinguish between males of different strains, but is generally not able to distinguish between individuals of the same strain. If the female is removed immediately after mating and placed with a strange male for a day, then returned to the stud male, pregnancy block will occur, as the female forms an olfactory imprint to the strange male during the postcoital critical period. Pregnancy block does not occur in anosmic females (Bruce & Parrott 1960). Pheromones mediating the pregnancy block selectively activate the accessory olfactory system via the vomeronasal organ (Lloyd-Thomas & Keverne 1982).

This behavioral expression of olfactory learning is also related to the release of norepinephrine by adrenergic neural input to the accessory bulb (from the locus coeruleus). If norepinephrine is depleted by microinfusion of 6-hydroxydopamine into the medial region of the olfactory tract during the 4-hr critical period, pregnancy block is prevented (Keverne & De La Riva 1982; Rosser & Keverne 1985). In some manner, then, the noradrenergic projection to the accessory olfactory bulb is activated during mating, possibly by cervical or vaginal stimulation, and remains active during the period of memory imprinting (Rosser & Keverne 1985). If norepinephrine is depleted after the critical period has passed and the stud male recognition imprint has formed, recognition of the stud male odor is retained, and pregnancy block is prevented. This implies that norepinephrine is required for formation of the memory, but is not necessary for the recall (Rosser & Keverne 1985). It has further been shown that blocking the α-adrenergic receptors, but not the β-adrenergic receptors (by infusing phentolamine into the accessory bulb immediately after mating), also prevents memory formation. The female retains the olfactory imprint of the stud male for 30–50 days after the original exposure (Brennan, Kaba, & Keverne 1990).

The memory for pregnancy block is stored in the accessory olfactory bulb, not in the amygdala, which is the next synaptic station for the vomeronasal system. Infusion of an anesthetic into the amygdala during the 4-hr postcoital period does not prevent formation of the recognition memory (Kaba, Rosser, & Keverne 1989). Infusion of the anesthetic into the female's accessory bulb, however, results in the stud male's pheromones blocking his own pregnancy (i.e., it interferes with formation of the memory protecting the female against pregnancy block) (Kaba et al. 1989). Lesions of the hippocampus, a part of the brain involved in learning and memory, are also without effect on olfactory memory formation in the context of pregnancy block (Selway & Keverne 1990).

Blocking the bulb's GABA receptors interferes with formation of the olfactory memory (Kaba et al. 1989). This may operate because the GABA receptors at the granule/mitral synapse are blocked; alternatively, if there are GABA receptors on noradrenergic terminals from the locus coeruleus (Gervais et al. 1988), the GABA blocker may work by blocking norepinephrine release. Local infusions of the excitatory amino acid receptor blocker γ-D-glutamylglycine into the accessory olfactory bulb prevents memory formation, but specific *N*-methyl-D-aspartate (NMDA) receptor antagonists have no effect (Brennan & Keverne 1989). It seems that NMDA receptors, involved in some forms of hippocampal memory formation, are not involved in the accessory bulb memory concerned with prevention of pregnancy block. The findings implicate both excitatory amino acid receptors and GABA receptors in memory formation to male urinary pheromones. Thus, the data point to the reciprocal dendrodendritic synapse between mitral and granule cells in the accessory olfactory bulb as the site involved in the formation of the olfactory memory to the stud male pheromone.

The mechanism for pregnancy block resulting from exposure of recently mated female mice to odors of strange males has been studied extensively. Exposure of female mice to male pheromones can result in suppression of prolactin (luteotropic hormone) secretion by the anterior lobe of the pituitary. Secretion of dopamine by arcuate nucleus cells in the hypothalamus inhibits the neuroendocrine pathway involved in pituitary secretion of prolactin. The hypothalamus is the next synaptic station in the vomeronasal system after the amygdala. Suppression of prolactin secretion interferes with the priming of the uterus for implantation of the embryo. Prolactin is required to maintain the corpus luteum, which in turn provides the hormones needed for preparation of the uterus. Thus, the cascade of events initiated by the male mouse pheromone stimulus ultimately works against implantation of the embryo. A simplified scheme is represented as follows:

$$\text{granule cells}$$
$$\text{excitation} \uparrow \quad \downarrow \text{inhibition}$$
$$\text{male pheromone} \rightarrow \text{VNO} \rightarrow \text{mitral cells} \rightarrow \text{amygdala} \rightarrow \text{hypothalamus}$$
$$\downarrow$$
$$\text{pregnancy block} \leftarrow \leftarrow \text{suppression of prolactin} \leftarrow \leftarrow \text{dopamine release}$$

Continuation of the pregnancy is dependent on successful formation of an olfactory imprint to the odor of the stud male during a critical period after coitus, thus preventing him from blocking his own pregnancy. The formation of the memory may involve a reduction of excitability of a subset of mitral cells responsive to the odor of the stud male pheromone. This interruption in the cascade, dependent at least in part on strengthening the granule/mitral inhibitory synapse, prevents the suppression of prolactin secretion, and implantation can occur. If a different subset of mitral cells is activated, that is, cells not involved in the specific memory formation and responsive to pheromones of a foreign (non-stud) male, prolactin secretion is suppressed, implantation does not occur, and the pregnancy is blocked.

The studies described in the preceding paragraphs, with mice and rats, and those with sheep, have resulted in the suggestion that a local increase in norepinephrine release within a specific region of the main or accessory olfactory bulb during biologically significant periods is closely associated with establishment of a memory to selective olfactory cues. In a recent review, Gervais et al. (1988) compared the findings regarding the mechanisms of norepinephrine action in the main and accessory olfactory bulbs and the findings from studies on plasticity in other parts of the brain: the visual cortex and the hippocampus. The common theme in these studies is that norepinephrine release is associated with the long-range retention of biologically meaningful information within the nervous system. In the olfactory system the reciprocal dendrodendritic synapse between mitral and granule cells is somehow involved. Cyclic AMP has been implicated. It has also been shown that blockade of protein kinase C prevents memory formation in the accessory olfactory bulb (Kaba et al. 1989). The precise cellular mechanism of this neural plasticity in the olfactory bulb remains to be determined.

6.6.3.4 Access to stored olfactory memory Acquisition of the kinds of olfactory memories described earlier can be confined to only one olfactory bulb if animals are trained in a particular way. In six-day-old rat pups it was found that the odor preference memory could be confined to one side if the contralateral naris was occluded during the training period. The training involved the pairing of a normally aversive odor with a reinforcing stimulus such as oral infusion of milk or sucrose (Kucharski, Johanson, & Hall 1986). Subsequent testing showed that the pups exhibited a preference for the odor when tested with the same naris that was open during training, but exhibited an aversion when tested with the contralateral (untrained) naris open (and the trained naris occluded).

When the same experiment was carried out in 12-day-old pups, the behavioral preference response was exhibited when either the trained or untrained naris was open (Kucharski & Hall 1988). This indicates that the learned odor preference of the older pups tested with only the untrained naris open depends on access to stored memories in the olfactory system on the trained side. This access is provided by the development of the connections passing by way of the anterior commissure from one side of the olfactory system to the other (Kucharski, Burka, & Hall 1990). These cross-projections develop during the second postnatal week in rats (Schwob & Price 1984).

These results are not consistent with those in the pregnancy block

experiments described earlier, which point to the granule/mitral reciprocal synapse in the accessory bulb as the place where olfactory memory
is stored. Rather, they suggest that memory storage in unilaterally
trained animals probably does not reside in the olfactory bulb itself, but
may be localized in the anterior olfactory nucleus or the anterior olfactory cortex, one or two synapses more central than the bulb (Kucharski
et al. 1990). It is clear that not all olfactory memories or recognitions
are formed in the same way. For example, the ability of female mice
to discriminate between normal and castrate male urine is carried by
the main olfactory system, not the vomeronasal pathway (Lloyd-Thomas
& Keverne 1982). The hippocampus is involved in the acquisition of
some learned olfactory discrimination tasks (Staubli et al. 1984), but
not the memory involved in preventing pregnancy block. It may be that
there are differences between neonatal animals and adults in the ways
memories are formed and recalled, as well as other differences depending on the relation between the olfactory cue and the learned
behavior.

7

Concluding Remarks

The preceding chapters have reviewed our current knowledge of olfactory cell biology. The word "current" is used with the awareness that by the time these words reach the reader, further advances will certainly have occurred. At this time, however, several gaps remain in our understanding of the olfactory system. In these concluding remarks, some of the major unanswered questions are discussed. This discussion is not meant to be comprehensive. The selection of topics reflects the author's bias and interest.

7.1 Odor receptor molecules

One unresolved matter is the identity of the olfactory receptor molecule(s). The issue was reviewed in Chapter 3, where it was noted that techniques in molecular biology are now being applied to identify these receptors. Some receptors for external chemicals have been identified and characterized in membranes of unicellular organisms, and their genes have been cloned (Burkholder & Hartwell 1985; Hagen et al. 1986; Klein et al. 1988). All of them have in common their linkage to a transduction system involving G-protein and a second messenger. These receptor molecules are all proteins with seven hydrophobic regions, thought to be transmembrane domains (see Figure 3.3). In the vertebrate olfactory epithelium, several candidate receptor molecules have been identified on the basis of their resemblance to other members of a superfamily of receptor proteins (in other tissues) that are linked to a G-protein-mediated transduction system and have seven transmembrane domains (Buck & Axel 1991). However, at this writing, there is no direct evidence that these are the functional olfactory receptor molecules. Now that powerful techniques are available, it is entirely likely that the identity of vertebrate olfactory receptors will be discovered during the 1990s, perhaps even before this book is published.

7.2 Coding in the olfactory system

The issue of olfactory coding has been addressed by many workers during the past three decades (reviewed by Kauer 1987). The basic question is, How is the sensory information coded and sent to the brain so that the brain can then interpret the information and respond? One way of addressing this issue is to determine if there are subtypes of sensory neurons, and, if so, how many. To what degree might this be species-dependent? Immunohistochemical data in rodents, rabbits, and amphibia suggest the possibility that there are subtypes both in the main olfactory sensory cell population (section 2.6.2) and in the vomeronasal organ, but we do not yet understand how or indeed whether the subtypes are related to cell specificity. If so, is the specificity related to the cell's responsiveness to a certain class of odors? For example, might there be pheromone-specific sensory cells in mice that are responsive to the stimulus or stimuli leading to pregnancy block? Might sensory cell subtypes be based on different combinations of receptor molecules or different constellations of signal transduction molecules? Could subtyping be related to the age of the cell? That is, does the cell synthesize one set of molecular "fingerprints" in its youth and a different one in its maturity? Is the immunological specificity related to topographical connections in the bulb, that is, the ability of the cell's axon to grow toward and establish synaptic connections with a certain region of the olfactory bulb?

A related question is, Are there subtypes of glomeruli or secondary neurons in the main olfactory bulb? Once again, there are immunological and physiological data suggesting that specificity does exist among regions in the olfactory bulb; that is, individual or small groups of glomeruli in certain regions of the bulb are responsive to particular odors. How is this topography represented on the olfactory cortex? What is the basis for the ability to discriminate among hundreds of different odorant stimuli? Does the analysis occur at the receptor, bulbar, or cortical level, or perhaps at all three levels? Does the analysis for the noncognitive aspects of olfactory behavior (those related to mating, alarm reactions, kin recognition, homing, etc.) occur in different regions of the brain than that for the hedonistic analysis of odors?

A promising approach to the study of coding is being made with some new imaging methods that permit examination of the evoked olfactory response of living epithelium or bulb in real time. This involves the use of voltage-sensitive dyes that are taken up by the responsive cells and then fluoresce when the voltage in the cell is changed by stimulating it with an odor (Kauer 1988; Brouwer et al. 1991; Gesteland, Brouwer, & Farmer 1991). Thus, one can examine a region of living epithelium, for example, and determine the spatial and temporal patterns of sensory cells responding to different odor stimuli, different concentrations of a given stimulus, and odor mixtures. This can be done optically, in real time, by relatively noninvasive methods.

7.3 Transduction

With respect to transduction of the odorant signal, monumental advances have been made during the latter half of the 1980s and the early 1990s, so that we know much more about intracellular transduction mechanisms than we do about the receptors or the basis of the cell subtypes. Work by several investigators has led to the identification of two major signal transduction mechanisms in vertebrate and crustacean olfactory cells, namely, the adenylate cyclase–cyclic AMP system and the phosphoinositide system (see Chapter 3). The latter (but not the adenylate cyclase system) seems to function in olfactory transduction in insects as well. The linkage between receptor and transduction mechanism is not clearly understood. It has been observed that some odors will cause hyperpolarization in some sensory cells and depolarization in others (Ache et al. 1989). This could mean that a given receptor molecule might be linked to two different transduction systems in different sensory cells, or there might be two receptors for the same stimulus, each linked to a different transduction system. It is possible that other signal transduction systems remain to be discovered.

7.4 Olfactory mucus and the role of supporting cells

The mucus bathing the responsive region of olfactory sensory neurons has received some attention in recent years (reviewed by T. Getchell et al. 1984; W. Carr, Gleeson, & Trapido-Rosenthal 1990b). Identification of olfactory binding proteins in vertebrates and in arthropods (section 3.2.1) has fueled interest in the composition and function of the perireceptor fluids. So, too, has the recognition of the continual need to cleanse the sensory epithelium by removing or degrading odor molecules (as well as toxic compounds). The glandular elements continually produce a watery mucus that now is known to contain a large variety of substances, including components of the immune barrier (M. Getchell & T. Getchell 1991). In humans who require long-term drug therapy, the mucus could also contain the drugs or their metabolites, either or both of which might affect receptors or the transduction mechanism, thereby modifying the olfactory response (Mania-Farnell, Bruch, & Farbman 1991). The complexity of the mucus is increased because it is in a state of flux; recent evidence suggests that at least some compounds are not homogeneously distributed (B. Menco & A. Farbman unpublished data). One can then imagine that local areas of the olfactory epithelial surface can be exposed to mucus of varying composition over a period of time. How might this affect olfactory function? What kinds of enzymes might be in the mucus, and how might they affect odor molecules? What precisely is the role of the supporting cells? Do they control ion flux into and out of the mucus? Do the supporting cells or Bowman's glands control the pH of the mucus, and, if so, how? Further, given that the olfactory sensory neuron is replaced within the epithelium so readily, why are not the supporting cells replaced at a similar rate?

7.5 Neurotransmitter

One would think that with all of the information we have about the sensory neuron, we would at least know the identity of its neurotransmitter. During the late 1970s and early 1980s the leading candidate for the olfactory neurotransmitter was the dipeptide carnosine, but there are conflicting data about the true function of this molecule. There has been little or no recent work directed to this problem; no other putative neurotransmitter has been proposed. Perhaps carnosine really is the neurotransmitter and this will ultimately be proved by new experiments.

7.6 Sensory cell replacement

Olfactory sensory neurons are unusual because they are continually produced throughout adult life. We have known for half a century that they are replaced in response to injury. More recently, it has become apparent that the neurons are produced under physiological conditions (section 2.7). We have yet to discover how genesis of these sensory neurons is regulated. If new neurons are continually made during adulthood, and the total size of the olfactory mucosa remains unchanged, then neurons must be continually dying. How is cell death in the neuronal population regulated?

The existence of factors regulating the rate of production of new neurons is implicit from the results of experiments showing that the rate of neurogenesis can be manipulated. In what might be called the wound-healing mode (i.e., in response to an injury), the rate of new neuron production is significantly increased, and the rate of neuron death is concomitantly increased (Carr & Farbman 1992). What factors are responsible for these phenomena, and where do they come from? Preliminary data from organ culture experiments suggest that epidermal growth factor can increase the rate of cell division in olfactory epithelium (A. Farbman unpublished data). No doubt other factors might do the same thing in an in vitro assay, but the mechanism in the living animal is not known.

The olfactory sensory neuron can serve as a model for the study of cell replacement in the central nervous system. If we knew why the globose basal cell is responsive to mitotis-inducing agents, this might lead us to the search for cells with similar properties in the brain. Adult cold-blooded animals have regenerative powers in the central nervous system. In fact, parts of their brains continue to make new neurons while the animals continue to grow. Indeed, in adults of some warm-blooded species, notably songbirds, there are cells in at least one part of the brain that can undergo proliferation and form new neurons (Paton & Nottebohm 1984). Granule cells in the olfactory bulb are added in small numbers during adult life as a result of continued mitosis at a low level (Kaplan & Hinds 1977; Forbes 1984; Kaplan et al. 1985). Are there other such cells in the adult mammalian brain for which we need to find the proper stimulus? Could we then promote the formation of new neurons in the mammalian central nervous system to replace those lost as a result of injury, for example, brain cells lost because of a stroke or illness (e.g., Alzheimer's disease), or those lost in the spinal cord as a result of trauma? Is it too optimistic to believe that it might be possible to use transplanted olfactory sensory neurons as replacement cells for lost neurons in the brain or spinal cord? The fact that olfactory neurons are always presynaptic and never postsynaptic suggests that they might not be able to function as interneurons, but they do make dopamine D2 receptors and acetylcholine receptors and perhaps others. Could they be induced to insert a sufficient number of neurotransmitter receptors in a localized region of membrane that could functionally serve as a postsynaptic site in a transplant?

7.7 Growth of olfactory nerves

Olfactory nerves have certain qualities not shared by other nerves that enter the central nervous system. They are not inhibited from growing from their peripheral location in the nasal cavity to the bulb, which is in the central nervous system. No other nerves retain an ability to grow from the periphery into the central nervous system in adulthood. The nonolfactory nerves are stopped at the glia limitans, the outer glial boundary of the central nervous system. Moreover, neurons in the central nervous system do not regenerate within the brain or spinal cord. It has been shown that central nervous system (CNS) myelin contains two proteins that inhibit growth of axons (Caroni & Schwab 1988), and these inhibitory molecules may explain, at least in part, why regeneration occurs to a limited extent or not at all in mammalian brain. If there are inhibitory molecules, it is implicit that receptors for these molecules must exist on the neurons that are not able to grow axons in the vicinity of CNS myelin. The interaction between the inhibitory molecules and the receptors on the membrane of the growing axon inhibits further growth in some manner. It is remarkable however, that CNS neurons can grow axons perfectly well on a substrate of peripheral nervous system myelin, which does not contain these inhibitory proteins.

How does this relate to olfactory nerves? Under physiological conditions, olfactory nerves never encounter myelin during their lifetime. Preliminary data indicate that olfactory axons can grow on CNS myelin (Farbman & Buchholz 1991). Olfactory axons can invade myelin-containing regions of the CNS and make synapses with foreign neurons (Graziadei et al. 1979). This suggests that these neurons, unlike other central and peripheral neurons, may not contain the receptors for the CNS myelin inhibitory proteins. If the olfactory axons lack these putative receptors, or if they are present and inactivated, further study may lead to the possibility of genetically or functionally altering other neurons to enable them to regenerate within the CNS.

When the olfactory axons find their target in the CNS, they elicit glomerular formation. They can do this even when they find inappropriate targets. What is the mechanism for this unusual phenomenon, and how much can we draw on the insect model (Tolbert & Oland 1989) to explain events in vertebrates? What role do the glia play in the formation of the glomeruli?

On a more microscopic level, what are the signals and the response mechanism for cessation of axonal growth and formation of a functional synapse? This is a fundamental question for developmental neurobiologists, for it applies widely in the growing nervous system.

Another problem that deserves further investigation is to what extent the olfactory nerve is a portal of entry to the nervous system for toxic or infectious agents. In section 2.8.2 the evidence for transport of certain viruses and other particulate agents into the CNS by way of the olfactory nerve was discussed. Little is known about the interactions between these particulates and the membrane of the olfactory sensory neuron that permits the cell to incorporate the particles. Once having incorporated the particles, why does the cell transport them along the axon instead of perhaps sequestering them and destroying them as a macrophage might do? Are these particulates also ingested by supporting cells? The answers to these questions may have important clinical implications.

7.8 Behavior

As indicated in the opening chapter, the ultimate reason for studying the biology of the olfactory system is to gain an understanding of olfactory-driven behavior. Human behavior is not so dependent on olfactory cues as is that of other vertebrates. Yet we are responsive to attractive and repulsive olfactory stimuli. Our responses to these stimuli do play an important role in certain aspects of our lives, particularly in our enjoyment of food and our ability to detect odors related to certain kinds of danger (e.g., smoke or other noxious elements). They may also play a noncognitive role in our attraction to members of the opposite sex, and in mother–infant bonding. Some olfactory stimuli evoke clear memories of past experiences.

In nonhuman vertebrates, some behavioral responses can be learned at various times in development (section 6.6.3). But other behaviors seem to be innate, such as nipple-search behavior in newborn mammals, or the response of the newly hatched snake to odorants in a food source. These behaviors are extremely complex because they involve binding of the odor stimulus to a receptor, transduction into a nerve response, excitation of the many neurons involved in a network (i.e., transmission of nerve impulses across several synapses in the CNS), and elicitation of directed motor behavior. Other response patterns are linked to the endocrine system. For example, in pregnancy block in mice, the hormonal balance of the pregnant female is altered (section 6.6.3.3); another example of olfactory-driven endocrine effects is seen in the courtship behavior of goldfish, in which the hormonal balance in males responding to female pheromones is altered, leading to formation and release of sperm (section 1.8.3).

The neural networks involved in many olfactory-driven behaviors are undoubtedly genetically determined. We can ask how these networks are constituted and regulated. What is the basis for the formation of the pathway that links certain sensory cells to the endocrine system, and other cells in the same animal to other efferent systems? Are there specific pathways (labeled lines?) from sensory cells to the CNS for stimuli related to predator avoidance? For stimuli related to kin recognition? For recognition of the male by the female, or vice versa? Do the CNS targets determine which receptors will be present in the sensory cells linked to them? Or do sensory cells first make certain receptors and then link themselves to pathways appropriate for the odors to which they are responsive?

In unicellular organisms, such as *Paramecium* or yeasts, stimulus binding, processing of the chemical signal, and behavioral response are all accomplished in the same cell. In vertebrates, coordinated networks of neuronal and nonneuronal cells are involved. For most behaviors, we have little evidence about the odor stimuli that drive them. We have little evidence of the network components and how they become linked into a network. In the wild, a mixture of odor stimuli could elicit different behaviors in different organisms. For example, the odors emanating from a female mouse in estrus would elicit one kind of behavior in a male mouse, but a very different behavior in a predator, such as a large snake. On the other hand, a rabbit might be completely indifferent to the mouse odorant cocktail. Do the snake and the male mouse respond to the same or different odor molecules in the mixture? Does the rabbit lack receptors for all the odorants in the cocktail, or if it can detect the odors, is its indifference based on the linkage of sensory cells to a different neural network?

7.9 Parallel sensory systems

Most of the fundamental questions pertaining to the biology of the olfactory system in the main nasal cavity apply as well to the vomeronasal and septal organs, because the sensory cells are similar. The other nerves in the nasal cavity, the trigeminal and terminal nerves, present other issues.

Branches of the trigeminal nerve terminate within the olfactory epithelium, essentially in direct contact with olfactory sensory neurons. Is there any communication between the trigeminal elements and the olfactory sensory neurons that might affect the physiological responses of either? For example, can the neuropeptides in the trigeminal nerve modulate the sensitivity of olfactory sensory neurons in any way? Can these peptides influence differentiation and maturation of neighboring sensory neurons?

The terminal nerve remains almost as mysterious now as it was when it was first described. Its presence in virtually all vertebrates, even toothed whales, which have essentially no olfactory system, indicates that it must have a function. Does it have an olfactory or olfactory-related (perhaps vomeronasal) function? What is the significance of its cells that contain luteinizing hormone-releasing hormone (LHRH)? Is this part of a yet-to-be-discovered olfactory–endocrine axis?

7.10 Final thoughts

Many of the exciting research findings in the olfactory system, particularly during the past decade, have come about as a result of the development of more sophisticated techniques and major advances in research by large numbers of dedicated scientists working on other biological systems. I hope it is clear that this book should be taken in the context of a progress report, a review of the current status. By no means do we have answers to many puzzling questions about how the olfactory system works. Those readers interested in research on this system need not be deterred, for many avenues must be traveled before the system is fully understood.

References

Ache, B.W. (1987) Chemoreception in invertebrates. In *Neurobiology of Taste and Smell*. Eds., T.E. Finger & W.L. Silver. Wiley, N.Y. pp. 39–64.

Ache, B.W., McClintock, T.S., Schmiedel-Jakob, I. & Michel, W.C. (1989) Evidence for multiple transduction pathways in lobster olfactory receptor neurons. In *ISOT X. Proceedings of the Tenth International Symposium on Olfaction and Taste*. Ed., K. Døving. GCS A/S, Oslo. pp. 101–105.

Adamek, G.D., Gesteland, R.C., Mair, R.G. & Oakley, B. (1984) Transduction physiology of olfactory receptor cells. *Brain Res.*, **310**,87–97.

Adams, D.A. & McFarland, L.Z. (1971) Septal olfactory organ in Peromyscus. *Comp. Biochem. Physiol.*, **40A**,971–974.

Adams, D.R. & Wiekamp, M.D. (1984) The canine vomeronasal organ. *J. Anat.*, **138**,771–787.

Adler, J. (1987) How motile bacteria are attracted and repelled by chemicals: an approach to neurobiology. *Biol. Chem. Hoppe-Seyler*, **368**,163–173.

Adrian, E.D. (1950) The electrical activity of the mammalian olfactory bulb. *Electroenceph. Clin. Neurophysiol.*, **2**,377–388.

Akeson, R.A. (1988) Primary olfactory neuron subclasses. In *Molecular Neurobiology of the Olfactory System*. Eds., F.L. Margolis & T.V. Getchell. Plenum, N.Y. pp. 297–318.

Akeson, R.A. & Haines, S.L. (1989) Rat olfactory cells and a central nervous system neuronal subpopulation share a cell surface antigen. *Brain Res.*, **488**,202–212.

Alberts, J.R. (1981) Ontogeny of olfaction: Reciprocal roles of sensation and behavior in the development of perception. In *Development of Perception. Vol. 1*. Eds., R.N. Aslin, J.R. Alberts & M.R. Petersen. Academic Press, N.Y. pp. 321–357.

Allen, W.K. & Akeson, R. (1985a) Identification of a cell surface glycoprotein family of olfactory receptor neurons with a monoclonal antibody. *J. Neuroscience*, **5**,284–296.

Allen, W.K. & Akeson, R. (1985b) Identification of an olfactory receptor neuron subclass: cellular and molecular analysis during development. *Devel. Biol.*, **109**,393–401.

Allison, A.C. (1953) The morphology of the olfactory system in the vertebrates. *Biol. Rev.*, **28**,195–244.

Allison, A.C. & Warwick, R.T.T. (1949) Quantitative observations on the olfactory system of the rabbit. *Brain*, **72**,186–197.

Alonso, J.R., Lara, J., Coveñas, R. & Aijón, J. (1988) Two types of mitral cells in the Teleostean olfactory bulb. *Neuroscience Res. Comm.*, **3**,113–118.

215

216   *References*

Alonso, J.R., Lara, J., Miguel, J.J. & Aijón, J. (1987) Ruffed cells in the olfactory bulb of freshwater teleosts. I. Golgi impregnation. *J. Anat.*, **155**,101–107.

Altman, J. (1966) Autoradiographic and histological studies of postnatal neurogenesis. II. A longitudinal investigation of the kinetics, migration and transformation of cells incorporating tritiated thymidine in infant rats, with special reference to postnatal neurogenesis in some brain regions. *J. Comp. Neurol.*, **128**,431–474.

Altman, J. (1969) Autoradiographic and histological studies of postnatal neurogenesis. IV. Cell proliferation and migration in the anterior forebrain, with special reference to persisting neurogenesis in the olfactory bulb. *J. Comp. Neurol.*, **137**,433–458.

Altner, H., Müller, W. & Brachner, I. (1970) The ultrastructure of the vomeronasal organ in Reptilia. *Z. Zellforsch.*, **105**,107–122.

Amoore, J.E. (1963a) Stereochemical theory of olfaction. *Nature*, **198**,271–272.

Amoore, J.E. (1963b) Stereochemical theory of olfaction. *Nature*, **199**,912–913.

Amoore, J.E. (1967a) Specific anosmia: a clue to the olfactory code. *Nature*, **214**,1095–1098.

Amoore, J.E. (1967b) Stereochemical theory of olfaction. In *Chemistry and Physiology of Flavors*. Eds. H.W. Schultz, E.A. Day & L.M. Libbey. Avi Publishing Co., Westport, CT. pp. 119–147.

Amoore, J.E. (1970) Computer correlation of molecular shape with odour: a model for structure-activity relationships. In *Taste and Smell in Vertebrates. A Ciba Foundation Symposium*. Eds., G.E.W. Wolstenholme & J. Knight. Churchill, London. pp. 293–306.

Amoore, J.E. (1977) Specific anosmia and the concept of primary odors. *Chem. Senses Flavour*, **2**,267–281.

Amoore, J.E. (1982) Odor theory and classification. In *Fragrance Chemistry*. Ed., E.T. Theimer. Academic Press, N.Y. pp. 28–76.

Amoore, J.E. & Forrester, L.J. (1976) Specific anosmia to trimethylamine: the fishy primary odor. *J. Chem. Ecol.*, **2**,49–56.

Amoore, J.E., Forrester, L.J. & Buttery, R.G. (1975) Specific anosmia to L-pyrroline: the spermous primary odor. *J. Chem. Ecol.*, **1**,299–310.

Amoore, J.E., Forrester, L.J. & Pelosi, P. (1976) Specific anosmia to isobutyraldehyde: the malty primary odor. *Chem. Senses Flavour*, **2**,17–25.

Amoore, J.E., Palmieri, G. & Wanke, E. (1967) Molecular shape and odour: pattern analysis by PAPA. *Nature*, **216**,1084–1087.

Amoore, J.E., Pelosi, P. & Forrester, J.L. (1977) Specific anosmias to 5α-androst–16-en–3-one and w-pentadecalactone: the urinous and musky primary odors. *Chem. Senses Flavour*, **2**,401–425.

Anasari, K.A. & Johnson, A. (1975) Olfactory function in patients with Parkinson's disease. *J. Chron. Dis.*, **28**,493–497.

Anderson, P.A.V. & Ache, B.W. (1985) Voltage- and current-clamp recordings of the receptor potential in olfactory receptor cells in situ. *Brain Res.*, **338**,273–280.

Anderson, P.A.V. & Hamilton, K.A. (1987) Intracellular recordings from isolated salamander olfactory receptor neurons. *Neuroscience*, **21**,167–173.

Andres, K.H. (1966) Der Feinbau der Regio olfactoria von Makrosmatikern. *Z. Zellforsch.*, **69**,140–154.

Andres, K.H. (1969) Der olfaktorisch Saum der Katze. *Z. Zellforsch.*, **96**,250–274.

Andres, K.H. (1970) Anatomy and ultrastructure of the olfactory bulb in fish, amphibia, reptiles, birds and mammals. In *Taste and Smell in Vertebrates. A Ciba Foundation Symposium.* Eds., G.E.W. Wolstenholme & J. Knight. Churchill, London. pp. 177–194.

Ånggard, A., Lundberg, J.M., Hökfelt, T., Nilsson, G., Fahrenkrug, J. & Said, S.J. (1979) Innervation of the cat nasal mucosa with special reference to relations between peptidergic and cholinergic neurons. *Acta Physiol. Scand. (Suppl.)*, **473**,50.

Anholt, R.R.H. (1987) Primary events in olfactory reception. *Trends Biochem. Sci.*, **12**,58–62.

Anholt, R.R.H. (1988) Functional reconstitution of the olfactory membrane: incorporation of the olfactory adenylate cyclase in liposomes. *Biochemistry*, **27**,6464–6468.

Anholt, R.R.H. (1989) Molecular physiology of olfaction. *Amer. J. Physiol.*, **257**,C1043–C1054.

Anholt, R.R.H., Aebi, U. & Snyder, S.H. (1986) A partially purified preparation of isolated chemosensory cilia from the olfactory epithelium of the bullfrog, *Rana catesbeiana. J. Neuroscience*, **6**,1962–1969.

Anholt, R.R.H., Mumby, S.M., Stoffers, D.A., Girard, P.R., Kuo, J.F. & Snyder, S.H. (1987) Transduction proteins of olfactory receptor cells: identification of guanine nucleotide binding proteins and protein kinase C. *Biochemistry*, **26**,788–795.

Anholt, R.R.H., Murphy, K.M.M., Mack, G.E. & Snyder, S.H. (1984) Peripheral-type benzodiazepine receptors in the central nervous system: localization to olfactory nerves. *J. Neuroscience*, **4**,593–603.

Anholt, R.R.H., Petro, A.E. & Rivers, A.M. (1990) Identification of a group of novel membrane proteins unique to chemosensory cilia of olfactory receptor cells. *Biochemistry*, **29**,3366–3373.

Anholt, R.R.H. & Rivers, A.M. (1990) Olfactory transduction: cross-talk between second-messenger systems. *Biochemistry*, **29**,4049–4054.

Ansari, K.A. & Johnson, A. (1975) Olfactory function in patients with Parkinson's disease. *J. Chron. Dis.*, **28**,493–497.

Apfelbach, R. (1986) Imprinting on prey odours in ferrets (*Mustela putorius F. Furo L.*) and its neural correlates. *Behav. Proc.*, **12**,363–381.

Apfelbach, R. & Weiler, E. (1985) Olfactory deprivation enhances normal spine loss in the olfactory bulb of developing ferrets. *Neuroscience Lett.*, **62**,169–173.

Ard, M.D., Morest, D.K. & Hauger, S.H. (1985) Trophic interactions between the cochleovestibular ganglion of the chick embryo and its synaptic targets in culture. *Neuroscience*, **16**,151–170.

Arey, L.B. (1974) *Developmental Anatomy.* Saunders, Philadelphia.

Astic, L., Le Pendu, J., Mollicone, R., Saucier, D. & Oriol, R. (1989) Cellular expression of H and B antigens in the rat olfactory system during development. *J. Comp. Neurol.*, **289**,386–394.

Astic, L. & Saucier, D. (1982) Ontogenesis of the functional activity of rat olfactory bulb: autoradiographic study with the 2-deoxyglucose method. *Devel. Brain Res.*, **2**,243–256.

Astic, L. & Saucier, D. (1986) Anatomical mapping of the neuroepithelial projection to the olfactory bulb in the rat. *Brain Res. Bull.*, **16**,445–454.

Astic, L. & Saucier, D. (1988) Topographical projection of the septal organ to the main olfactory bulb in rats: ontogenetic study. *Devel. Brain Res.*, **42**,297–303.

Astic, L., Saucier, D. & Holley, A. (1987) Topographical relationships between olfactory receptor cells and glomerular foci in the rat olfactory bulb. *Brain Res.*, **424,**144–152.

Bailey, M.S., Poston, M.R. & Shipley, M.T. (1989) Unique morphology of olfactory bulb radial glia: role in glomerular development. *Soc. Neuroscience Abstr.*, **15,**690.

Bakalyar, H.A. & Reed, R.R. (1990) Identification of a specialized adenylyl cyclase that may mediate odorant detection. *Science*, **250,**1403–1406.

Baker, H. (1986a) Species differences in the distribution of substance P and tyrosine hydroxylase immunoreactivity in the olfactory bulb. *J. Comp. Neurol.*, **252,**206–226.

Baker, H. (1986b) Substance P and tyrosine hydroxylase are localized in different neurons of the hamster olfactory bulb. *Exp. Brain Res.*, **65,**245–249.

Baker, H. & Farbman, A.I. (1989a) Tyrosine hydroxylase (TH) and olfactory marker protein (OMP) expression in rat olfactory bulb of embryos and in culture. *Soc. Neuroscience Abstr.*, **15,**1121.

Baker, H. & Farbman, A.I. (1989b) Tyrosine hydroxylase expression in olfactory bulb in vitro is dependent on receptor afferent innervation. *Chem. Senses*, **14,**683.

Baker, H., Kawano, T., Albert, V., Joh, T.H., Reis, D.J. & Margolis, F.L. (1984) Olfactory bulb dopamine neurons survive deafferentation-induced loss of tyrosine hydroxylase. *Neuroscience*, **11,**605–15.

Baker, H., Kawano, T., Margolis, F.L. & Joh, T.H. (1983) Transneuronal regulation of tyrosine hydroxylase expression in olfactory bulb of mouse and rat. *J. Neuroscience*, **3,**69–78.

Baker, H. & Spencer, R.F. (1986) Transneuronal transport of peroxidase-conjugated wheat germ agglutinin (WGA-HRP) from the olfactory epithelium to the brain of the adult rat. *Exp. Brain Res.*, **63,**461–473.

Baker, H., Towle, A.C. & Margolis, F.L. (1988) Differential afferent regulation of dopaminergic and GABAergic neurons in the mouse main olfactory bulb. *Brain Res.*, **450,**69–80.

Balthazart, J. & Schoffeniels, E. (1979) Pheromones are involved in the control of sexual behavior in birds. *Naturwissensch.*, **66,**55–56.

Bang, B.G. (1971) Functional anatomy of the olfactory system in 23 orders of birds. *Acta Anat. (Suppl.)*, **79,**1–76.

Bannister, L.H. (1965) The fine structure of the olfactory surface of teleostean fishes. *Quart. J. Microsc. Sci.*, **106,**333–342.

Barber, P.C. (1981a) Axonal growth by newly-formed vomeronasal neurosensory cells in the normal adult mouse. *Brain Res.*, **216,**229–237.

Barber, P.C. (1981b) Regeneration of vomeronasal nerves into the main olfactory bulb in the mouse. *Brain Res.*, **216,**239–251.

Barber, P.C. (1989) *Ulex europeus* agglutinin I binds exclusively to primary olfactory neurons in the rat nervous system. *Neuroscience*, **30,**1–9.

Barber, P.C., Jensen, S. & Zimmer, J. (1982) Differentiation of neurons containing olfactory marker protein in adult rat olfactory epithelium transplanted to the anterior chamber of the eye. *Neuroscience*, **7,**2687–2695.

Barber, P.C. & Lindsay, R.M. (1982) Schwann cells of the olfactory nerves contain glial fibrillary acidic protein and resemble astrocytes. *Neuroscience*, **7,**3077–3090.

Barber, P.C. & Raisman, G. (1974) An autoradiographic investigation of the projection of the vomeronasal organ to the accessory olfactory bulb in the mouse. *Brain Res.*, **81,**21–30.

Barber, P.C. & Raisman, G. (1978) Cell division in the vomeronasal organ of the adult mouse. *Brain Res.*, **141,**57–66.

Bargmann, C.I. & Horvitz, H.R. (1991) Control of larval development by chemosensory neurons in *Caenorhabditis elegans. Science,* **251,**1243–1246.

Baron, G., Frahm, H.D., Bhatnagar, K.P. & Stephan, H. (1983) Comparison of brain structure volumes in Insectivora and Primates. III. Main olfactory bulb (MOB). *J. Hirnforsch.,* **24,**551–568.

Barron, D.H. (1943) The early development of the motor cells and columns in the spinal cord of the sheep. *J. Comp. Neurol.,* **78,**1–26.

Barron, D.H. (1946) Observations on the early differentiation of the motor neuroblasts in the spinal cord of the chick. *J. Comp. Neurol.,* **85,** 149–169.

Barron, D.H. (1948) Some effects of amputation of the chick wing bud on the early differentiation of the motor neuroblasts in the associated segments of the spinal cord. *J. Comp. Neurol.,* **88,**93–127.

Bassett, J.C., Shipley, M.T. & Foote, S.L. (1992) Localization of corticotropin-releasing factor-like immunoreactivity in monkey olfactory bulb and secondary olfactory areas. *J. Comp. Neurol.,* **316,**348–362.

Bayer, S.A. (1983) ^{3}H-thymidine-radiographic studies of neurogenesis in the rat olfactory bulb. *Exp. Brain Res.,* **50,**329–340.

Beauchamp, G.K., Yamazaki, K. & Boyse, E.A. (1985) The chemosensory recognition of genetic individuality. *Sci. Amer.,* **253,**86–92.

Bedini, C., Fiaschi, V. & Lanfranchi, A. (1976) Olfactory nerve reconstitution in the homing pigeon after resection: ultrastructural and electrophysiological data. *Arch. Ital. Biol.,* **114,**1–22.

Beets, M.G.J. (1978) *Structure-Activity Relationships in Human Chemoreception.* Applied Science Publishers, Ltd., London.

Beets, M.G.J. & Theimer, E.T. (1970) Odour similarity between structurally unrelated odorants. In *Taste and Smell in Vertebrates. A Ciba Foundation Symposium.* Eds., G.E.W. Wolstenholme & J. Knight. Churchill, London. pp. 313–321.

Beidler, L.M. & Tucker, D. (1955) Response of nasal epithelium to odor stimulation. *Science,* **122,**76.

Benowitz, L.I. & Routtenberg, A. (1987) A membrane phosphoprotein associated with neural development, axonal regeneration, phospholipid metabolism and synaptic plasticity. *Trends Neuroscience,* **10,**527–532.

Benson, T.E., Burd, G., Greer, C.A., Landis, D.M.D. & Shepherd, G.M. (1985) High-resolution 2-deoxyglucose autoradiography in quick-frozen slabs of neonatal rat olfactory bulb. *Brain Res.,* **339,**67–78.

Benson, T.E., Ryugo, D.K. & Hinds, J.W. (1984) Effects of sensory deprivation on the developing mouse olfactory system: a light and electron microscopic morphometric analysis. *J. Neuroscience,* **4,**638–653.

Beroza, M. & Knipling, E.F. (1972) Gypsy moth control with the sex attractant pheromone. *Science,* **177,**19–27.

Bertmar, G. (1981) Evolution of vomeronasal organs in vertebrates. *Evolution,* **35,**359–366.

Bhatnagar, K.P. (1977) Olfactory receptor-glomerular ratio versus olfactory acuity in mammals. In *Olfaction and Taste VI.* Eds., J. LeMagnen & P. MacLeod. Information Retrieval, Ltd., London. p. 186.

Bhatnagar, K.P., Kennedy, R.C., Baron, G. & Greenberg, R.A. (1987) Number of mitral cells and the bulb volume in the aging human olfactory bulb: a quantitative morphological study. *Anat. Rec.,* **218,**73–87.

Bhatnagar, K.P., Matulionis, D.H. & Breipohl, W. (1982) Fine structure of the vomeronasal neuroepithelium of bats: a comparative study. *Acta Anat.,* **112,**158–177.

Biffo, S., Goren, T., Khew-Goodall, Y.S., Miara, J. & Margolis, F.L. (1991) Expression of calmodulin mRNA in rat olfactory epithelium. *Mol. Brain Res.*, **10,**13–21.

Bignetti, E., Cavaggioni, A., Pelosi, P., Persaud, K.C., Sorbi, R.T. & Tirindelli, R. (1985a) Purification and characterization of an odorant-binding protein from cow nasal tissue. *Eur. J. Biochem.*, **149,**227–231.

Bignetti, E., Tirindelli, R., Rossi, G.L., Bolognesi, M., Coda, A. & Gatti, G. (1985b) Crystallization of an odorant-binding protein from cow nasal mucosa. *J. Mol. Biol.*, **186,**211–212.

Blakely, R.D., Ory-Lavollee, L., Grzanna, R., Koller, K.J. & Coyle, J.T. (1987) Selective immunocytochemical staining of mitral cells in rat olfactory bulb with affinity purified antibodies against *N*-acetyl-aspartyl-glutamate. *Brain Res.*, **402,**373–378.

Bland, K.P. & Cottrell, D.F. (1989) The nervous control of intraluminal pressure in the vomeronasal organ of the domestic ram. *Quart. J. Exp. Physiol.*, **74,**813–824.

Bloom, G. (1954) Studies on the olfactory epithelium of the frog and the toad with the aid of light and electron microscopy. *Z. Zellforsch. Mikroskop. Anat.*, **41,**89–100.

Bodian, D. & Howe, H.A. (1941a) Experimental studies on intraneural spread of poliomyelitis virus. *Bull. Johns Hopkins Hosp.*, **68,**248–267.

Bodian, D. & Howe, H.A. (1941b) The rate of progression of poliomyelitis virus in nerves. *Bull. Johns Hopkins Hosp.*, **69,**79–85.

Boekhoff, I., Strotmann, J., Raming, K., Tareilus, E. & Breer, H. (1990a) Odorant-sensitive phospholipase C in insect antennae. *Cell. Signalling*, **2,**49–56.

Boekhoff, I., Tareilus, E., Strotmann, J. & Breer, H. (1990b) Rapid activation of alternative second messenger pathways in olfactory cilia from rats by different odorants. *EMBO J.*, **9,**2453–2458.

Boelens, H. (1976) In *Structure-Activity Relationships in Chemoreception*. Ed., G. Benz. IRL, London. pp. 197–210.

Boelens, H. (1983) Structure-activity relationships in chemoreception by human olfaction. *Trends Pharmacol. Sci.*, **4,**421–426.

Bogan, N., Brecha, N., Gall, C. & Karten, H.J. (1982) Distribution of enkephalin-like immunoreactivity in the rat main olfactory bulb. *Neuroscience*, **7,**895–906.

Bojsen-Møller, F. (1975) Demonstration of terminalis, olfactory, trigeminal and perivascular nerves in the rat nasal septum. *J. Comp. Neurol.*, **159,**245–256.

Booth, W.S., Baldwin, B.A., Poynder, T.M., Bannister, L.H. & Gower, D.B. (1981) Degeneration and regeneration of the olfactory epithelium after olfactory bulb ablation in the pig: a morphological and electrophysiological study. *Quart. J. Exp. Physiol.*, **66,**533–540.

Bossy, Y. (1980) Development of olfactory and related structures in staged human embryos. *Anat. Embryol.*, **161,**225–236.

Bouvet, J.F., Delaleu, J.C. & Holley, A. (1987a) Olfactory receptor cell function is affected by trigeminal nerve activity. *Neuroscience Lett.*, **77,**181–186.

Bouvet, J.F., Delaleu, J.C. & Holley, A. (1988) The activity of olfactory receptor cells is affected by acetylcholine and substance P. *Neuroscience Res.*, **5,**214–223.

Bouvet, J.F., Godinot, F., Croze, S. & Delaleu, J.C. (1987b) Trigeminal substance P-like immunoreactive fibres in the frog olfactory mucosa. *Chem. Senses*, **12**,499–505.

Boyse, E.A., Beauchamp, G.K., Yamazaki, K., Bard, J. & Thomas, L. (1982) A new aspect of the major histocompatibility complex and other genes in the mouse. *Oncodevel. Biol., Med.*, **4**,101–116.

Bradford, H.F. & Richards, C.D. (1976) Specific release of endogenous glutamate from piriform cortex stimulated in vitro. *Brain Res.*, **105**,168–172.

Breer, H. & Boekhoff, I. (1991) Odorants of the same odor class activate different second messenger pathways. *Chem. Senses*, **16**,19–29.

Breer, H., Boekhoff, I. & Tareilus, E. (1990) Rapid kinetics of second messenger formation in olfactory transduction. *Nature*, **345**,65–68.

Breipohl, W., Bhatnagar, K.P., Blank, M. & Mendoza, A.S. (1981) Intraepithelial blood vessels in the vomeronasal neuroepithelium of the rat. A light and electron microscopic study. *Cell Tissue Res.*, **215**,465–473.

Breipohl, W., Bhatnagar, K.P. & Mendoza, A. (1979) Fine structure of the receptor-free epithelium in the vomeronasal organ of the rat. *Cell Tissue Res.*, **200**,383–395.

Breipohl, W. & Fernandez, M. (1977) Scanning electron microscopic investigations of olfactory epithelium in the chick embryo. *Cell Tissue Res.*, **183**,105–114.

Breipohl, W., Mackay-Sim, A., Grandt, D., Rehn, B. & Darrelmann, C. (1986) Neurogenesis in the vertebrate main olfactory epithelium. In *Ontogeny of Olfaction*. Ed., W. Breipohl. Springer-Verlag, Berlin. pp. 21–33.

Breipohl, W., Moulton, D., Ummels, M. & Matulionis, D.H. (1982) Spatial pattern of sensory cell terminals in the olfactory sac of the tiger salamander. I. A scanning electron microscope study. *J. Anat.*, **134**,757–769.

Breipohl, W., Naguro, T. & Miragall, F. (1983) Morphology of the Masera organ in NMRI mice: combined morphometric, freeze-fracture, light- and scanning electron microscopic investigations. *Verh. Anat. Ges.*, **77**,741–743.

Breipohl, W. & Ohyama, M. (1981) Comparative and developmental SEM studies on olfactory epithelia in vertebrates. *Biomed. Res. (Suppl.)*, **2**,437–448.

Brennan, P., Kaba, H. & Keverne, E.B. (1990) Olfactory recognition: a simple memory system. *Science*, **250**,1223–1226.

Brennan, P.A. & Keverne, E.B. (1989) Impairment of olfactory memory by local infusions of non-selective excitatory amino acid receptor antagonists into the accessory olfactory bulb. *Neuroscience*, **33**,463–468.

Brennells, A.B. (1974) Spontaneous and neurally evoked release of labelled noradrenaline from rabbit olfactory bulb *in vivo*. *J. Physiol. (Lond.)*, **240**,279–293.

Breucker, H., Zeiske, E. & Melinkat, R. (1979) Development of the olfactory organ in the rainbow fish *Nematocentris macculochi (Atheriniformes, Melanotaeniidae). Cell Tissue Res.*, **200**,53–68.

Broadwell, R.D. (1975a) Olfactory relationships of the telencephalon and diencephalon in the rabbit. I. An autoradiographic study of the efferent connections of the main and accessory olfactory bulbs. *J. Comp. Neurol.*, **163**,329–346.

Broadwell, R.D. (1975b) Olfactory relationships of the telencephalon and diencephalon in the rabbit. II. An autoradiographic and horseradish peroxidase study of the efferent connections of the anterior olfactory nucleus. *J. Comp. Neurol.*, **164**,389–410.

Broadwell, R.D. & Jacobowitz, D.M. (1976) Olfactory relationships of the telencephalon and diencephalon in the rabbit. III. The ipsilateral centrifugal fibers to the olfactory bulbar and retrobulbar formations. *J. Comp. Neurol.,* **170,**321–346.

Broman, I. (1920) Das Organon vomero-nasale Jacobsoni – ein Wassergeruchorgane. *Anat. Hefte,* **58,**137–191.

Bronshtein, A.A. & Minor, A.V. (1977) The regeneration of olfactory flagella and restoration of electroolfactogram after the treatment of the olfactory mucosa with Triton X–100. *Cytology,* **19,**33–39.

Brouwer, J.N., Farmer, P., Gesteland, R.C. & Drazba, J. (1991) Voltage-sensitive dye confocal microscopy of living olfactory epithelia. *Chem. Senses,* **16,**504–505.

Brown, D., Garcia-Segura, L.-M. & Orci, L. (1984) Carbonic anhydrase is present in olfactory receptor cells. *Histochemistry,* **80,**307–309.

Brown, R.E. (1988) Individual odors of rats are discriminable independently of changes in gonadal hormone levels. *Physiol. Behav.,* **43,**359–363.

Brown, R.E., Roser, B. & Singh, P.B. (1989) Class I and Class II regions of the major histocompatibility complex both contribute to individual odors in congenic inbred strains of rats. *Behav. Genet.,* **19,**659–674.

Brown, R.E., Singh, P.B. & Roser, B. (1987) The major histocompatibility complex and the chemosensory recognition of individuality in rats. *Physiol. Behav.,* **40,**65–73.

Brown, S.B. & Hara, T.J. (1981) Accumulation of chemostimulatory amino acids by a sedimentable fraction isolated from olfactory rosettes of rainbow trout (*Salmo gairdneri*). *Biochim. Biophys. Acta,* **675,**149–162.

Bruce, H.M. (1959) An exteroceptive block to pregnancy in the mouse. *Nature,* **184,**105.

Bruce, H.M. (1960) A block to pregnancy in the mouse caused by proximity to strange males. *J. Reprod. Fertil.,* **1,**96–103.

Bruce, H.M. & Parrott, D.M.V. (1960) Role of olfactory sense in pregnancy block by strange males. *Science,* **131,**1526.

Bruch, R.C. (1990a) Signal transduction in olfaction and taste. In *G-Proteins.* Eds., R. Iyengar & L. Birnbaumer. Academic Press, N.Y. pp. 411–428.

Bruch, R.C. (1990b) Signal transducing GTP-binding proteins in olfaction. *Comp. Biochem. Physiol.,* **95A,**27–29.

Bruch, R.C. & Carr, V.M. (1991) Rat olfactory neurons express a 200 kDa neurofilament. *Brain Res.,* **550,**133–136.

Bruch, R.C. & Gold, G.H. (1990) G-protein-mediated signalling in olfaction. In *G-Proteins as Mediators of Cellular Signalling Processes.* Eds., M.D. Houslay & G. Milligan. Wiley, N.Y. pp. 113–124.

Bruch, R.C. & Kalinoski, D.L. (1987) Interaction of GTP-binding regulatory proteins with chemosensory receptors. *J. Biol. Chem.,* **262,**2401–2404.

Bruch, R.C. & Rulli, R.D. (1988) Ligand binding specificity of a neutral L-amino acid olfactory receptor. *Comp. Biochem. Physiol. B,* **91,**535–540.

Bruch, R.C. & Teeter, J.H. (1989) Second messenger signalling mechanisms in olfaction. In *Chemical Senses: Receptor Events and Transduction in Taste and Olfaction.* Eds., J.G. Brand, J.H. Teeter, R.H. Cagan & M.R. Kare. Dekker, N.Y. pp. 283–298.

Bruch, R.C. & Teeter, J.H. (1990) Cyclic AMP links amino acid chemoreceptors to ion channels in olfactory cilia. *Chem. Senses,* **15,**419–430.

Brunjes, P.C. (1983) Olfactory bulb maturation in *Acomys cahirinus:* is neural growth similar in precocial and altricial murids. *Devel. Brain Res.,* **8,**335–341.

Brunjes, P.C. (1985) Unilateral odor deprivation: time course of changes in laminar volume. *Brain Res. Bull.*, **14**,233–237.

Brunjes, P.C. & Alberts, J.R. (1979) Olfactory stimulation induces filial huddling preferences in pups. *J. Comp. Physiol. Psychol.*, **93**,548–555.

Brunjes, P.C. & Borror, M.J. (1983) Unilateral odor deprivation: Differential effects due to time of treatment. *Brain Res. Bull.*, **11**,501–503.

Brunjes, P.C. & Frazier, L.L. (1986) Maturation and plasticity in the olfactory system of vertebrates. *Brain Res. Rev.*, **11**,1–45.

Brunjes, P.C., Schwark, H.D. & Greenough, W.T. (1982) Olfactory granule cell development in normal and hyperthyroid rats. *Devel. Brain Res.*, **5**,149–159.

Brunjes, P.C., Smith-Crafts, L.K. & McCarty, R. (1985) Unilateral odor deprivation: effects on the development of olfactory bulb catecholamines and behavior. *Devel. Brain Res.*, **22**,1–6.

Buchner, K., Seitz-Tutter, D., Schönitzer, K. & Weiss, D. (1987) A quantitative study of anterograde and retrograde axonal transport of exogenous proteins in olfactory nerve C-fibers. *Neuroscience*, **22**,697–707.

Buck, L. & Axel, R. (1991) A novel multigene family may encode odorant receptors: a molecular basis for odor recognition. *Cell*, **65**,175–187.

Buhl, E.H. & Oelschläger, H.A. (1986) Ontogenetic development of the nervus terminalis in toothed whales. *Anat. Embryol.*, **173**,285–294.

Buonviso, N. & Chaput, M.A. (1990) Response similarity to odors in olfactory bulb output cells presumed to be connected to the same glomerulus: electrophysiological study using simultaneous single-unit recordings. *J. Neurophysiol.*, **63**,447–454.

Burd, G.D., Davis, B.J. & Macrides, F. (1982a) Ultrastructural identification of substance P immunoreactive neurons in the main olfactory bulb of the hamster. *Neuroscience*, **7**,2697–2704.

Burd, G.D., Davis, B.J., Macrides, F., Grillo, M. & Margolis, F.L. (1982b) Carnosine in primary afferents of the olfactory system: an autoradiographic and biochemical study. *J. Neuroscience*, **2**,244–255.

Burk, D., Sadler, T.W. & Langman, J. (1979) Distribution of surface coat material on nasal folds of mouse embryos as demonstrated by Concanavalin A staining. *Anat. Rec.*, **193**,185–196.

Burkholder, A.C. & Hartwell, L.H. (1985) The yeast α-factor receptor: structural properties deduced from the sequence of the STE2 gene. *Nucleic Acids Res.*, **13**,8463–8473.

Burns, S.M., Mitchell, J.A., Getchell, M.L. & Getchell, T.V. (1981). Functional correlates of degeneration and renewal of cilia and knobs of olfactory receptor neurons in the frog. *Chem. Senses*, **6**,307–315.

Burr, H.S. (1916) The effects of the removal of the nasal pits in Amblystoma embryos. *J. Exp. Zool.*, **20**,27–57.

Burr, H.S. (1924) Some experiments on the transplantation of the olfactory placode in Amblystoma. I. An experimentally produced aberrant cranial nerve. *J. Comp. Neurol.*, **37**,455–479.

Burr, H.S. (1930) Hyperplasia in the brain of Amblystoma. *J. Exp. Zool.*, **55**,171–191.

Burton, P.R. (1984) Luminal material in microtubules of frog olfactory axons: structure and distribution. *J. Cell Biol.*, **99**,520–528.

Burton, P.R. (1987) Microtubules of frog olfactory axons: their length and number/axon. *Brain Res.*, **409**,71–78.

Burton, P.R. & Laveri, L.A. (1985) The distribution, relationships to other organelles, and calcium-sequestering ability of smooth endoplasmic reticulum in frog olfactory axons. *J. Neuroscience*, **5**,3047–3060.

Burton, P.R. & Paige, J.L. (1981) Polarity of axoplasmic microtubules in the olfactory nerve of the frog. *Proc. Natl. Acad. Sci. USA*, **78**,3269–3273.

Butler, A., Graziadei, P.P.C., Monti Graziadei, G.A. & Slotnick, B.M. (1984) Neonatally bulbectomized rats with new olfactory-neocortical connections are anosmic. *Neuroscience Lett.*, **48**,247–254.

Byrd, C.A. & Burd, G.D. (1991) Development of the olfactory bulb in the clawed frog, *Xenopus laevis:* a morphological and quantitative analysis. *J. Comp. Neurol.*, **314**,79–90.

Cagan, R.H. & Zeiger, W.N. (1978) Biochemical studies of olfaction: binding specificity of radioactively labeled stimuli to an isolated olfactory preparation from rainbow trout (*Salmo gairdneri*). *Proc. Natl. Acad. Sci. USA*, **75**,4679–4683.

Cain, W.S. (1974) Contribution of the trigeminal nerve to perceived odor magnitude. *Ann. N.Y. Acad. Sci.*, **237**,28–34.

Cain, W.S. (1976) Olfaction and the common chemical sense: some psychophysical contrasts. *Sensory Proc.*, **1**,57–67.

Cain, W.S. & Murphy, C. (1980) Interaction between chemoreceptive modalities of odour and irritation. *Nature*, **284**,255–257.

Calof, A.L. & Chikaraishi, D.M. (1989) Analysis of neurogenesis in a mammalian neuroepithelium: proliferation and differentiation of an olfactory neuron precursor *in vitro*. *Neuron*, **3**,115–127.

Càmara, C.G. & Harding, J.W. (1984) Thymidine incorporation in the olfactory epithelium of mice: early exponential response induced by olfactory neurectomy. *Brain Res.*, **308**,63–68.

Cancalon, P. (1979) Subcellular and polypeptide distributions of slowly transported proteins in the garfish olfactory nerve. *Brain Res.*, **161**,115–130.

Cancalon, P. (1982) Slow flow in axons detached from their perikarya. *J. Cell Biol.*, **95**,989–992.

Cancalon, P. (1983a) Slow flow in regenerating C-fibers. *Devel. Brain Res.*, **6**,197–200.

Cancalon, P. (1983b) Regeneration of three populations of olfactory axons as a function of temperature. *Devel. Brain Res.*, **9**,265–278.

Cancalon, P. (1983c) Influence of temperature on slow flow in populations of regenerating axons with different elongation velocities. *Devel. Brain Res.*, **9**,279–289.

Cancalon, P. (1983d) Proximodistal degeneration of C-fibers detached from their perikarya. *J. Cell Biol.*, **97**,6–14.

Cancalon, P. (1983e) Receptor cells of the catfish olfactory mucosa. *Chem. Senses*, **8**,203–209.

Cancalon, P.F. (1987) Survival and subsequent regeneration of olfactory neurons after a distal axonal lesion. *J. Neurocytol.*, **16**,829–841.

Cancalon, P.F. (1988) Axonal transport in the garfish optic nerve: comparison with the olfactory system. *J. Neurochem.*, **51**,266–276.

Cancalon, P. & Beidler, L.M. (1975) Distribution along the axon and into various subcellular fractions of molecules labeled with [³H]leucine and rapidly transported in the garfish olfactory nerve. *Brain Res.*, **89**,225–244.

Cancalon, P., Brady, S.T. & Lasek, R.J. (1988) Slow transport in a nerve with embryonic characteristics, the olfactory nerve. *Devel. Brain Res.*, **38**,275–285.

Cancalon, P. & Elam, J.S. (1980a) Study of regeneration in the garfish olfactory nerve. *J. Cell Biol.*, **84**,779–794.

Cancalon, P. & Elam, J.S. (1980b) Rate of movement and composition of rapidly transported proteins in regenerating olfactory nerve. *J. Neurochem.*, **35**,889–897.

Caprio, J. (1977) Electrophysiological distinctions between the taste and smell of amino acids in catfish. *Nature*, **266**,850–851.

Caprio, J. (1978) Olfaction and taste in the channel catfish: an electrophysiological study of the responses to amino acids and derivatives. *J. Comp. Physiol. A*, **123**,357–371.

Caprio, J. (1982) High sensitivity and specificity of olfactory and gustatory receptors of catfish to amino acids. In *Chemoreception in Fishes*. Ed., T.J. Hara. Elsevier, Amsterdam. pp. 109–134.

Caprio, J. & Byrd, R.P., Jr. (1984) Electrophysiological evidence of acidic, basic, and neutral amino acid olfactory receptor sites in catfish. *J. Gen. Physiol.*, **84**,403–422.

Caroni, P. & Schwab, M.E. (1988) Two membrane protein fractions from rat central myelin with inhibitory properties for neurite growth and fibroblast spreading. *J. Cell Biol.*, **106**,1281–1288.

Carr, V.M. & Farbman, A.I. (1989) [³H]-thymidine labeling of degenerating cells in the olfactory epithelium of bulbectomized rats. *Soc. Neuroscience Abstr.*, **15**,444.

Carr, V.M. & Farbman, A.I. (1990) Nongenetic regulation of proliferation and cell death in the olfactory epithelium (OE). *Soc. Neuroscience Abstr.*, **16**,837.

Carr, V.M. & Farbman, A.I. (1991) Heat shock protein HSP70 in control and bulbectomized rat olfactory epithelium. *Chem. Senses*, **16**,507.

Carr, V.M. & Farbman, A.I. (1992) Ablation of the olfactory bulb up-regulates the rate of neurogenesis and induces precocious cell death in olfactory epithelium. *Exper. Neurol.*, **115**,55–59.

Carr, V.M., Farbman, A.I., Colletti, L.M. & Morgan, J.I. (1991) Identification of a new nonneuronal cell type in rat olfactory epithelium. *Neuroscience*, **45**,433–449.

Carr, V.M., Farbman, A.I., Lidow, M.S., Colletti, L.M., Hempstead, J.L. & Morgan, J.I. (1989) Developmental expression of reactivity to monoclonal antibodies generated against olfactory epithelia. *J. Neuroscience*, **9**,1179–1198.

Carr, V.M. & Simpson, S.B., Jr. (1982) Rapid appearance of labeled degenerating cells in the dorsal root ganglia after exposure of chick embryos to tritiated thymidine. *Devel. Brain Res.*, **2**,157–162.

Carr, W.E.S. (1967) Chemoreception in the mud snail, *Nassarius obsoletus*. II. Identification of stimulatory substances. *Bio. Bull.*, **133**,106–127.

Carr, W.E.S. (1990) Chemical signaling systems in lower organisms: a prelude to the evolution of chemical communication in the nervous system. In *Evolution of the First Nervous Systems*. Ed., P.A.V. Anderson. Plenum Press, N.Y. pp. 81–94.

Carr, W.E.S., Gleeson, R.A. & Trapido-Rosenthal, H.G. (1990a) The role of degradative enzymes in chemosensory processes. *Chem. Senses*, **15**,181–190.

Carr, W.E.S., Gleeson, R.A. & Trapido-Rosenthal, H.G. (1990b) The role of perireceptor events in chemosensory processes. *Trends Neuroscience*, **13**,212–215.

Carson, K.A. (1984) Localization of acetylcholinesterase-positive neurons projecting to the mouse main olfactory bulb. *Brain Res. Bull.*, **12**,635–639.

Carson, K.A. & Burd, G.D. (1980) Localization of acetylcholinesterase in the main and accessory olfactory bulbs of the mouse by light and electron microscopic histochemistry. *J. Comp. Neurol.,* **191,**353–374.

Caza, P.A. & Spear, N.E. (1984) Short-term exposure to an odor increases its subsequent preferences in preweanling rats: a descriptive profile of the phenomenon. *Devel. Psychobiol.,* **17,**407–422.

Chacko, G.K., Goldman, D.E., Malhotra, H.C. & Dewey, M.M. (1974) Isolation and characterization of plasma membrane fractions from garfish *Lepisosteus osseus* olfactory nerve. *J. Cell Biol.,* **62,**831–843.

Chase, N.T. & Kopin, I.J. (1968) Stimulus-induced release of substances from olfactory bulb using the push-pull cannula. *Nature,* **217,**466–467.

Chase, R. (1986) Lessons from snail tentacles. *Chem. Senses,* **11,**411–426.

Chase, R. & Rieling, J. (1986) Autoradiographic evidence for receptor cell renewal in the olfactory epithelium of a snail. *Brain Res.,* **384,**232–239.

Chase, R. & Tolloczko, B. (1986) Synaptic glomeruli in the olfactory system of a snail, *Achatina fulica. Cell Tissue Res.,* **246,**567–573.

Chen, Z. & Lancet, D. (1984) Membrane proteins unique to vertebrate olfactory cilia: candidates for sensory receptor molecules. *Proc. Natl. Acad. Sci. USA,* **81,**1859–1863.

Chen, Z., Pace, U., Heldman, J., Shapira, A. & Lancet, D. (1986a) Isolated frog olfactory cilia: a preparation of dendritic membranes from chemosensory neurons. *J. Neuroscience,* **6,**2146–2154.

Chen, Z., Pace, U., Ronen, D. & Lancet, D. (1986b) Polypeptide gp95. A unique glycoprotein of olfactory cilia with transmembrane receptor properties. *J. Biol. Chem.,* **261,**1299–1305.

Chiszar, D., Melcer, T., Lee, R., Radcliffe, C.W. & Duvall, D. (1990) Chemical cues used by prairie rattlesnakes (*Crotalus viridis*) to follow trails of rodent prey. *J. Chem. Ecol.,* **16,**79–86.

Chuah, M.I. & Au, C. (1991) Olfactory Schwann cells are derived from precursor cells in the olfactory epithelium. *J. Neuroscience Res.,* **29,**172–180.

Chuah, M.I. & Farbman, A.I. (1983) Olfactory bulb increases marker protein in olfactory receptor cells. *J. Neuroscience,* **3,**2197–2205.

Chuah, M.I. & Farbman, A.I. (1986) Mitral cell differentiation and synaptogenesis of rat presumptive olfactory bulb in organ culture. *Cell Tissue Res.,* **243,**359–365.

Chuah, M.I., Farbman, A.I. & Menco, B. P. M. (1985) Influence of olfactory bulb on dendritic knob density of rat olfactory receptor neurons in vitro. *Brain Res.,* **338,**259–266.

Ciges, M., Labella, T., Gayoso, M. & Sanchez, G. (1977) Ultrastructure of the organ of Jacobson and comparative study with olfactory mucosa. *Acta Otolaryngol.* **83,**47–58.

Clancy, A.N., Macrides, F., Singer, A.G. & Agosta, W.C. (1984) Male hamster copulatory responses to a high molecular weight fraction of vaginal discharge: effects of vomeronasal organ removal. *Physiol. Behav.,* **33,**653–660.

Clark, L. & Smeraski, C.A. (1990) Seasonal shifts in odor acuity by starlings. *J. Exp. Zool.,* **255,**22–29.

Claude, P. & Goodenough, D.A. (1973) Fracture faces of zonulae occludentes from "tight" and "leaky" epithelia. *J. Cell Biol.,* **58,**390–400.

Cochard, P. & Paulin, D. (1984) Initial expression of neurofilaments and vimentin in the central and peripheral nervous system of the mouse embryo in vivo. *J. Neuroscience,* **4,**2080–2094.

Coggins, P.J. & Zwiers, H. (1991) B–50 (GAP–43): biochemistry and functional neurochemistry of a neuron-specific phosphoprotein. *J. Neurochem.*, **56**,1095–1106.

Collins, G.G.S. (1979) Evidence of a neurotransmitter role for aspartate and gamma-aminobutyric acid in the rat olfactory cortex. *J. Physiol. (Lond.)*, **291**,51–60.

Collins, G.G.S. & Probett, G.A. (1981) Aspartate and not glutamate is the likely transmitter of the rat lateral olfactory tract fibres. *Brain Res.*, **209**,231–234.

Cooper, W.E. (1990) Chemical detection of predators by a lizard, the broad-headed skink (*Eumeces laticeps*). *J. Exp. Zool.*, **256**,162–167.

Cooper, W.E. & Burghardt, G.M. (1990) Vomerolfaction and vomodor. *J. Chem. Ecol.*, **16**,103–104.

Coopersmith, R. & Leon, M. (1984) Enhanced neural response to familiar olfactory cues. *Science*, **225**,849–851.

Cornwell-Jones, C.A. & Bollers, H.R. (1983) Neonatal 6-hydroxy-dopamine alters conspecific odor investigation by male rats. *Brain Res.*, **268**,291–294.

Costanzo, R.M. (1984) Comparison of neurogenesis and cell replacement in the hamster olfactory system with and without a target (olfactory bulb). *Brain Res.*, **307**,295–301.

Costanzo, R.M. & Graziadei, P.P.C. (1983) A quantitative analysis of changes in the olfactory epithelium following bulbectomy in hamster. *J. Comp. Neurol.*, **215**,370–381.

Costanzo, R.M. & Graziadei, P.P.C. (1987) Development and plasticity of the olfactory system. In *Neurobiology of Taste and Smell*. Eds., T.E. Finger & W.L. Silver. Wiley, N.Y. pp. 233–250.

Costanzo, R.M. & Morrison, E.E. (1989) Three-dimensional scanning electron microscopic study of the normal hamster olfactory epithelium. *J. Neurocytol.*, **18**,381–391.

Costanzo, R.M. & Mozell, M.M. (1976) Electrophysiological evidence for a topographical projection of the nasal mucosa onto the olfactory bulb of the frog. *J. Gen. Physiol.*, **68**,297–312.

Costanzo, R.M. & O'Connell, R.J. (1978) Spatially organized projections of hamster olfactory nerves. *Brain Res.*, **139**,327–332.

Costanzo, R.M. & O'Connell, R.J. (1980) Receptive fields of second-order neurons in the olfactory bulb of the hamster. *J. Gen. Physiol.*, **76**,53–68.

Coughlin, M.D., Dibner, M.D., Boyer, D.M. & Black, I.B. (1978) Factors regulating development of an embryonic mouse sympathetic ganglion. *Devel. Biol.*, **66**,513–528.

Couly, G.F. & Le Douarin, N.M. (1985) Mapping of the early neural primordium in quail-chick chimeras. I. Developmental relationships between placodes, facial ectoderm and prosencephalon. *Devel. Biol.*, **110**,422–439.

Cowan, W.M. & Wenger, E. (1967) Cell loss in the trochlear nucleus of the chick during normal development and after radical extirpation of the optic vesicle. *J. Exp. Zool.*, **164**,513–528.

Crowe, M.J. & Pixley, S.K. (1991) Monoclonal antibody to carnosine synthetase identifies a subpopulation of frog olfactory receptor neurons. *Brain Res.*, **538**,147–151.

Cullinan, W.E. & Brunjes, P.C. (1987) Unilateral odor deprivation: effects on the development of staining for olfactory bulb succinate dehydrogenase and cytochrome oxidase. *Devel. Brain Res.*, **35**,35–42.

Cuschieri, A. & Bannister, L.H. (1975a) The development of the olfactory mucosa in the mouse: light microscopy. *J. Anat.*, **119**,277–286.

Cuschieri, A. & Bannister, L.H. (1975b) The development of the olfactory mucosa in the mouse: electron microscopy. *J. Anat.,* **119,**471–498.

Dahl, A.R. (1988) The effect of cytochrome P–450-dependent metabolism and other enzyme activities on olfaction. In *Molecular Neurobiology of the Olfactory System.* Eds., F.L. Margolis & T.V. Getchell. Plenum Press, N.Y. pp. 51–70.

Dahl, A.R., Hadley, W.M., Hahn, F.F., Benson, J.M. & McClellan, R.O. (1982) Cytochrome P–450-dependent monooxygenases in olfactory epithelium of dogs; possible role in tumorigenicity. *Science,* **216,**57–59.

Daston, M.M., Adamek, G.D., & Gesteland, R.C. (1990) Ultrastructural organization of receptor cell axons in frog olfactory nerve. *Brain Res.,* **537,**69–75.

Dau, B., Menco, B.P.M., Bruch, R.C., Danho, W. & Farbman, A. (1991) Appearance of the transduction proteins $G_s\alpha$, $G_{olf}\alpha$ and adenylate cyclase in the olfactory epithelium of rats occurs on different prenatal days. *Chem. Senses,* **16,**511–512.

Davies, J.T. (1971) Olfactory theories. In *Handbook of Sensory Physiology. Vol. IV. Chemical Senses, 1. Olfaction.* Ed., L.M. Beidler. Springer-Verlag, Berlin. pp. 322–350.

Davis, B.J., Burd, G.D. & Macrides, F. (1982) Localization of methionine-enkephalin, substance P, and somatostatin immunoreactivities in the main olfactory bulb of the hamster. *J. Comp. Neurol.,* **204,**377–383.

Davis, B.J. & Macrides, F. (1981) The organization of centrifugal projections from the anterior olfactory nucleus, ventral hippocampal rudiment and piriform cortex to the main olfactory bulb in the hamster: an autoradiographic study. *J. Comp. Neurol.,* **203,**475–493.

Davis, B.J. & Macrides, F. (1983) Tyrosine hydroxylase immunoreactive neurons and fibers in the olfactory system of the hamster. *J. Comp. Neurol.,* **214,**427–440.

Davis, B.J., Macrides, F., Youngs, W.M., Schneider, S.P. & Rosene, D.L. (1978) Efferents and centrifugal afferents of the main and accessory olfactory bulbs in the hamster. *Brain Res. Bull.,* **3,**59–72.

Dawley, E.M. & Bass, A.H. (1988) Organization of the vomeronasal organ in a *Plethodontid* salamander. *J. Morphol.,* **198,**243–255.

Dawley, E.M. & Bass, A.H. (1989) Chemical access to the vomeronasal organs of a *Plethodontid* salamander. *J. Morphol.,* **200,**163–174.

Dawson, W.W. (1962) Chemical stimulation of the peripheral trigeminal nerve. *Nature,* **196,**341–345.

DeLong, R.E. & Getchell, T.V. (1987) Nasal respiratory function – vasomotor and secretory regulation. *Chem. Senses,* **12,**3–36.

DeLorenzo, A.J. (1970) The olfactory neuron and the blood-brain barrier. In *Taste and Smell in Vertebrates. A Ciba Foundation Symposium.* Eds., G.E.W. Wolstenholme & J. Knight. Churchill, London. pp. 151–173.

Demski, L.S. & Northcutt, R.G. (1983) The terminal nerve: a new chemosensory system in vertebrates? *Science,* **220,**435–437.

de Olmos, J., Hardy, H. & Heimer, L. (1978) The afferent connections of the main and the accessory olfactory bulb formations in the rat: an experimental HRP study. *J. Comp. Neurol.,* **181,**213–244.

Dhallan, R.S., Yau, K.-W., Schrader, K.A. & Reed, R.R. (1990) Primary structure and functional expression of a cyclic nucleotide-activated channel from olfactory neurons. *Nature,* **347,**184–187.

Ding, X. & Coon, M.J. (1990) Induction of cytochrome P–450 and isozyme 3a

(P-450IIE1) in rabbit olfactory mucosa by ethanol and acetone. *Drug Metabol. Disposition*, **18**,742–745.

Dionne, V.E. (1986) Membrane conductance mechanisms in dissociated cells from the *Necturus* olfactory epithelium. *Chem. Senses*, **11**,594–595.

Disse, J. (1897) Die erste Entwicklung des Riechnerven. *Anat. Hefte*, **9**,257–300.

Ditraglia, G.M., Press, D.S., Butters, N., Jernigan, T.L., Cermak, L.S., Velin, R.A., Shear, P.K., Irwin, M. & Schuckit, M. (1991) Assessment of olfactory deficits in detoxified alcoholics. *Alcohol*, **8**,109–115.

Dodd, G. & Persaud, K. (1981) Biochemical mechanisms in vertebrate primary olfactory neurons. In *Biochemistry of Taste and Olfaction*. Eds., R.H. Cagan & M.R. Kare. Academic Press, N.Y. pp. 333–357.

Doty, R.L. (1975) Intranasal trigeminal detection of chemical vapors by humans. *Physiol. Behav.*, **14**,855–859.

Doty, R.L. (1990) Olfaction. In *Handbook of Neuropsychology, Vol. 4*. Eds., F. Boller & J. Grafman. Elsevier, Amsterdam. pp. 213–228.

Doty, R.L., Brugger, W.E., Jurs, P.C., Orndorff, M.A., Snyder, P.F. & Lowry, L.D. (1978) Intranasal trigeminal stimulation from odorous volatiles: psychometric responses from anosmic and normal humans. *Physiol. Behav.*, **20**,175–187.

Doty, R.L., Deems, D.A. & Stellar, S. (1988) Olfactory dysfunction in Parkinsonism: a general deficit unrelated to neurologic signs, disease stage, or disease duration. *Neurology*, **38**,1237–1244.

Doty, R.L., Reyes, P.F. & Gregor, T. (1987) Presence of both odor identification and detection deficits in Alzheimer's disease. *Brain Res. Bull.*, **18**,597–600.

Doty, R.L., Riklan, M., Deems, D.A., Reynolds, C. & Stellar, S. (1989) The olfactory and cognitive deficits of Parkinson's disease: evidence for independence. *Ann. Neurol.*, **25**,166–171.

Doty, R.L., Shaman, P., Applebaum, S.L., Giberson, R., Siksorski, L. & Rosenberg, L. (1984) Smell identification ability: changes with age. *Science*, **226**,1441–1443.

Doucette, J.R. (1984) The glial cells in the nerve fiber layer of the rat olfactory bulb. *Anat. Rec.*, **210**,385–391.

Doucette, J.R. (1989) Development of the nerve fiber layer in the olfactory bulb of mouse embryos. *J. Comp. Neurol.*, **285**,514–527.

Doucette, J.R. (1990) Glial influences on axonal growth in the primary olfactory system. *Glia*, **3**,433–449.

Doucette, J.R., Kiernan, J.A. & Flumerfelt, B.A. (1983a) Two different patterns of retrograde degeneration in the olfactory epithelium following transection of primary olfactory axons. *J. Anat.*, **136**,673–689.

Doucette, J.R., Kiernan, J.A. & Flumerfelt, B.A. (1983b) The re-innervation of olfactory glomeruli following transection of primary olfactory axons in the central or peripheral nervous system. *J. Anat.*, **137**,1–19.

Douek, E. (1974) *The Sense of Smell and Its Abnormalities*. Churchill Livingstone, Edinburgh.

Døving, K.B. (1964) Studies of the relation between the frog's electroolfactogram (EOG) and single unit activity in the olfactory bulb. *Acta Physiol. Scand.*, **60**,150–153.

Døving, K.B. & Holmberg, K. (1974) A note on the function of the olfactory organ of the hagfish *Myxine glutinosa*. *Acta Physiol. Scand.*, **75**,111–123.

Døving, K.B. & Pinching, A.J. (1973) Selective degeneration of neurones in the olfactory bulb following prolonged odour exposure. *Brain Res.*, **52**,115–129.

Døving, K.B. & Selset, R. (1980) Behavior patterns in cod released by electrical stimulation of olfactory tract bundlets. *Science*, **207**,559–560.

Døving, K.B. & Thommesen, G. (1977) Some properties of the fish olfactory system. In *Olfaction and Taste VI*, Eds., J. LeMagnen & P. MacLeod. Information Retrieval Ltd., London. pp. 175–183.

Dubois-Dauphin, M., Tribollet, E. & Dreifuss, J.J., (1981) Relations somatotopiques entre la muquese olfactive et le bulbe olfactif chez le triton. *Brain Res.*, **219**,269–287.

Duchamp-Viret, P., Duchamp, A. & Vigoroux, M. (1989) Amplifying role of convergence in olfactory system. A comparative study of receptor cell and second-order neuron sensitivities. *J. Neurophysiol.*, **61**,1085–1094.

Dulka, J.G., Stacey, N.E., Sorensen, P.W. & Van Der Kraak, G.J. (1987) Sex steroid pheromone synchronizes male-female spawning readiness in the goldfish. *Nature*, **325**,251–253.

Duncan, H.J., Nickell, W.T., Shipley, M.T. & Gesteland, R.C. (1990) Organization of projections from olfactory epithelium to olfactory bulb in the frog, *Rana pipiens*. *J. Comp. Neurol.*, **299**,299–311.

Easton, D.M. (1971) Garfish olfactory nerve: easily accessible source of numerous long, homogeneous, nonmyelinated axons. *Science*, **172**,952–955.

Eccles, R. (1982) Autonomic innervation of the vomeronasal organ of the cat. *Physiol. Behav.*, **28**,1011–1015.

Ehrlich, M.E., Grillo, M., Joh, T.H., Margolis, F.L. & Baker, H. (1990) Transneuronal regulation of neuronal specific gene expression in the mouse olfactory bulb. *Mol. Brain Res.*, **7**,115–122.

Eisthen, H.L., Wysocki, C.J. & Beauchamp, G.K. (1987) Behavioral responses of male guinea pigs to conspecific chemical signals following neonatal vomeronasal organ removal. *Physiol. Behav.*, **41**,445–449.

Elam, J.S. (1982) Composition and subcellular distribution of glycoproteins and glycosaminoglycans undergoing axonal transport in garfish olfactory nerves. *J. Neurochem.*, **39**,1220–1229.

Elam, J.S. & Peterson, N.W. (1979) Axonal transport of glycoproteins in the garfish olfactory nerve: isolation of high molecular weight glycopeptides labeled with [^{3}H]fucose and [^{3}H]glucosamine. *J. Neurochem.*, **33**,571–573.

Emerson, H.S. (1944) Embryonic grafts in regenerating tissue. III. The development of dorsal and ventral ectoderm of *Rana pipiens* larvae. *J. Exp. Zool.*, **97**,1–19.

Emerson, H.S. (1945) The development of late gastrula explants of *Rana pipiens* in salt solution. *J. Exp. Zool.*, **100**,497–521.

Emery, D.G. (1975) The histology and fine structure of the olfactory organ of the squid *Lolliguncula brevis* Blainville. *Tissue & Cell*, **7**,357–367.

Emery, D.G. (1992) Fine structure of olfactory epithelia in gastropods. *Micro. Res. Technique.* (in press)

Erickson, J.R. & Caprio, J. (1984) The spatial distribution of ciliated and microvillous olfactory receptor neurons in the channel catfish is not matched by a differential specificity to amino acid and bile salt stimuli. *Chem. Senses*, **9**,127–141.

Esiri, M.M. (1982) Viruses and Alzheimer's disease. *J. Neurol. Neurosurg. Psychiatry*, **45**,759.

Esiri, M.M. & Tomlinson, A.H. (1984) Herpes simplex encephalitis. *J. Neurol. Sci.*, **64**,213–217.

Esiri, M.M. & Wilcock, G.K. (1984) The olfactory bulbs in Alzheimer's disease. *J. Neurol. Neurosurg. Psychiatry*, **47**,56–60.

Estes, R.D. (1972) The role of the vomeronasal organ in mammalian reproduction. *Mammalia*, **36**,315–341.

Farbman, A.I. (1977) Differentiation of olfactory receptor cells in organ culture. *Anat. Rec.*, **189**,187–200.

Farbman, A.I. (1986) Prenatal development of mammalian olfactory receptor cells. *Chem. Senses*, **11**,3–18.

Farbman, A.I. (1988) Cellular interactions in the development of the vertebrate olfactory system. In *Molecular Neurobiology of the Olfactory System*. Eds., F.L. Margolis & T.V. Getchell. Plenum, N.Y. pp. 319–332.

Farbman, A.I. (1990) Olfactory neurogenesis: genetic or environmental controls? *Trends Neuroscience*, **13**,362–365.

Farbman, A.I. (1991) Developmental neurobiology of the olfactory system. In *Smell and Taste in Health and Disease*. Eds., T.V. Getchell, R.L. Doty, L.M. Bartoshuk & J.B. Snow, Jr. Raven Press, N.Y. pp. 19–33.

Farbman, A.I., Brunjes, P.C., Rentfro, L., Michas, J. & Ritz, S. (1988) The effect of unilateral naris occlusion on cell dynamics in the developing rat olfactory epithelium. *J. Neuroscience*, **8**,3290–3295.

Farbman, A.I. & Buchholz, J.A. (1991) The growth of olfactory nerves in explant cultures is not inhibited by CNS myelin. *Soc. Neuroscience Abstr.*, **17**,634.

Farbman, A.I. & Gesteland, R.C. (1974) Fine structure of olfactory epithelium in the mudpuppy, *Necturus maculosus. Amer. J. Anat.*, **139**,227–244.

Farbman, A.I. & Gesteland, R.C. (1975) Development and electrophysiological studies of olfactory mucosa in organ culture. In *International Symposium on Olfaction and Taste, Vol. V*. Eds., D.A. Denton & J.P. Coghlan. Academic Press, N.Y. pp. 107–110.

Farbman, A.I. & Margolis, F.L. (1980) Olfactory marker protein during ontogeny: immunohistochemical localization. *Devel. Biol.*, **74**,205–215.

Farbman, A.I. & Menco, B.P.M. (1986) Development of olfactory epithelium in the rat. In *Ontogeny of Olfaction*. Ed., W. Breipohl. Springer-Verlag, Berlin. pp. 45–56.

Farbman, A.I. & Squinto, L.M. (1985) Early development of olfactory receptor cell axons. *Devel. Brain Res.*, **19**,205–213.

Ferriero, D. & Margolis, F.L. (1975) Denervation in the primary olfactory pathway of mice. II. Effects on carnosine and other amine compounds. *Brain Res.*, **94**,75–86.

Fesenko, E.E., Novoselov, V.I. & Bystrova, M.F. (1987) The subunits of specific odor-binding glycoproteins from rat olfactory epithelium. *FEBS Lett.*, **219**,224–226.

Fesenko, E.E., Novoselov, V.I. & Bystrova, M.F. (1988) Properties of odour-binding glycoproteins from rat olfactory epithelium. *Biochim. Biophys. Acta*, **937**,369–378.

Fesenko, E.E., Novoselov, V.I. & Krapivinskaya, L.D. (1979) Molecular mechanisms of olfactory reception. IV. Some biochemical characteristics of the camphor receptor from rat olfactory epithelium. *Biochim. Biophys. Acta*, **587**,424–433.

Fesenko, E.E., Novoselov, V.I., Krapivinskaya, L.D., Mjasoedov, N.F. & Zolotarev, J.A. (1983) Molecular mechanisms of odor sensing. VI. Some biochemical characteristics of a possible receptor for amino acids from the olfactory epithelium of the skate *Dasyatis pastinaca* and carp *Cyprinus carpio. Biochim. Biophys. Acta,* **759,**250–256.

ffrench-Mullen, J.M., Koller, K., Zaczek, R., Coyle, J.T., Hori, N. & Carpenter, D.O. (1985) *N*-acetylaspartylglutamate: possible role as the neurotransmitter of the lateral olfactory tract. *Proc. Natl. Acad. Sci. USA,* **82,**3897–3900.

Finger, T.E., St. Jeor, V.L., Kinnamon, J.C. & Silver, W.L. (1990) Ultrastructure of substance P- and CGRP-immunoreactive nerve fibers in the nasal epithelium of rodents. *J. Comp. Neurol.,* **294,**293–305.

Firestein, S., Darrow, B. & Shepherd, G.M. (1991) Activation of the sensory current in salamander olfactory receptor neurons depends on a G protein-mediated cAMP second messenger system. *Neuron,* **6,**825–835.

Firestein, S. & Shepherd, G.M. (1989) Olfactory transduction is mediated by the direct action of cAMP. *Soc. Neuroscience Abstr.,* **15,**749.

Firestein, S., Shepherd, G.M. & Werblin, F.S. (1990) Time course of the membrane current underlying sensory transduction in salamander olfactory receptor neurones. *J. Physiol.,* **430,**135–158.

Firestein, S. & Werblin, F. (1987) Gated currents in isolated olfactory receptor neurons of the larval tiger salamander. *Proc. Natl. Acad. Sci. USA,* **84,**6292–6296.

Firestein, S. & Werblin, F. (1989) Odor-induced membrane currents in vertebrate olfactory receptor neurons. *Science,* **244,**79–82.

Forbes, W.B. (1984) Aging-related morphological changes in the main olfactory bulb of the Fischer 344 rat. *Neurobiol. Aging,* **5,**93–99.

Fraher, J.P. (1982) The ultrastructure of sheath cells in developing rat vomeronasal nerve. *J. Anat.,* **134,**149–168.

Frahm, H.D. (1981) Volumetric comparison of the accessory olfactory bulb in bats. *Acta Anat.,* **109,**173–183.

Frahm, H.D., Stephan, H. & Baron, G. (1984) Comparison of accessory olfactory bulb volumes in the common tree shrew (*Tupaia glis*). *Acta Anat.,* **119,**129–135.

Frazier, L.L. & Brunjes, P.C. (1988) Unilateral odor deprivation: early postnatal changes in olfactory bulb cell density and number. *J. Comp. Neurol.,* **269,**355–370.

Frazier-Cierpial, L.L. & Brunjes, P.C. (1989a) Early postnatal cellular proliferation and survival in the olfactory bulb and rostral migratory stream of normal and unilaterally odor-deprived rats. *J. Comp. Neurol.,* **289,**481–492.

Frazier-Cierpial, L.L. & Brunjes, P.C. (1989b) Early postnatal differentiation of granule cell dendrites in the olfactory bulbs of normal and unilaterally odor-deprived rats. *Devel. Brain Res.,* **47,**129–136.

Friedman, L. & Miller, J.G. (1971) Odor incongruity and chirality. *Science,* **172,**1044–1046.

Frings, S. & Lindemann, B. (1988) Odorant response of isolated olfactory receptor cells is blocked by amiloride. *J. Membrane Biol.,* **105,**233–243.

Frings, S. & Lindemann, B. (1990) Single unit recording from olfactory cilia. *Biophys. J.,* **57,**1091–1094.

Frings, S. & Lindemann, B. (1991) Current recording from sensory cilia of olfactory receptor cells *in situ.* I. The neuronal response to cyclic nucleotides. *J. Gen. Physiol.,* **97,**1–16.

Frisch, D. (1967) Ultrastructure of the mouse olfactory mucosa. *Amer. J. Anat.,* **121,**87–120.

Frosch, M.P. & Dichter, M.A. (1984) Physiology and pharmacology of olfactory bulb neurons in dissociated cell culture. *Brain Res.*, **290**,321–332.

Frye, R.E., Schwartz, B.S. & Doty, R.L. (1990) Dose-related effects of cigarette smoking on olfactory function. *J. Amer. Med. Assoc.*, **263**,1233–1236.

Fujita, I., Satou, M. & Ueda, K. (1985) Ganglion cells of the terminal nerve: morphology and electrophysiology. *Brain Res.*, **335**,148–152.

Fujita, I., Satou, M. & Ueda, K. (1988) Morphology of physiologically identified mitral cells in the carp olfactory bulb: a light microscopic study after intracellular staining with horseradish peroxidase. *J. Comp. Neurol.*, **267**,253–268.

Fujita, S.C., Mori, K., Imamura, K. & Obata, K. (1985) Subclasses of olfactory receptor cells and their segregated central projections demonstrated by a monoclonal antibody. *Brain Res.*, **326**,192–196.

Fuller, T.A. & Price, J.L. (1988) Putative glutamatergic and/or aspartatergic cells in the main and accessory olfactory bulbs of the rat. *J. Comp. Neurol.*, **276**,209–218.

Gall, C., Seroogy, K.B. & Brecha, N. (1986) Distribution of VIP- and NPY-like immunoreactivities in rat main olfactory bulb. *Brain Res.*, **374**,389–394.

Gall, C.M., Hendry, S.H.C., Seroogy, K.B., Jones, E.G. & Haycock, J.W. (1987) Evidence for coexistence of GABA and dopamine in neurons of the rat olfactory bulb. *J. Comp. Neurol.*, **266**,307–318.

Garcia-Verdugo, J.M., Llahi, S., Ferrer, I. & Lopez-Garcia, C. (1989) Postnatal neurogenesis in the olfactory bulbs of a lizard. A tritiated thymidine autoradiographic study. *Neuroscience Lett.*, **98**,247–252.

Garrosa, M., Coca, S. & Mora, O.A. (1986) Histological development of the vomeronasal complex in the pre- and postnatal rat. *Acta Otolaryngol. (Stockh.)*, **102**,291–301.

Garstka, W.R. & Crews, D. (1981) Female sex pheromone in the skin and circulation of a garter snake. *Science*, **214**,681–683.

Gasser, H.S. (1956) Olfactory nerve fibers. *J. Gen. Physiol.*, **39**,473–496.

Gasser, H.S. (1958) Comparison of the structure, as revealed with the electron microscope, and the physiology of the unmedullated fibers in the skin nerves and in the olfactory nerves. *Exper. Cell Res. (Suppl.)*, **5**,3–17.

Gemne, G. & Døving, K.B. (1969) Ultrastructural properties of primary olfactory neurons in fish (*Lota lota L.*). *Amer. J. Anat.*, **126**,457–476.

Gennings, J.N., Gower, D.B. & Bannister, L.H. (1977) Studies on the receptors to 5-androst–16-en–3-one and 5-androst–16-en–13-ol in sow nasal mucosa. *Biochim. Biophys. Acta*, **496**,547–566.

Gerisch, G. (1982) Chemotaxis in *Dictyostelium*. *Ann. Rev. Physiol.*, **44**,535–542.

Gervais, R. (1987) Local GABAergic modulation of noradrenaline release in rat olfactory bulb measured on superfused slices. *Brain Res.*, **400**,151–154.

Gervais, R., Holley, A. & Keverne, B. (1988) The importance of central noradrenergic influences on the olfactory bulb in the processing of learned olfactory cues. *Chem. Senses*, **13**,3–12.

Gervais, R. & Pager, J. (1983) Olfactory bulb excitability selectively modified in behaving rats after local 6-hydroxydopamine treatment. *Behav. Brain Res.*, **9**,165–179.

Gesteland, R.C. (1964) Initial events of the electro-olfactogram. *Ann. N.Y. Acad. Sci.*, **116**,440–447.

Gesteland, R.C. (1971) Neural coding in olfactory receptor cells. In *Handbook of Sensory Physiology. Vol. IV. Chemical Senses 1, Olfaction.* Ed., L.M. Beidler. Springer-Verlag, Berlin. pp. 132–150.

Gesteland, R.C. (1986) Speculations on receptor cells as analyzers and filters. *Experientia,* **42,**287–291.

Gesteland, R.C., Brouwer, J. & Farmer, P. (1991) Voltage-sensitive dyes localize transduction events within olfactory receptor neurons. *Soc. Neuroscience Abstr.,* **17,**1103.

Gesteland, R.C., Lettvin, J.Y. & Pitts, W.H. (1965) Chemical transmission in the nose of the frog. *J. Physiol. (Lond.),* **181,**525–559.

Gesteland, R.C., Yancey, R.A. & Farbman, A.I. (1982) Development of olfactory receptor neuron selectivity in the rat fetus. *Neuroscience,* **7,**3127–3136.

Getchell, M.L., Bouvet, J.F., Finger, T.E., Holley, A. & Getchell, T.V. (1989) Peptidergic regulation of secretory activity in the olfactory mucosa of the tiger salamander: immunocytochemistry and pharmacology. *Cell Tissue Res.,* **256,**381–389.

Getchell, M.L. & Gesteland, R.C. (1972) The chemistry of olfactory reception: stimulus-specific protection from sulphydryl reagent inhibition. *Proc. Natl. Acad. Sci. USA,* **69,**1494–1498.

Getchell, M.L. & Getchell, T.V. (1984) β-Adrenergic regulation of the secretory granule content of acinar cells in olfactory glands of the salamander. *J. Comp. Physiol. A,* **155,**435–443.

Getchell, M.L. & Getchell, T.V. (1991) Immunohistochemical localization of components of the immune barrier in the olfactory mucosae of salamanders and rats. *Anat. Rec.,* **231,**358–374.

Getchell, M.L., Rafols, J.A. & Getchell, T.V. (1984) Histological and histochemical studies of the secretory components of the salamander olfactory mucosa: effects of isoproterenol and olfactory nerve section. *Anat. Rec.,* **208,**553–565.

Getchell, M.L., Zielinski, B. & Getchell, T.V. (1988) Odorant and autonomic regulation of secretion in the olfactory mucosa. In *Molecular Neurobiology of the Olfactory System.* Eds., F.L. Margolis & T.V. Getchell. Plenum, N.Y. pp. 71–98.

Getchell, T.V. (1973) Analysis of unitary spikes recorded extracellularly from frog olfactory receptor cells and axons. *J. Physiol.,* **234,**533–551.

Getchell, T.V. (1974a) Unitary responses in frog olfactory epithelium to sterically related molecules at low concentrations. *J. Gen. Physiol.,* **64,**241–261.

Getchell, T.V. (1974b) Electrogenic sources of slow voltage transients recorded from frog olfactory epithelium. *J. Neurophysiol.,* **37,**1115–1130.

Getchell, T.V. (1977) Analysis of intracellular recordings from salamander olfactory epithelium. *Brain Res.,* **123,**275–286.

Getchell, T.V. (1986) Functional properties of vertebrate olfactory receptor neurons. *Physiol. Rev.,* **66,**772–818.

Getchell, T.V. & Getchell, M.L. (1974) Signal-detecting mechanisms in the olfactory epithelium: molecular discrimination. *Ann. N.Y. Acad. Sci.,* **237,**62–71.

Getchell, T.V. & Getchell, M.L. (1987) Peripheral mechanisms of olfaction: biochemistry and neurophysiology. In *Neurobiology of Taste and Smell.* Eds., T.E. Finger & W.L. Silver. Wiley, N.Y. pp. 91–123.

Getchell, T.V. & Getchell, M.L. (1990) Regulatory factors in the vertebrate olfactory mucosa. *Chem. Senses,* **15,**223–231.

Getchell, T.V., Heck, G.L., DeSimone, J.A. & Price, S. (1980) The location of olfactory receptor sites. Inferences from latency measurements. *Biophys. J.*, **29**,397–412.

Getchell, T.V., Margolis, F.L. & Getchell, M.L. (1984) Perireceptor and receptor events in vertebrate olfaction. *Progr. Neurobiol.*, **23**,317–345.

Getchell, T.V. & Shepherd, G.M. (1978a) Responses of olfactory receptor cells to step pulses of odour at different concentrations in the salamander. *J. Physiol.*, **282**,521–540.

Getchell, T.V. & Shepherd, G.M. (1978b) Adaptive properties of olfactory receptors analysed with odour pulses of varying durations. *J. Physiol.*, **282**,541–560.

Giroud, A., Martinet, M. & Deluchat, C. (1965) Mécanisme de développement du bulbe olfactif. *Arch. Anat., Histol. Embryol.* **48**,203–217.

Godfrey, D.A., Ross, C.D., Carter, J.A., Lowry, O.H. & Matschinsky, F.H. (1980) Effect of intervening lesions on amino acid distributions in rat olfactory cortex and olfactory bulb. *J. Histochem. Cytochem.*, **28**,1157–1169.

Gold, G.H. & Nakamura, T. (1987) Cyclic nucleotide-gated conductances: a new class of ion channels mediates visual and olfactory transduction. *Trends Pharmacol. Sci.*, **8**,312–316.

Goldberg, S.J., Turpin, J., & Price, S. (1979) Anisole binding protein from dog olfactory epithelium: evidence for a role in transduction. *Chem. Senses*, **4**,207–214.

Gomez-Pinilla, F., Guthrie, K.M., Leon, M., & Nieto-Sampedro, M. (1989) NGF receptor increase in the olfactory bulb of the rat after early unilateral deprivation. *Devel. Brain Res.*, **48**,161–165.

Gonzales, F., Farbman, A.I. & Gesteland, R.C. (1985) Cell and explant culture of olfactory chemoreceptor cells. *J. Neuroscience Meth.*, **14**,77–90.

González-Estrada, M.T. & Freeman, W.J. (1980) Effects of carnosine on olfactory bulb EEG, evoked potentials and DC potentials. *Brain Res.*, **202**,373–386.

Goto, N., Hirano, N., Aiuchi, M., Hayashi, T. & Fujiwara, K. (1977) Nasoencephalopathy of mice infected intranasally with a mouse hepatitis virus, JHM strain. *Japan. J. Exp. Med.*, **47**,59–70.

Gozzo, S., & Fülöp, Z. (1984) Transneuronal degeneration in different inbred strains of mice – a preliminary study of olfactory bulb events after olfactory nerve lesion. *Int. J. Neuroscience*, **23**,187–194.

Grafe, M.R. & Leonard, C.M. (1982) Developmental changes in the topographical distribution of cells contributing to the lateral olfactory tract. *Devel. Brain Res.*, **3**,387–400.

Graves, B.M. & Halpern, M. (1989) Chemical access to the vomeronasal organs of the lizard, *Chalcides ocellatus. J. Exp. Zool.*, **249**,150–157.

Graziadei, P.P.C. (1971) The olfactory mucosa of vertebrates. In *Handbook of Sensory Physiology. Vol. IV. Chemical Senses 1, Olfaction.* Ed., L.M. Beidler. Springer-Verlag, Berlin. pp. 27–58.

Graziadei, P.P.C. (1973a) Cell dynamics in the olfactory mucosa. *Tissue & Cell*, **5**,113–131.

Graziadei, P.P.C. (1973b) The ultrastructure of vertebrates olfactory mucosa. In *The Ultrastructure of Sensory Organs.* Ed., I. Friedmann. North Holland, Amsterdam. pp. 267–305.

Graziadei, P.P.C. (1977) Functional anatomy of the mammalian chemoreceptor system. In *Chemical Senses in Vertebrates.* Eds., D. Müller-Schwarze & M.M. Mozell. Plenum Press, N.Y. pp. 435–454.

Graziadei, P.P.C. & Bannister, L.H. (1967) Some observations on the fine structure of the olfactory epithelium in the domestic duck. Z. Zellforsch., **80**,220–228.

Graziadei, P.P.C. & Gagne, H.T. (1973) Extrinsic innervation of olfactory epithelium. Z. Zellforsch., **138**,315–326.

Graziadei, P.P.C. & Kaplan, M.S. (1980) Regrowth of olfactory sensory axons into transplanted neural tissue. 1. Development of connections with the occipital cortex. Brain Res., **201**,39–44.

Graziadei, P.P.C., Karlan, M.S., Monti Graziadei, G.A. & Bernstein, J.J. (1980) Neurogenesis of sensory neurons in the primate olfactory system after section of the fila olfactoria. Brain Res., **186**,289–300.

Graziadei, P.P.C., Levine, R.R. & Monti Graziadei, G.A. (1978) Regeneration of olfactory axons and synapse formation in the forebrain after bulbectomy in neonatal mice. Proc. Natl. Acad. Sci. USA, **75**,5320–5324

Graziadei, P.P.C., Levine, R.R. & Monti Graziadei, G.A. (1979) Regeneration into the forebrain following bulbectomy in the neonatal mouse. Neuroscience, **4**,713–727.

Graziadei, P.P.C. & Metcalf, J.F. (1971) Autoradiographic and ultrastructural observations on the frog's olfactory mucosa. Z. Zellforsch., **116**,305–318.

Graziadei, P.P.C. & Monti Graziadei, G.A. (1976) Olfactory epithelium of Necturus maculosus and Ambystoma tigrinum. J. Neurocytol., **5**,11–32.

Graziadei, P.P.C. & Monti Graziadei, G.A. (1978) Continuous nerve cell renewal in the olfactory system. In Handbook of Sensory Physiology, Vol. IX. Ed., M. Jacobson. Springer-Verlag, Berlin. pp. 55–82.

Graziadei, P.P.C. & Monti Graziadei, G.A. (1979) Neurogenesis and neuron regeneration in the olfactory system of mammals. I. Morphological aspects of differentiation and structural organization of the olfactory sensory neurons. J. Neurocytol., **8**,1–18.

Graziadei, P.P.C. & Monti Graziadei, G.A. (1980a) Neurogenesis and neuron regeneration in the olfactory system of mammals. III. Deafferentation and reinnervation of the olfactory bulb following section of the fila olfactoria in rat. J. Neurocytol., **9**,145–162.

Graziadei, P.P.C. & Monti Graziadei, G.A. (1980b) Plasticity of connections in the olfactory sensory pathway: transplantation studies. In Olfaction and Taste VII. Ed., H. Van der Starre. IRL Press, London. pp. 155–158.

Graziadei, P.P.C. & Monti Graziadei, G.A. (1983) Regeneration in the olfactory system of vertebrates. Amer. J. Otolaryngol., **4**,228–233.

Graziadei, P.P.C. & Monti Graziadei, G.A. (1986) Principles of organization of the vertebrate olfactory glomerulus: an hypothesis. Neuroscience, **19**,1025–1035.

Graziadei, P.P.C. & Okano, M. (1979) Neuronal degeneration and regeneration in the olfactory epithelium of pigeon following transection of the first cranial nerve. Acta Anat., **104**,220–236.

Graziadei, P.P.C. & Samanen, D.W. (1980) Ectopic glomerular structures in the olfactory bulb of neonatal and adult mice. Brain Res., **187**,467–472.

Graziadei, P.P.C. & Tucker, D. (1970) Vomeronasal receptors in turtles. Z. Zellforsch., **105**,498–514.

Greer, C.A. (1987) Golgi analyses of dendritic organization among denervated olfactory bulb granule cells. J. Comp. Neurol., **257**,442–452.

Greer, C.A. & Halász, N. (1987) Plasticity of dendrodendritic microcircuits following mitral cell loss in the olfactory bulb of the murine mutant Purkinje cell degeneration. J. Comp. Neurol., **256**,284–298.

Greer, C.A. & Shepherd, G.M. (1982) Mitral cell degeneration and sensory function in the neurological mutant mouse PCD. *Brain Res.*, **235**,156–161.

Greer, C.A., Stewart, W.B., Kauer, J.S. & Shepherd, G.M. (1981) Topographical and laminar localization of 2-deoxyglucose uptake in rat olfactory bulb induced by stimulation of olfactory nerves. *Brain Res.*, **217**,279–293.

Greer, C.A., Stewart, W.B., Teicher, M.H. & Shepherd, G.M. (1982) Functional development of the olfactory bulb and a unique glomerular complex in the neonatal rat. *J. Neuroscience*, **2**,1744–1759.

Gross, G.W. & Beidler, L.M. (1973) Fast axonal transport in the C-fibers of the garfish olfactory nerve. *J. Neurobiol.*, **4**,413–428.

Gross, G.W. & Beidler, L.M. (1975) A quantitative analysis of isotope concentration profiles and rapid transport velocities in the C-fibers of the garfish olfactory nerve. *J. Neurobiol.*, **6**,213–232.

Gross, G.W. & Kreutzberg, G.W. (1978) Rapid axoplasmic transport in the olfactory nerve of the pike: I. Basic transport parameters for proteins and amino acids. *Brain Res.*, **139**,65–76.

Guthrie, K.M. & Leon, M. (1989) Induction of tyrosine hydroxylase expression in rat forebrain neurons. *Brain Res.*, **497**,117–131.

Guthrie, K.M., Wilson, D.A. & Leon, M. (1990) Early unilateral deprivation modifies olfactory bulb function. *J. Neuroscience*, **10**,3402–3412.

Gyorgyi, T.K., Roby-Shemkowitz, A.J. & Lerner, M.R. (1988) Characterization and cDNA cloning of the pheromone binding protein from the tobacco hornworm, *Manduca sexta. Proc. Natl. Acad. Sci. USA*, **85**,9851–9855.

Hagen, D.C., McCaffrey, G. & Sprague, G.F., Jr. (1986) Evidence the yeast STE3 gene encodes a receptor for the peptide pheromone a factor: gene sequence and implications for the structure of the presumed receptor. *Proc. Natl. Acad. Sci. USA*, **83**,1418–1422.

Haggis, A.J. (1956) Analysis of the determination of the olfactory placode in *Amblystoma punctatum. J. Embryol. Exper. Morphol.*, **4**,120–139.

Halász, N. (1990) *The Vertebrate Olfactory System.* Akadémiai Kiadó, Budapest. 281 pp.

Halász, N., Johanssen, O., Hökfelt, T., Ljungdahl, Å. & Goldstein, M. (1981a) Immunohistochemical identification of two types of dopamine neurons in the rat olfactory bulb as seen by serial sectioning. *J. Neurocytol.*, **10**,251–259.

Halász, N., Ljungdahl, Å. & Hökfelt, T. (1978) Transmitter histochemistry of the rat olfactory bulb. II. Fluorescence histochemical, autoradiographic and electron microscopic localization of monoamines. *Brain Res.*, **154**,253–271.

Halász, N., Ljungdahl, Å. & Hökfelt, T. (1979) Transmitter histochemistry of the rat olfactory bulb. III. Autoradiographic localization of [^{3}H]GABA. *Brain Res.*, **167**,221–240.

Halász, N., Ljungdahl, Å., Hökfelt, T., Johansson, O., Goldstein, M., Park, D. & Biberfeld, P. (1977) Transmitter histochemistry of the rat olfactory bulb. I. Immunohistochemical localization of monoamine synthesizing enzymes. Support for intrabulbar, periglomerular dopamine neurons. *Brain Res.*, **126**,455–474.

Halász, N., Nowycky, M.C. & Shepherd, G.M. (1983) Autoradiographic analysis of [^{3}H]dopamine and [^{3}H]DOPA uptake in the turtle olfactory bulb. *Neuroscience*, **8**,705–715.

Halász, N., Parry, D.M., Blackett, N.M., Ljungdahl, Å. & Hökfelt, T. (1981b) [^{3}H]γ-aminobutyrate autoradiography of the rat olfactory bulb: hypothetical grain analysis of the distribution of silver grains. *Neuroscience*, **6**,473–479.

238   *References*

Halász, N. & Shepherd, G.M. (1983) Neurochemistry of the vertebrate olfactory bulb. *Neuroscience*, **10,**579–619.
Hallberg, E., Eloffson, R. & Johansson, K.U.I. (1992) The aesthetasc concept: structural variations of putative olfactory receptor cell complexes in *Crustacea. Micro. Res. Technique.* (in press)
Halpern, M. (1987) The organization and function of the vomeronasal system. *Ann. Rev. Neuroscience*, **10,**325–362.
Halpern, M. & Kubie, J.L. (1980) Chemical access to the vomeronasal organs of garter snakes. *Physiol. Behav.,* **24,**367–371.
Halpern, M., Schulman, N. & Kirschenbaum, D.M. (1986) Characteristics of earthworm washings detected by the vomeronasal system of snakes. In *Chemical Signals in Vertebrates. Vol. 4, Ecology, Evolution and Comparative Biology.* Eds., D. Duvall, D. Muller-Schwarze & R. M. Silverstein. Plenum Press, N.Y., pp. 63–77.
Hamburger, V. (1934) The effects of wing bud extirpation on the development of the central nervous system in chick embryos. *J. Exp. Zool.,* **68,**449–494.
Hamburger, V. (1958) Regression versus peripheral control of differentiation in motor hypoplasia. *Amer. J. Anat.,* **102,**365–410.
Hamburger, V. & Keefe, E.L. (1944) The effects of peripheral factors on the proliferation and differentiation in the spinal cord of the chick embryo. *J. Exp. Zool.,* **96,**223–242.
Hamburger, V. & Levi-Montalcini, R. (1949) Proliferation, differentiation and degeneration in the spinal ganglia of the chick embryo under normal and experimental conditions. *J. Exp. Zool.,* **111,**457–501.
Hamill, O.P., Marty, A., Neher, E., Sakmann, B. & Sigworth, F.J. (1981) Improved patch-clamp techniques for high-resolution current recording from cells and cell-free membrane patches. *Pflügers Arch.,* **391,**85–100.
Hammerschlag, R. & Brady, S.T. (1989) Axonal transport and the neuronal cytoskeleton. In *Basic Neurochemistry*, 4th ed. Eds., G. Siegel, B. Agranoff, R.W. Albers & P. Molinoff. Raven Press, N.Y. pp. 457–478.
Hara, T.J. (1973) Olfactory responses to amino acids in rainbow trout, *Salmo gairdneri. Comp. Biochem. Physiol.,* **44A,**407–416.
Hara, T.J. (1975) Olfaction in fish. *Prog. Neurobiol.,* **5,**271–335.
Hara, T.J. (1982) Structure-activity relationships of amino acids as olfactory stimuli. In *Chemoreception in Fishes.* Ed., T.J. Hara. Elsevier, Amsterdam. pp. 109–134.
Hara, T.J. & Zielinski, B. (1989) Structural and functional development of the olfactory organ in Teleosts. *Trans. Amer. Fisheries Soc.,* **118,**183–194.
Harding, J., Graziadei, P.P.C., Monti Graziadei, G.A. & Margolis, F.L. (1977) Denervation in the primary olfactory pathway of mice. IV. Biochemical and morphological evidence for neuronal replacement following nerve section. *Brain Res.,* **132,**11–28.
Harding, J. & Margolis, F.L. (1976) Denervation in the primary olfactory pathway of mice. III. Effect on enzymes of carnosine metabolism. *Brain Res.,* **110,**351–360.
Harding, J. & Wright, J.W. (1979) Reversible effects of olfactory nerve section on behavior and biochemistry in mice. *Brain Res. Bull.,* **4,**17–22.
Harvey, J.A., Scholfield, L.T., Graham, L.T. & Aprison, M.H. (1975) Putative transmitters in denervated olfactory cortex. *J. Neurochem.,* **24,**445–449.
Harvey, P.H. & Krebs, J.R. (1990) Comparing brains. *Science*, **249,**140–146.
Hasler, A.D., Scholz, A.T. & Horrall, R.M. (1978) Olfactory imprinting and homing in salmon. *Amer. Scientist,* **66,**347–355.

Healy, S. & Guilford, T. (1990) Olfactory-bulb size and nocturnality in birds. *Evolution,* **44,**339–346.

Heckroth, J.A., Monti Graziadei, G.A. & Graziadei, P.P.C. (1983) Intraocular transplants of olfactory neuroepithelium in rat. *Int. J. Devel. Neuroscience,* **1,**273–287.

Hedlund, B., Masukawa, L.M. & Shepherd, G.M. (1987) Excitable properties of olfactory receptor neurons. *J. Neuroscience,* **7,**2338–2343.

Hedlund, B. & Shepherd, G.M. (1983) Biochemical studies on muscarinic receptors in the salamander olfactory epithelium. *FEBS Lett.,* **162,**428–431.

Heimer, L. & Larsson, K. (1967) Impairment of mating behavior in male rats following lesions in the preoptic-anterior hypothalamic continuum. *Brain Res.,* **3,**248–263.

Hempstead, J.L. & Morgan, J.I. (1983) Monoclonal antibodies to the rat olfactory sustentacular cell. *Brain Res.,* **288,**289–295.

Hempstead, J.L. & Morgan, J.I. (1985) Monoclonal antibodies reveal novel aspects of the biochemistry and organization of olfactory neurons following unilateral olfactory bulbectomy. *J. Neuroscience,* **5,**2382–2387.

Herrick, C.J. (1921) The connections of the vomeronasal nerve, accessory olfactory bulb and amygdala in amphibia. *J. Comp. Neurol.,* **33,**213–280.

Hinds, J.W. (1968a) Autoradiographic study of histogenesis in the mouse olfactory bulb. I. Time of origin of neurons and neuroglia. *J. Comp. Neurol.,* **134,**287–304.

Hinds, J.W. (1968b) Autoradiographic study of histogenesis in the mouse olfactory bulb. II. Cell proliferation and migration. *J. Comp. Neurol.,* **134,**305–322.

Hinds, J.W. (1972a) Early neuron differentiation in the mouse olfactory bulb. I. Light microscopy. *J. Comp. Neurol.,* **146,**233–252.

Hinds, J.W. (1972b) Early neuron differentiation in the mouse olfactory bulb. II. Electron microscopy. *J. Comp. Neurol.,* **146,**253–276.

Hinds, J.W. & Hinds, P.L. (1976a) Synapse formation in the mouse olfactory bulb. I. Quantitative studies. *J. Comp. Neurol.,* **169,**15–40.

Hinds, J.W. & Hinds, P.L. (1976b) Synapse formation in the mouse olfactory bulb. II. Morphogenesis. *J. Comp. Neurol.,* **169,**41–62.

Hinds, J.W., Hinds, P.L. & McNelly, N.A. (1984) An autoradiographic study of the mouse olfactory epithelium: evidence for long-lived receptors. *Anat. Rec.,* **210,**375–383.

Hinds, J.W. & McNelly, N.A. (1977) Aging of the rat olfactory bulb: growth and atrophy of constituent layers and changes in size and number of mitral cells. *J. Comp. Neurol.,* **171,**345–368.

Hinds, J.W. & McNelly, N.A. (1981) Aging in the rat olfactory system: correlation of changes in the olfactory epithelium and olfactory bulb. *J. Comp. Neurol.,* **203,**441–453.

Hinds, J.W. & Ruffett, T.L. (1973) Mitral cell development in the mouse olfactory bulb: reorientation of the perikaryon and maturation of the axon initial segment. *J. Comp. Neurol.,* **151,**281–306.

Hirsch, J.D., Grillo, M. & Margolis, F.L. (1978) Ligand binding studies in the mouse olfactory bulb: identification and characterization of a L-[³H]carnosine binding site. *Brain Res.,* **158,**407–422.

Hirsch, J.D. & Margolis, F.L. (1979) L-[³H]carnosine binding in the olfactory bulb. II. Biochemical and biological studies. *Brain Res.,* **174,**81–94.

Hirsch, J.D. & Margolis, F.L. (1981) Isolation, separation, and analysis of cells from olfactory epithelium. In *Biochemistry of Taste and Olfaction.* Eds., R.H. Cagan & M.R. Kare. Academic Press, N.Y. pp. 311–332.

Hofer, M.A., Shair, H. & Singh, P. (1976) Evidence that maternal ventral skin substances promote suckling in infant rats. *Physiol. Behav.,* **17,**131–136.

Hofmann, M.H. & Meyer, D.L. (1989) The nervus terminalis in larval and adult *Xenopus laevis. Brain Res.,* **497,**167–169.

Hökfelt, T., Halász, N., Ljungdahl, Å., Johansson, O., Goldstein, M. & Park, D. (1975) Histochemical support for a dopaminergic mechanism in the dendrites of certain periglomerular cells of the rat olfactory bulb. *Neuroscience Lett.,* **1,**85–90.

Holl, A. (1981) Marking of olfactory axons of fishes by intravital staining with Procion Brilliant Yellow. *Stain Technol.,* **56,**67–70.

Holley, A., Duchamp, A., Revial, M.-F., Juge, A. & MacLeod, P. (1974) Qualitative and quantitative discrimination in the frog olfactory receptors: analysis from electrophysiological data. *Ann. N.Y. Acad. Sci.,* **237,**102–114.

Holmgren, N. (1920) Zur Anatomie und Histologie des Vorder- und Zwischenhirns der Knochenfische. Hauptsächlich nach Untersuchungen an *Osmerus eperlanus. Acta Zool.* (Stockh.), **1:**137–315.

Holtzman, D.A. & Halpern, M. (1990) Embryonic and neonatal development of the vomeronasal and olfactory systems in garter snakes (*Thamnophis* spp.) *J. Morphol.,* **203,**123–140.

Hoppe, P.C. (1975) Genetic and endocrine studies of the pregnancy-blocking pheromone of mice. *J. Reprod. Fertil.,* **45,**109–115.

Hori, N., Auker, C.R., Braitman, D.J. & Carpenter, D.O. (1981) Lateral olfactory tract transmitter: glutamate, aspartate or neither? *Cell. Mol. Neurobiol.,* **1,**115–120.

Hornung, D.E., Lansing, R.D. & Mozell, M.M. (1975) Distribution of butanol molecules along bullfrog olfactory mucosa. *Nature,* **254,**617–618.

Hornung, D.E. & Mozell, M.M. (1977) Factors influencing the differential sorption of odorant molecules across the olfactory mucosa. *J. Gen. Physiol.,* **69,**343–361.

Hornung, D.E. & Mozell, M.M. (1980) Tritiated odorants to monitor retention in the olfactory and vomeronasal organs. *Brain Res.,* **181,**488–492.

Hornung, D.E. & Mozell, M.M. (1981) Accessibility of odorant molecules to the receptors. In *Biochemistry of Taste and Olfaction.* Eds., R.H. Cagan & M.R. Kare. Academic Press, N.Y. pp. 33–45.

Hornykiewicz, O. (1983) Parkinson's disease: from brain homogenate to treatment. *Fed. Proc.,* **32,**183–190.

Houston, D.C. (1987) Scavenging efficiency of turkey vultures in tropical forests. *Condor,* **88,**318–323.

Hubel, D.H. & Wiesel, T.N. (1977) Functional architecture of macaque monkey visual cortex. *Proc. Roy. Soc. London B.,* **198,**1–59.

Hudson, R. & Distel, H. (1983) Nipple location by newborn rabbits: behavioural evidence for pheromonal guidance. *Behaviour,* **85,**260–275.

Hudson, R. & Distel, H. (1984) Nipple search pheromone in rabbits: dependence on season and reproductive state. *J. Comp. Physiol. A,* **155,**13–17.

Hudson, R. & Distel, H. (1986) Olfactory guidance of nipple-search behaviour in newborn rabbits. In *Ontogeny of Olfaction.* Ed., W. Breipohl. Springer-Verlag, Berlin. pp. 243–254.

Humphrey, T. (1940) The development of the olfactory and the accessory olfactory formations in human embryos and fetuses. *J. Comp. Neurol.,* **73,**431–468.

Huque, T. & Bruch, R.C. (1986) Odorant- and guanine nucleotide-stimulated phosphoinositide turnover in olfactory cilia. *Biochem. Biophys. Res. Comm.,* **137,**36–42.

Hurwitz, T., Kopala, L., Clark, C. & Jones, B. (1988) Olfactory deficits in schizophrenia. *Biol. Psychiatry,* **23,**123–128.

Hutchison, L.V., Wenzel, B.M., Stager, K.E. & Tedford, B.L. (1984) Further evidence for olfactory foraging by sooty shearwaters and northern fulmars. In *Marine Birds: Their Feeding Ecology and Commercial Fisheries Relationships.* Eds., D.N. Nettleship, G.A. Sanger & P.F. Springer. Can. Wildl. Serv., Spec. Publ., Ottawa. pp. 72–77.

Imaki, T., Nahon, J.L., Sawchenko, P.E. & Vale, W. (1989) Widespread expression of corticotropin-releasing factor messenger RNA and immunoreactivity in the rat olfactory bulb. *Brain Res.,* **496,**35–44.

Imamura, K., Mori, K., Fujita, S.C. & Obata, K. (1985) Immunochemical identification of subgroups of vomeronasal nerve fibers and their segregated terminations in the accessory olfactory bulb. *Brain Res.,* **328,**362–366.

Itaya, S.K. (1987) Anterograde transsynaptic transport of WGA-HRP in rat olfactory pathways. *Brain Res.,* **409,**205–214.

Jackowski, A., Parnavelas, J.G., & Lieberman, A.R. (1978) The reciprocal synapse in the external plexiform layer of the mammalian olfactory bulb. *Brain Res.,* **159,**17–28.

Jackson, R.T., Tigges, J. & Arnold, W. (1979) Subarachnoid space of the CNS, nasal mucosa, and lymphatic system. *Arch. Otolaryngol.,* **105,**180–184.

Jacobson, A.G. (1963) The determination and positioning of the nose, lens and ear. I. Interactions within the ectoderm and underlying tissues. *J. Exp. Zool.,* **154,**273–283.

Jafek, B.W. (1983) Ultrastructure of human nasal mucosa. *Laryngoscope,* **93,**1576–1599.

Jafek, B.W., Eller, P.M., Esses, B.A. & Moran, D.T. (1989) Post-traumatic anosmia. Ultrastructural correlates. *Arch. Neurol.,* **46,**300–304.

Jafek, B.W., Hartman, D., Eller, P.M., Johnson, E.W., Strahan, R.C. & Moran, D.T. (1990) Postviral olfactory dysfunction. *Amer. J. Rhinol.,* **4,**91–100.

Jafek, B.W., Moran, D.T., Eller, P.M., Rowley, J.C. & Jafek, T.B. (1987) Steroid-dependent anosmia. *Arch. Otolaryngol.,* **113,**547–549.

Jahr, C.E. & Nicoll, R.A. (1980) Dendrodendritic inhibition: demonstration with intracellular recording. *Science,* **207,**1473–1475.

Jahr, C.E. & Nicoll, R.A. (1982a) Noradrenergic modulation of dendrodendritic inhibition in the olfactory bulb. *Nature,* **297,**227–229.

Jahr, C.E. & Nicoll, R.A. (1982b) An intracellular analysis of dendrodendritic inhibition in the turtle *in vitro* olfactory bulb. *J. Physiol.,* **326,**213–234.

Jastreboff, P.J., Pedersen, P.E., Greer, C.A., Stewart, W.B., Kauer, J.S., Benson, T.E. & Shepherd, G.M. (1984) Specific olfactory receptor populations projecting to identified glomeruli in the rat olfactory bulb. *Proc. Natl. Acad. Sci. USA,* **81,**5250–5254.

Jennes, L. (1986) The olfactory gonadotropin-releasing hormone immunoreactive system in mouse. *Brain Res.,* **386,**351–363.

Jiang, X.C., Inouchi, J., Wang, D. & Halpern, M. (1990) Purification and characterization of a chemoattractant from electric shock-induced earthworm secretion, its receptor binding, and signal transduction through the vomeronasal system of garter snakes. *J. Biol. Chem.,* **265,**8736–8744.

Johanson, I.B. & Hall, W.G. (1982) Appetitive conditioning in neonatal rats: conditioned orientation to a novel odor. *Devel. Psychobiol.,* **15,**379–397.

Johnson, A., Josephson, R. & Hawke, M. (1985) Clinical and histological evidence for the presence of the vomeronasal (Jacobson's) organ in adult humans. *J. Otolaryngol.*, **14**,71–79.

Johnston, M.C. & Sulik, K.K. (1980) Development of face and oral cavity. In *Orban's Oral Histology and Embryology*, 9th ed. Ed., S.N. Bhaskar. C.V. Mosby, St. Louis. pp. 1–23.

Jones, B.P., Butters, N., Moskowitz, H.R. & Montgomery, K. (1978) Olfactory and gustatory capacities of alcoholic Korsakoff patients. *Neuropsychologia*, **16**,323–336.

Jones, D.T. (1990) Distribution of the stimulatory GTP-binding proteins, G_s and G_{olf}, within olfactory neuroepithelium. *Chem. Senses*, **15**,333–340.

Jones, D.T., Barbosa, E. & Reed, R.R. (1989) Expression of G-protein alpha subunits in rat olfactory neuroepithelium: candidates for olfactory signal transduction. *Cold Spring Harbor Symp. Quant. Biol.*, **53**,349–353.

Jones, D.T. & Reed, R.R. (1987) Molecular cloning of five GTP-binding protein cDNA species from rat olfactory neuroepithelium. *J. Biol. Chem.*, **262**,14241–14249.

Jones, D.T. & Reed, R.R. (1989) G_{olf}: an olfactory neuron specific G-protein involved in odorant signal transduction. *Science*, **244**,790–795.

Jourdan, F. (1975) Ultrastructure de l'épithélium olfactif du rat: polymorphisme des récepteurs. *Comp. Rend. Acad. Sci., Paris*, **280**,443–446.

Jourdan, F., Duveau, A., Astic, L. & Holley, A. (1980) Spatial distribution of [^{14}C]2-deoxyglucose uptake in the olfactory bulbs of rats stimulated with two different odours. *Brain Res.*, **188**,139–154.

Kaba, H. & Keverne, E.B. (1988) The effect of microinfusions of drugs into the accessory olfactory bulb on the olfactory block to pregnancy. *Neuroscience*, **25**,1007–1011.

Kaba, H., Rosser, A. & Keverne, B. (1989) Neural basis of olfactory memory in the context of pregnancy block. *Neuroscience*, **32**,657–662..

Kaissling, K.-E. (1986) Chemo-electrical transduction in insect olfactory receptors. *Ann. Rev. Neuroscience*, **9**,121–145.

Kaitz, M., Good, A., Rokem, A.M. & Eidelman, A.I. (1987) Mother's recognition of their newborns by olfactory cues. *Devel. Psychobiol.*, **20**,587–591.

Kalinoski, D.L., Bruch, R.C. & Brand, J.G. (1987) Differential interaction of lectins with chemosensory receptors. *Brain Res.*, **418**,34–40.

Kaplan, M.S. & Hinds, J.W. (1977) Neurogenesis in adult rat: electron microscopic analysis of light autoradiographs. *Science*, **197**,1092–1094.

Kaplan, M.S., McNelly, N.A. & Hinds, J.W. (1985) Population dynamics of adult-formed granule neurons of the rat olfactory bulb. *J. Comp. Neurol.*, **239**,117–125.

Kashiwayanagi, M. & Kurihara, K. (1984) Neuroblastoma cell as model for olfactory cell: mechanism of depolarization in response to odorants. *Brain Res.*, **293**,251–258.

Kashiwayanagi, M. & Kurihara, K. (1985) Evidence for non-receptor odor discrimination using neuroblastoma cells as a model for olfactory cells. *Brain Res.*, **359**,97–103.

Kashiwayanagi, M., Sai, K. & Kurihara, K. (1987) Cell suspensions from porcine olfactory mucosa. *J. Gen. Physiol.*, **89**,443–457.

Kashiwayanagi, M., Suenaga, A., Enomoto, S. & Kurihara, K. (1990) Membrane fluidity changes of liposomes in response to various odorants. *Biophys. J.*, **58**,887–895.

Katz, S. & Merzel, J. (1977) Distribution of epithelia and glands of the nasal septum mucosa in the rat. *Acta Anat.,* **99,**58–66.

Kauer, J.S. (1974) Response patterns of amphibian olfactory bulb neurones to odour stimulation. *J. Physiol.,* **243,**695–715.

Kauer, J.S. (1981) Olfactory receptor cell staining using horseradish peroxidase. *Anat. Rec.,* **200,**331–336.

Kauer, J.S. (1987) Coding in the olfactory system. In *Neurobiology of Taste and Smell.* Eds., T.E. Finger & W.S. Silver. Wiley, N.Y. pp. 205–231.

Kauer, J.S. (1988) Real-time imaging of evoked activity in local circuits of the salamander olfactory bulb. *Nature,* **331,**166–168.

Kauer, J.S. (1991) Contributions of topography and parallel processing to odor coding in the vertebrate olfactory pathway. *Trends Neuroscience,* **14,**79–85.

Kauer, J.S. & Moulton, D.G. (1974) Responses of olfactory bulb neurones to odour stimulation of small nasal areas in the salamander. *J. Physiol.,* **243,**717–737.

Kauer, J.S. & Shepherd, G.M. (1977) Analysis of the onset phase of olfactory bulb unit responses to odour pulses in the salamander. *J. Physiol.,* **272,**495–516.

Kawano, T. & Margolis, F.L. (1982) Transsynaptic regulation of olfactory bulb catecholamines in mice and rats. *J. Neurochem.,* **39,**342–348.

Kendrick, K.M., Keverne, E.B., Chapman, C. & Baldwin, B.A. (1988) Microdialysis measurement of oxytocin, aspartate, γ-aminobutyric acid and glutamate release from the olfactory bulb of the sheep during vaginocervical stimulation. *Brain Res.,* **442,**171–174.

Kerjaschki, D. (1977) Some freeze-etching data on the olfactory epithelium. In *Olfaction and Taste VI.* Eds., J. Le Magnen & P. MacLeod. Information Retrieval, Ltd., London. pp. 75–85.

Kerjaschki, D. & Hörandner, H. (1976) The development of mouse olfactory vesicles and their cell contacts: a freeze-etching study. *J. Ultrastruct. Res.,* **54,**420–444.

Kesbeke, F., Van Haastert, J.M., De Wit, R.J.W. & Snaar-Jagalska, B.E. (1990) Chemotaxis to cyclic AMP and folic acid is mediated by different G proteins in *Dictyostelium discoideum. J. Cell Sci.,* **96,**669–673.

Keverne, E.B. & De La Riva, C. (1982) Pheromones in mice: reciprocal interaction between the nose and the brain. *Nature,* **296,**148–150.

Keverne, E.B., Lévy, F., Poindron, P. & Lindsay, D.R. (1983) Vaginal stimulation: an important determinant of maternal bonding in sheep. *Science,* **219,**81–83.

Key, B. & Akeson, R.A. (1990a) Immunochemical markers for the frog olfactory neuroepithelium. *Devel. Brain Res.,* **57,**103–117.

Key, B. & Akeson, R.A. (1990b) Olfactory neurons express a unique glycosylated form of the neural cell adhesion molecule (N-CAM). *J. Cell Biol.,* **110,**1729–1743.

Key, B. & Giorgi, P.P. (1986) Soybean agglutinin binding to the olfactory systems of the rat and mouse. *Neuroscience Lett.,* **69,**131–136.

Kimelberg, H.K. & Norenberg, M.D. (1989) Astrocytes. *Sci. Amer.,* **260,**66–76.

Kinnamon, S. (1988) Taste transduction: a diversity of mechanisms. *Trends Neuroscience,* **11,**491–496.

Kishi, K.K., Mori, K. & Ojima, H. (1984) Distribution of local axon collaterals of mitral, displaced mitral, and tufted cells in the rabbit olfactory bulb. *J. Comp. Neurol.*, **225**,511–526.

Kishi, K., Peng, J.Y., Kakuta, S., Murakami, K., Kuroda, M., Yokota, S., Hayakawa, S., Kuge, T. & Asayama, T. (1990) Migration of bipolar subependymal cells, precursors of the granule cells of the rat olfactory bulb, with reference to the arrangement of the radial glial fibers. *Arch. Histol. Cytol.*, **53**,219–226.

Kiyohara, S. & Tucker, D. (1978) Activity of new receptors after transection of the primary olfactory nerve in pigeons. *Physiol. Behav.*, **21**,987–994.

Kleene, S.J. (1986) Bacterial chemotaxis and vertebrate olfaction. *Experientia*, **24**,241–250.

Kleene, S.J. & Gesteland, R.C. (1983) Dissociation of frog olfactory epithelium. *J. Neuroscience Meth.*, **9**,173–183.

Klein, P., Sun, T.J., Saxe, C.L., Kimmel, A.R., Johnson, R.L. & Devreotes, P.N. (1988) A chemoattractant receptor controls development in *Dictyostelium discoideum. Science*, **241**,1467–1472.

Koch, A.L., Carr, A. & Ehrenfeld, D.W. (1969) The problem of open-sea navigation: the migration of the green turtle to Ascension Island. *J. Theoret. Biol.*, **22**,163–179.

Kolesnikov, S.S., Zhainazarov, A.B. & Kosolapov, A.V. (1990) Cyclic nucleotide-activated channels in the frog olfactory receptor plasma membrane. *FEBS Lett.*, **266**,96–98.

Kolnberger, I. (1971) Vergleichende Untersuchungen am Riechepithel, insbesondere des Jacobsonschen Organs von Amphibien, Reptilien und Säugetieren. *Z. Zellforsch.*, **122**,53–67.

Kolnberger, I. & Altner, H. (1971) Ciliary-structure precursor bodies as stable constituents in the sensory cells of the vomeronasal organ of reptiles and mammals. *Z. Zellforsch.*, **118**,254–262.

Kosaka, T. (1980) Ruffed cell: a new type of neuron with a distinctive initial unmyelinated portion of the axon in the olfactory bulb of the goldfish (*Carassius auratus*). II. Fine structure of the ruffed cell. *J. Comp. Neurol.*, **193**,119–145.

Kosaka, T. & Hama, K. (1979) Ruffed cell: a new type of neuron with a distinctive initial unmyelinated portion of the axon in the olfactory bulb of the goldfish (*Carassius auratus*). I. Golgi impregnation and serial sectioning studies. *J. Comp. Neurol.*, **186**,301–320.

Kosaka, T. & Hama, K. (1982) Structure of the mitral cell in the olfactory bulb of the goldfish *(Carassius auratus). J. Comp. Neurol.*, **212**,365–384.

Kosaka, T., Hataguchi, Y., Hama, K., Nagatsu, I. & Wu, J.-Y. (1985) Coexistence of immunoreactivities for glutamate decarboxylase and tyrosine hydroxylase in some neurons in the periglomerular region of the rat main olfactory bulb: possible coexistence of gamma-aminobutyric acid (GABA) and dopamine. *Brain Res.*, **343**,166–171.

Kosaka, T., Kosaka, K., Hama, K., Wu, J.-Y. & Nagatsu, I. (1987) Differential effect of functional olfactory deprivation on the GABAergic and catecholaminergic traits in the rat main olfactory bulb. *Brain Res.*, **413**,197–203.

Koss, E., Weiffenbach, J.M., Haxby, J.V. & Friedland, R.P. (1987) Olfactory detection and recognition in Alzheimer's disease. *Lancet*, **1**,622.

Koyama, N. & Kurihara, K. (1972) Effect of odorants on lipid monolayers from bovine olfactory epithelium. *Nature*, **236**,402–404.

Kracke, G.R. & Chacko, G.K. (1979) Polypeptide components of the $(Na + K +)$-ATPase of garfish olfactory nerve axon plasma membrane. *Life Sci.*, **25**,2125–2129.

Kratzing, J.E. (1970) The olfactory mucosa of the sheep. *Austral. J. Biol. Sci.*, **23**,447–458.
Kratzing, J.E. (1971a) The fine structure of the sensory epithelium in suckling rats. *Austral. J. Biol. Sci.*, **24**,787–796.
Kratzing, J.E. (1971b) The structure of the vomeronasal organ in the sheep. *J. Anat.*, **108**,247–260.
Kratzing, J.E. (1972) The structure of olfactory cilia in a lizard. *J. Ultrastruct. Res.*, **39**,295–300.
Kratzing, J.E. (1975) The fine structure of the olfactory and vomeronasal organs of a lizard *(Tiliqua scinocoides scincoides)*. *Cell Tissue Res.*, **156**,239–252.
Kratzing, J.E. (1978) The olfactory apparatus of the bandicoot *(Isoodon macrourus)*: fine structure and presence of a septal olfactory organ. *J. Anat.*, **125**,601–613.
Kratzing, J.E. (1984) The anatomy and histology of the nasal cavity of the koala *(Phascolarctos cinereus)*. *J. Anat.*, **138**,55–65.
Kream, R.M., Davis, B.J., Kawano, T., Margolis, F.L. & Macrides, F. (1984) Substance P and catecholaminergic expression in neurons of the hamster main olfactory bulb. *J. Comp. Neurol.*, **222**,140–154.
Kreutzberg, G.W. & Gross, G.W. (1977) General morphology and axonal ultrastructure of the olfactory nerve of the pike, *Esox lucius*. *Cell Tissue Res.*, **181**,443–457.
Kreutzer, E.W. & Jafek, B.W. (1980) The vomeronasal organ of Jacobson in the human embryo and fetus. *Otolaryngol. Head Neck Surg.*, **88**,119–123.
Kristensson, K. & Olsson, Y. (1971) Uptake of exogenous proteins in mouse olfactory cells. *Acta Neuropath.*, **19**,145–154.
Kubie, J., Mackay-Sim, A. & Moulton, D.G. (1980) Inherent spatial patterning of responses to odorants in the salamander olfactory epithelium. In *Olfaction and Taste VII*. Ed., H. van der Starre. IRL Press, Ltd., London. pp. 163–166.
Kubie, J.L., Vagvolgyi, A. & Halpern, M. (1978) The roles of the vomeronasal and olfactory systems in the courtship behavior of male garter snakes. *J. Comp. Physiol. Psychol.*, **92**,627–641.
Kucharski, D., Burka, N. & Hall, W.G. (1990) The anterior limb of the anterior commissure is an access route to contralaterally stored olfactory preference memories. *Psychobiology*, **18**,195–204.
Kucharski, D. & Hall, W.G. (1988) Developmental change in the access to olfactory memories. *Behav. Neuroscience*, **102**,340–348.
Kucharski, D., Johanson, I.B. & Hall, W.G. (1986) Unilateral olfactory conditioning in 6-day-old rat pups. *Behav. Neural Biol.*, **46**,472–490.
Kurahashi, T. (1989) Activation by odorants of cation-selective conductance in the olfactory receptor cell isolated from the newt. *J. Physiol.*, **419**,177–192.
Kurahashi, T. (1990) The response induced by intracellular cyclic AMP in isolated olfactory receptor cells of the newt. *J. Physiol.*, **430**,355–371.
Kurahashi, T. & Kaneko, A. (1991) High density cAMP-gated channels at the ciliary membrane in the olfactory receptor cell. *NeuroReport*, **2**,5–8.
Kurihara, K. & Koyama, N. (1972) High activity of adenyl cyclase in olfactory and gustatory organs. *Biochem. Biophys. Res. Comm.*, **48**,30–34.
Kurihara, K., Miyake, M. & Yoshii, K. (1981) Molecular mechanisms of transduction in chemoreception. In *Biochemistry of Taste and Olfaction*. Eds., R.H. Cagan & M.R. Kare. Academic Press, N.Y. pp. 249–285.
Kurihara, K., Yoshii, K. & Kashiwayanagi, M. (1986) Transduction mechanisms in chemoreception. *Comp. Biochem. Physiol.*, **85A**,1–22.

Labarca, P. & Bacigalupo, J. (1988) Ion channels from chemosensory olfactory neurons. *J. Bioenergetics Biomembranes*, **20**,551–569.

Labarca, P., Simon, S.A. & Anholt, R.R.H. (1988) Activation by odorants of a multistate cation channel from olfactory cilia. *Proc. Natl. Acad. Sci. USA*, **85**,944–947.

Laffort, P. & Dravnieks, A. (1974) An approach to a physico-chemical model of olfactory stimulation in vertebrates by single compounds. *J. Theoret. Biol.*, **38**,335–345.

Laing, D.G. (1984) The effect of environmental odours on the sense of smell. In *Animal Models in Psychopathology*. Ed., N.W. Bond. Academic Press, Sydney. pp. 59–98.

Laing, D.G. & Panhuber, H. (1978) Neural and behavioral changes in rats following continuous exposure to an odor. *J. Comp. Physiol. A*, **124**,259–265.

Laing, D.G. & Panhuber, H. (1980) Olfactory sensitivity of rats reared in an odorous or deodorized environment. *Physiol. Behav.*, **25**,555–558.

Laing, D.G., Panhuber, H., Pittman, E.A., Willcox, M.E. & Eagleson, G.K. (1985) Prolonged exposure to an odor or deodorized air alters the size of mitral cells in the olfactory bulb. *Brain Res.*, **336**,81–87.

LaMantia, A.-S. & Purves, D. (1989) Development of glomerular pattern visualized in the olfactory bulbs of living mice. *Nature*, **341**,646–649.

Lambert, M.P., Megerian, T., Garden, G. & Klein, W.L. (1988) Soluble proteins from rat olfactory bulb promote the survival and differentiation of cultured basal forebrain cells. *Devel. Brain Res.*, **41**,263–276.

Lancet, D. (1984) Molecular view of olfactory reception. *Trends Neuroscience*, **7**,35–36.

Lancet, D. (1986) Vertebrate olfactory reception. *Ann. Rev. Neuroscience*, **9**,329–355.

Lancet, D. (1988) Molecular components of olfactory reception and transduction. In *Molecular Neurobiology of the Olfactory System*. Eds., F.L. Margolis & T.V. Getchell. Plenum Press, N.Y. pp. 25–50.

Lancet, D., Greer, C.A., Kauer, J.S. & Shepherd, G.M. (1982) Mapping of odor-related neuronal activity in the olfactory bulb by high-resolution 2-deoxyglucose localization. *Proc. Natl. Acad. Sci. USA*, **79**,670–674.

Lancet, D. & Pace, U. (1987) The molecular basis of odor recognition. *Trends Biochem. Sci.*, **12**,63–66.

Land, L.J. (1973) Localized projections of olfactory nerves to rabbit olfactory bulb. *Brain Res.*, **63**,153–166.

Land, L.J. & Shepherd, G.M. (1974) Autoradiographic analysis of olfactory receptor projections in the rabbit. *Brain Res.*, **70**,506–510.

Landmesser, L. & Pilar, G. (1974) Synaptic transmission and cell death during normal ganglionic development. *J. Physiol. (Lond.)*, **241**,738–750.

Large, T.H., Lambert, M.P., Gremillion, M.A. & Klein, W.L. (1986) Parallel postnatal development of choline acetyltransferase activity and muscarinic acetylcholine receptors in the rat olfactory bulb. *J. Neurochem.*, **46**,671–680.

Lazard, D., Zupko, K., Heldman, J., Nef, P. & Lancet, D. (1989) Molecular cloning of olfactory-specific cytochrome P450 and UDP glucuronosyl transferase: candidate signal termination and odorant clearance enzymes. *Chem. Senses*, **14**,721.

Lazard, D., Zupko, K., Poria, Y., Nef, P., Lazarovits, J., Horn, S., Khen M. & Lancet, D. (1991) Odorant signal termination by olfactory UDP glucuronosyl transferase. *Nature*, **49**,790–793.

Leblond, C.P. & Walker, B.E. (1956) Renewal of cell populations. *Physiol. Rev.*, **36**,255–275.

Le Gros Clark, W.E. (1951) The projection of the olfactory epithelium on the olfactory bulb in the rabbit. *J. Neurol. Neurosurg. Psychiatry*, **14**,1–10.

Le Gros Clark, W. (1957) Inquiries into the anatomical basis of olfactory discrimination. *Proc. Roy. Soc. (Ser. B)*, **146**,299–319.

Leon, M. (1974) Maternal pheromone. *Physiol. Behav.*, **13**,441–453.

Leon, M. (1983) Chemical communication in mother-young interactions. In *Pheromones and Reproduction in Mammals*. Academic Press, N.Y. pp. 39–77.

Leon, M., Coopersmith, R., Ulibarri, C., Porter, R.H. & Powers, B., (1984) Development of olfactory bulb organization in precocial and altricial rodents. *Devel. Brain Res.*, **12**,45–53.

Leon, M., Galef, B.G. & Behse, J. (1977) Establishment of pheromonal bonds and diet choice in young rats by odor preexposure. *Physiol. Behav.*, **18**,387–391.

Lerner, M.R., Reagan, J., Gyorgyi, T. & Roby, A. (1988) Olfaction by melanophores: What does it mean? *Proc. Natl. Acad. Sci. USA*, **85**,261–264.

Leveteau, J. & MacLeod, P. (1966) Olfactory discrimination in the rabbit olfactory glomerulus. *Science*, **153**,175–176.

Lévy, F., Gervais, R., Kindermann, U., Orgeur, P. & Piketty, V. (1990) Importance of β-adrenergic receptors in the olfactory bulb of sheep for recognition of lambs. *Behav. Neuroscience*, **104**,464–469.

Licht, G. & Meredith, M. (1987) Convergence of main and accessory olfactory pathways onto single neurons in the hamster amygdala. *Exp. Brain Res.*, **69**,7–18.

Lidow, M.S., Gesteland, R.C., Shipley, M.T. & Kleene, S.J. (1987) Comparative study of immature and mature olfactory receptor cells in adult frogs. *Devel. Brain Res.*, **31**,243–258.

Lidow, M.S. & Menco, B.P.M. (1984) Observations on axonemes and membranes of olfactory and respiratory cilia in frogs and rats using tannic acid-supplemented fixation and photographic rotation. *J. Ultrastruct. Res.*, **86**,18–30.

Lloyd-Thomas, A. & Keverne, E.B. (1982) Role of the main and accessory olfactory system in the block to pregnancy in mice. *Neuroscience*, **7**,907–913.

Lomas, D.E. & Keverne, E.B. (1982) Role of the vomeronasal organ and prolactin in the acceleration of puberty in female mice. *J. Reprod. Fertil.*, **66**,101–107.

Loo, S.K. & Kanagasuntheram, R. (1972) The vomeronasal organ in tree shrew and slow loris. *J. Anat.*, **112**,165–172.

Lowe, G., Nakamura, T. & Gold, G.H. (1989) Adenylate cyclase mediates olfactory transduction for a wide variety of odorants. *Proc. Natl. Acad. Sci. USA*, **86**,5641–5645.

Ludwig, J., Margalit, T., Eisman, E., Lancet, D. & Kaupp. U. B. (1990) Primary structure of cAMP-gated channel from bovine olfactory epithelium. *FEBS Lett.*, **270**,24–29.

Lundblad, L., Lundberg, J.M., Brodin, E. & Ånggard, A. (1983) Origin and distribution of capsaicin-sensitive substance P-immunoreactive nerves in the nasal mucosa. *Acta Otolaryngol.*, **96**,485–493.

Lundh, B., Kristensson, K. & Norrby, E. (1987) Selective infections of olfactory and respiratory epithelium by vesicular stomatitis and Sendai viruses. *Neuropath. Appl. Neurobiol.*, **13**,111–122.

Luskin, M.B. & Price, J.L. (1983a) The topographic organization of associational fibers of the olfactory system in the rat, including centrifugal fibers to the olfactory bulb. *J. Comp. Neurol.,* **216,**264–291.

Luskin, M.B. & Price, J.L. (1983b) The laminar distribution of intracortical fibers originating in the olfactory cortex of the rat. *J. Comp. Neurol.,* **216,**292–302.

Lynch, J.W. & Barry, P.H. (1989) Action potentials initiated by single channels opening in a small neuron (rat olfactory receptor). *Biophys. J.,* **55,**755–768.

McClintock, T.S. & Ache, B.W. (1989a) Ionic currents and ion channels of lobster olfactory receptor neurons. *J. Gen. Physiol.,* **94,**1085–1099.

McClintock, T.S. & Ache, B.W. (1989b) Hyperpolarizing receptor potentials in lobster olfactory receptor cells: implications for transduction and mixture suppression. *Chem. Senses,* **14,**637–647.

McClintock, T.S., Schütte, K. & Ache, B.W. (1989) Failure to implicate cAMP in transduction in lobster olfactory cells. *Chem. Senses,* **14,**817–827.

McCotter, R.E. (1912) The connection of the vomeronasal nerves with the accessory olfactory bulb in the opossum and other mammals. *Anat. Rec.,* **6,**299–318.

MacFarlane, A. (1975) Olfaction in the development of social preferences in the human neonate. In *Parent–Infant Interaction. Ciba Foundation Symposium 33.* Ed., M.A. Hofer. Associated Scientific Publishers, Amsterdam. pp. 103–117.

Mack, A. & Wolburg, H. (1986) Heterogeneity of glial membranes in the rat olfactory system as revealed by freeze-fracturing. *Neuroscience Lett.,* **65,**117–122.

Mackay-Sim, A. & Beard, M.D. (1987) Hypothyroidism disrupts neural development in the olfactory epithelium of adult mice. *Devel. Brain Res.,* **36,**190–198.

Mackay-Sim, A., Breipohl, W. & Kremer, M. (1988) Cell dynamics in the olfactory epithelium of the tiger salamander: a morphometric analysis. *Exp. Brain Res.,* **71,**189–198.

Mackay-Sim, A. & Kittel, P.W. (1991) On the life span of olfactory receptor neurons. *Eur. J. Neuroscience,* **3,**209–215.

Mackay-Sim, A. & Kubie, J.L. (1981) The salamander nose: a model system for the study of spatial coding of olfactory quality. *Chem. Senses,* **6,**249–257.

Mackay-Sim, A. & Patel, U. (1984) Regional differences in cell density and cell genesis in the olfactory epithelium of the salamander, *Ambystoma tigrinum. Exp. Brain Res.,* **57,**99–106.

Mackay-Sim, A. & Shaman, P. (1984) Topographic coding of odorant quality is maintained at different concentrations in the salamander olfactory epithelium. *Brain Res.,* **297,**207–216.

Mackay-Sim, A., Shaman, P. & Moulton, D.G. (1982) Topographic coding of olfactory quality: odorant-specific patterns of epithelial responsivity in the salamander. *J. Neurophysiol.,* **48,**584–596.

McLean, J.H. & Shipley, M.T. (1987a) Serotonergic afferents to the rat olfactory bulb: I. Origins and laminar specificity of serotonergic inputs in the adult rat. *J. Neuroscience,* **7,**3016–3028.

McLean, J.H. & Shipley, M.T. (1987b) Serotonergic afferents to the rat olfactory bulb: II. Changes in fiber distribution during development. *J. Neuroscience,* **7,**3029–3039.

McLean, J.H. & Shipley, M.T. (1988) Postmitotic, postmigrational expression of tyrosine hydroxylase in olfactory bulb dopaminergic neurons. *J. Neuroscience,* **8,**3658–3669.

McLean, J.H. & Shipley, M.T. (1992) Neuroanatomical substrates of olfaction. In *The Science of Olfaction.* Eds., M. Serby & K. Chobor. Springer-Verlag, Berlin. pp. 126–171.

McLean, J.H., Shipley, M.T. & Bernstein, D.I. (1989a) Golgi-like, transneuronal retrograde labelling with CNS injections of *Herpes simplex* virus type I. *Brain Res. Bull.,* **22,**867–881.

McLean, J.H., Shipley, M.T., Nickell, W.T., Aston-Jones, G. & Reyher, C.K.H. (1989b) Chemoanatomical organization of the noradrenergic input from locus coeruleus to the olfactory bulb of the adult rat. *J. Comp. Neurol.,* **285,**339–349.

McLennan, H. (1971) The pharmacology of inhibition of mitral cells in the olfactory bulb. *Brain Res.,* **29,**177–184.

MacLeod, N. (1978) Is carnosine (β-ala-L-his) a neurotransmitter in the primary olfactory pathway? In *Iontophoresis and Transmitter Mechanisms in the Mammalian Central Nervous System.* Eds., R.W. Ryall & J.S. Kelly. Elsevier/North Holland, Amsterdam. pp. 117–119.

MacLeod, N.K. & Straughan, D.W. (1979) Responses of olfactory bulb neurons to the dipeptide carnosine. *Exp. Brain Res.,* **34,**183–188.

Macrides, F. & Davis, B.J. (1983) The olfactory bulb. In *Chemical Neuroanatomy.* Ed., P.C. Emson. Raven Press, N.Y. pp. 391–426.

Macrides, F. & Schneider, S.P. (1982) Laminar organization of mitral and tufted cells in the main olfactory bulb of the adult hamster. *J. Comp. Neurol.,* **208,**419–430.

Macrides, F., Schoenfeld, T.A., Marchand, J.E., & Clancy, A.N. (1985) Evidence for morphologically, neurochemically and functionally heterogeneous classes of mitral and tufted cells in the olfactory bulb. *Chem. Senses,* **10,**175–202.

Magrassi, L. & Graziadei, P.P.C. (1985) Interaction of the transplanted olfactory placode with the optic stalk and the diencephalon in *Xenopus laevis* embryos. *Neuroscience,* **15,**903–921.

Mair, R.G. (1982) Response properties of rat olfactory bulb neurones. *J. Physiol.,* **326,**341–359.

Mair, R.G., Doty, R.L., Kelly, K.M., Wilson, C.S., Langlais, P.J., McEntee, W.J. & Vollmecke, T.A. (1986) Multimodal sensory discrimination deficits in Korsakoff's psychosis. *Neuropsychologia,* **24,**831–839.

Mair, R.G. & Gesteland, R.C. (1982) Response properties of mitral cells in the olfactory bulb of the neonatal rat. *Neuroscience,* **7,**3117–3125.

Mair, R.G., Gesteland, R.C. & Blank, D.L. (1982) Changes in olfactory receptor cilia morphology and physiology during development. *Neuroscience,* **7,**3091–3103.

Males, J.L., Townsend, J.L. & Schneider, R.A. (1973) Hypogonadotropic hypogonadism with anosmia – Kallmann's syndrome. *Arch. Int. Med.,* **131,**501–507.

Mania-Farnell, B., Bruch, R.C. & Farbman, A.I. (1991) Reduction of olfactory neuron adenylate cyclase activity by amitriptyline *in vitro. Soc. Neuroscience Abstr.,* **17,**119.

Mania-Farnell, B. & Farbman, A.I. (1988) Olfactory bulb influence on G-protein expression in rat olfactory epithelium. *Soc. Neuroscience Abstr.,* **14,**428.

Mania-Farnell, B. & Farbman, A.I. (1990) Immunohistochemical localization of guanine nucleotide-binding proteins in rat olfactory epithelium during development. *Devel. Brain Res.,* **51,**103–112.

Manson, M.D. (1990) Introduction to bacterial motility and chemotaxis. *J. Chem. Ecol.*, **16,**107–113.

Margolis, F.L. (1972) A brain protein unique to the olfactory bulb. *Proc. Natl. Acad. Sci. USA,* **69,**1221–1224.

Margolis, F.L. (1974) Carnosine in the primary olfactory pathway. *Science,* **184,**909–911.

Margolis, F.L. (1980a) Neurotransmitter biochemistry of the mammalian olfactory bulb. In *Biochemistry of Taste and Olfaction.* Eds., R.H. Cagan & M.R. Kare. Academic Press, N.Y. pp. 369–394.

Margolis, F.L. (1980b) Carnosine: an olfactory neuropeptide. In *Role of Peptides in Neuronal Function.* Eds., J.L. Barker & T. Smith. Marcel Dekker, N.Y. pp. 545–572.

Margolis, F.L. (1980c) A marker protein for the olfactory chemoreceptor neuron. In *Proteins of the Nervous System.* Eds., R.A. Bradshaw & D. Schneider. Raven, N. Y. pp. 59–84.

Margolis, F.L. (1982) Olfactory marker protein (OMP). *Scand. J. Immunol.,* **15(Suppl. 9),** 181–199.

Margolis, F.L. (1988) Molecular cloning of olfactory-specific gene products. In *Molecular Neurobiology of the Olfactory System.* Eds., F.L. Margolis & T.V. Getchell. Plenum, N.Y. pp. 237–265.

Margolis, F.L. & Getchell, T.V. (1991) Receptors: current status and future directions. In *Perfumes: Art, Science and Technology.* Eds., P. Muller & D. Lamparsky. Elsevier Applied Science, Essex, England. pp. 481–498.

Margolis, F.L. & Grillo, M. (1977) Axoplasmic transport of carnosine (β-alanyl-L-histidine) in the mouse olfactory pathway. *Neurochem. Res.,* 2,507–519.

Margolis, F.L., Grillo, M., Grannot-Reisfeld, N. & Farbman, A.I. (1983) Purification, characterization and immunocytochemical localization of mouse kidney carnosinase. *Biochim. Biophys. Acta,* **744,**237–248.

Margolis, F.L., Grillo, M., Kawano, T. & Farbman, A.I. (1985) Carnosine synthesis in olfactory tissue during ontogeny: effect of exogenous β-alanine. *J. Neurochem.,* **44,**1459–1464.

Margolis, F.L., Kawano, T. & Grillo, M. (1986) Ontogeny of carnosine, olfactory marker protein and neurotransmitter enzymes in olfactory bulb and mucosa of the rat. In *Ontogeny of Olfaction.* Ed., W. Breipohl. Springer-Verlag, Berlin. pp. 107–116.

Marin-Padilla, M. & Amieva B., M.R. (1989) Early neurogenesis of the mouse olfactory nerve: Golgi and electron microscopic studies. *J. Comp. Neurol.,* **288,**339–352.

Marshall, D.A. & Maruniak, J.A. (1986) Masera's organ responds to odorants. *Brain Res.,* **366,**329–332.

Maruniak, J.A., Silver, W.L. & Moulton, D.G. (1982) Olfactory receptors respond to blood-borne odorants. *Brain Res.,* **265,**312–316.

Mason, J.R., Clark, L. & Morton, T.H. (1984) Selective deficits in the sense of smell caused by chemical modification of the olfactory epithelium. *Science,* **226,**1092–1094.

Mason, J.R., Greenspon, J.M. & Silver, W.L. (1987) Capsaicin and its effects on olfaction and trigeminal chemoreception. *Acta Physiol. Hungar.,* **69,**469–479.

Mason, R.T., Fales, H.M., Jones, T.H., Pannell, L.K., Chinn, J.W. & Crews, D. (1989) Sex pheromones in snakes. *Science,* **245,**290–293.

Mason, R.T. & Gutzke, W.H.N. (1990) Sex recognition in the leopard gecko, *Eublepharis macularius* (Sauria: Gekkonidae). Possible mediation by skin-derived semiochemicals. *J. Chem. Ecol.,* **16,**27–36.

Masson, C., Kouprach, S., Giachetti, I. & MacLeod, P. (1978) Relation between intramembranous particle density of frog olfactory cilia and EOG response. In *Olfaction and Taste VI*. Eds., J. LeMagnen & P. MacLeod. Information Retrieval, London. p. 195.

Masukawa, L.M., Hedlund, B. & Shepherd, G.M. (1985a) Morpholgical and electrophysiological correlations of identified cells in the in vitro olfactory epithelium of the tiger salamander. *J. Neuroscience*, **5**,128–135.

Masukawa, L.M., Hedlund, B. & Shepherd, G.M. (1985b) Changes in the electrical properties of olfactory epithelial cells in the tiger salamander after olfactory nerve transection. *J. Neuroscience*, **5**,136–141.

Masukawa, L.M., Kauer, J.S. & Shepherd, G.M. (1983) Intracellular recordings from two cell types in an in vitro preparation of the salamander olfactory epithelium. *Neuroscience Lett.*, **35**,59–64.

Math, F. & Davrainville, J.L. (1980) Electrophysiological study on the postnatal development of mitral cell activity in the rat olfactory bulb. *Brain Res.*, **190**,243–247.

Mathews, D.F. (1972) Response patterns of single neurons in the tortoise olfactory epithelium and olfactory bulb. *J. Gen. Physiol.*, **60**,166–180.

Matsui, S. & Yamamoto, C. (1975) Release of radioactive glutamic acid from thin sections of guinea-pig olfactory cortex in vitro. *J. Neurochem.*, **24**,245–250.

Matsutani, S., Senba, E. & Tohyama, M. (1988) Neuropeptide- and neurotransmitter-related immunoreactivities in the developing rat olfactory bulb. *J. Comp. Neurol.*, **272**,331–342.

Matulionis, D.H. (1975) Ultrastructural study of mouse olfactory epithelium following destruction by $ZnSO_4$ and its subsequent regeneration. *Amer. J. Anat.*, **142**,67–90.

Matulionis, D.H. (1976) Light and electron microscopic study of the degeneration and early regeneration of olfactory epithelium in the mouse. *Amer. J. Anat.*, **145**,79–100.

Maue, R.A. & Dionne, V.E. (1987) Patch-clamp studies of isolated mouse olfactory receptor neurons. *J. Gen. Physiol.*, **90**,95–125.

Meisami, E. (1976) Effects of olfactory deprivation on postnatal growth of the rat olfactory bulb utilizing a new method for production of neonatal unilateral anosmia. *Brain Res.*, **107**,437–444.

Meisami, E. (1979) The developing rat olfactory bulb: prospects of a new model system in developmental neurobiology. In *Neural Growth and Differentiation*. Eds., E. Meisami & M.A.B. Brazier. Raven Press, N.Y. pp. 183–206.

Meisami, E. (1989) A proposed relationship between increases in the number of olfactory receptor neurons, convergence ratio and sensitivity in the developing rat. *Devel. Brain Res.*, **46**,9–19.

Meisami, E., Louie, J., Hudson, R. & Distel, H. (1990) A morphometric comparison of the olfactory epithelium of newborn and weanling rabbits. *Cell Tissue Res.*, **262**,89–97.

Meisami, E. & Mousavi, R. (1982) Lasting effects of early olfactory deprivation on the growth, DNA, RNA and protein content, and Na-K-ATPase and AChE activity of the rat olfactory bulb. *Devel. Brain Res.*, 2.⌐ ⌐–229.

Meisami, E. & Noushinfar, E. (1986) Early olfactory deprivatior ⌐ mitral cells of the olfactory bulb: a Golgi study. *Int. J. Devel. N⌐ .ce*, **4**,431–434.

Meisami, E. & Safari, L. (1981) A quantitative study ⌐ ⌐ effects of early unilateral olfactory deprivation on the number and distribution of mitral and tufted cells and of glomeruli in the rat olfactory bulb. *Brain Res.*, **221**,81–107.

Melese-d'Hospital, P.Y. & Hart, B.L. (1985) Vomeronasal organ cannulation in male goats: evidence for transport of fluid from the oral cavity to the vomeronasal organ during flehmen. *Physiol. Behav.*, **35**,941–944.

Menco, B.P.M. (1977) A qualitative and quantitative investigation of olfactory and nasal respiratory mucosal surfaces of cow and sheep based on various ultrastructural and biochemical techniques. *Commun. Agricult. Univ. Wageningen*, **77–13**,1–157.

Menco, B.P.M. (1980a) Qualitative and quantitative freeze-fracture studies on olfactory and nasal respiratory structures of frog, ox, rat, and dog. I. A general survey. *Cell Tissue Res.*, **207**,183–209.

Menco, B.P.M. (1980b) Qualitative and quantitative freeze-fracture studies on olfactory and nasal respiratory epithelial surfaces of frog, ox, rat, and dog. II. Cell apices, cilia, and microvilli. *Cell Tissue Res.*, **211**,5–30.

Menco, B.P.M. (1980c) Qualitative and quantitative freeze-fracture studies on olfactory and nasal respiratory epithelial surfaces of frog, ox, rat, and dog. III. Tight junctions. *Cell Tissue Res.*, **211**,361–373.

Menco, B.P.M. (1980d) Qualitative and quantitative freeze-fracture studies on olfactory and nasal respiratory epithelial surfaces of frog, ox, rat, and dog. IV. Ciliogenesis and ciliary necklaces (including high-voltage observations). *Cell Tissue Res.*, **212**,1–16.

Menco, B.P.M. (1983) The ultrastructure of olfactory and nasal respiratory epithelium surfaces. In *Nasal Tumors in Animals and Man, Anatomy, Physiology and Epidemiology, Vol. 1.* Eds., G. Reznik & S.F. Stinson. CRC Press Inc., Boca Raton, Florida. pp. 45–102.

Menco, B.P.M. (1984) Ciliated and microvillous structures of rat olfactory and nasal respiratory epithelia. A study using ultrarapid cryo-fixation followed by freeze-substitution or freeze-etching. *Cell Tissue Res.*, **235**,225–341.

Menco, B.P.M. (1987) A freeze-fracture study on the prenatal development of ciliated surfaces in rat olfactory epithelia. *Ann. N.Y. Acad. Sci.*, **510**,491–493.

Menco, B.P.M. (1988a) Pre-natal development of rat nasal epithelia. IV. Freeze-fracturing on apices, microvilli and primary and secondary cilia of olfactory and respiratory epithelial cells, and on olfactory axons. *Anat. Embryol.*, **178**,309–326.

Menco, B.P.M. (1988b) Pre-natal development of rat nasal epithelia. V. Freeze-fracturing on necklaces of primary and secondary cilia of olfactory and respiratory cells. *Anat. Embryol.*, **178**,381–388.

Menco, B.P.M. (1988c) Tight-junctional strands first appear in regions where three cells meet in differentiating olfactory epithelium: a freeze-fracture study. *J. Cell Sci.*, **89**,495–505.

Menco, B.P.M. (1989) Electron-microscopic demonstration of olfactory-marker protein with protein G-gold in freeze-substituted, Lowicryl K11M-embedded rat olfactory-receptor cells. *Cell Tissue Res.*, **256**,275–281.

Menco, B.P.M. (1991) Ultrastructural localization of the transduction apparatus in the rat's olfactory epithelium. *Chem. Senses*, **16**,555.

Menco, B.P.M. (1992) Lectins bind differentially to cilia and microvilli of major and minor cell populations in olfactory and nasal respiratory epithelia. *Micro. Res. Technique.* (in press)

Menco, B.P.M. & Benos, D.J. (1989) Freeze-substitution and freeze-etch cytochemistry on cilia and microvilli of the rat's olfactory epithelium. *J. Cell Biol.*, **109**,254a.

Menco, B.P.M., Dodd, G.H., Davey, M. & Bannister, L.H. (1976) Presence of membrane particles in freeze-etched bovine olfactory cilia. *Nature*, **263**,597–599.

Menco, B.P.M. & Farbman, A.I. (1985a) Genesis of cilia and microvilli of rat nasal epithelia during pre-natal development. I. Olfactory epithelium, qualitative studies. *J. Cell Sci.*, **78**,283–310.

Menco, B.P.M. & Farbman, A.I. (1985b) Genesis of cilia and microvilli of rat nasal epithelia during pre-natal development. II. Olfactory epithelium, a morphometric analysis. *J. Cell Sci.*, **78**,311–336.

Menco, B.P.M., Minner, E.W. & Farbman, A.I. (1988) Preliminary observations on rapidly-frozen, freeze-fractured and deep-etched rat olfactory cilia rotary-replicated with tantalum/tungsten. *J. Elect. Micro. Res. Technique*, **8**,441–442.

Mendoza, A.S. (1986) The mouse vomeronasal glands: a light and electron microscopical study. *Chem. Senses*, **11**,541–555.

Mendoza, A.S. & Breipohl, W. (1983) The cell coat of the olfactory epithelium proper and vomeronasal neuroepithelium of the rat as revealed by means of the Ruthenium-red reaction. *Cell Tissue Res.*, **230**,139–146.

Mendoza, A.S., Breipohl, W. & Miragall, F. (1982) Cell migration from the chick olfactory placode: a light and electron microscopic study. *J. Embryol. Exper. Morphol.*, **69**,47–59.

Mendoza, A.S. & Kühnel, W. (1987) Morphological evidence for a direct innervation of the mouse vomeronasal glands. *Cell Tissue Res.*, **247**,457–459.

Mendoza, A.S. & Kühnel, W. (1989) Das vomeronasale Organ (VNO) von neugeborenen Meerschweinchen ist funktionstüchtig. *Anat. Anz., Jena*, **168**,279–281.

Mendoza, A.S. & Szabo, K. (1988) Developmental studies on the rat vomeronasal organ: vascular pattern and neuroepithelial differentiation. II. Electron microscopy. *Devel. Brain Res.*, **39**,259–268.

Menevse, A., Dodd, G. & Poynder, T.M. (1977) Evidence for the specific involvement of cyclic AMP in the olfactory transduction mechanism. *Biochem. Biophys. Comm.*, **77**,671–677.

Meredith, M. (1983) Sensory physiology of pheromone communication. In *Pheromones and Reproduction in Mammals*. Ed., J.G. Vandenbergh. Academic Press, N.Y. pp. 199–252.

Meredith, M. (1986) Patterned response to odor in mammalian olfactory bulb: the influence of intensity. *J. Neurophysiol.*, **56**,572–597.

Meredith, M. (1991) Sensory processing in the main and accessory olfactory systems: comparisons and contrasts. *J. Steroid Biochem. Mol. Biol.*, **39**,601–614.

Meredith, M., Marques, D.M., O'Connell, R.J. & Stern, F.L. (1980) Vomeronasal pump: significance for male hamster sexual behavior. *Science*, **207**,1224–1226.

Meredith, M. & Moulton, D.G. (1978) Patterned response to odor in single neurones of goldfish olfactory bulb: influence of odor quality and other stimulus parameters. *J. Gen. Physiol.*, **71**,615–643.

Meredith, M. & O'Connell, R.J. (1979) Efferent control of stimulus access to the hamster vomeronasal organ. *J. Physiol.*, **286**,301–316.

Meredith, M. & O'Connell, R.J. (1988) HRP uptake by olfactory and vomeronasal receptor neurons: use as an indicator of incomplete lesions and relevance for non-volatile chemoreception. *Chem. Senses*, **13**,487–515.

Merkel, K.H.H. & Maibach, E.A. (1984) Experimental herpes simplex encephalitis in rats after intranasal inoculation. An immunohistologic study. *Histochem. J.*, **16**,467–469.

Michel, W.C., McClintock, T.S. & Ache, B.W. (1991) Inhibition of lobster olfactory receptor cells by an odor-activated potassium conductance. *J. Neurophysiol.*, **65**,446–453.

Miragall, F., Breipohl, W. & Bhatnagar, K.P. (1979) Ultrastructural investigation on the cell membranes of the vomeronasal organ in the rat: a freeze-etching study. *Cell Tissue Res.*, **200**,397–408.

Miragall, F., Breipohl, W. & Mendoza, A.S. (1981) Morphological investigations on the rat vomeronasal olfactory sensory epithelia in rodents. A freeze-fracture study. *Verh. Anat. Ges.*, **75**,967–968.

Miragall, F., Breipohl, W., Naguro, T. & Voss-Wermbter, G. (1984) Freeze-fracture study of the plasma membranes of the septal olfactory organ of Masera. *J. Neurocytol.*, **13**,111–125.

Miragall, F., Kadmon, G. & Schachner, M. (1989) Expression of L1 and N-CAM cell adhesion molecules during development of the mouse olfactory system. *Devel. Biol.*, **135**,272–286.

Miragall, F. & Mendoza, A.S. (1982) Intercellular junctions in the rat vomeronasal neuroepithelium. *J. Submicrosc. Cytol.*, **14**,597–605.

Miragall, F. & Monti Graziadei, G.A. (1982) Experimental studies on the olfactory marker protein. II. Appearance of the olfactory marker protein during differentiation of the olfactory sensory neurons of mouse: an immunohistochemical and autoradiographic study. *Brain Res.*, **329**,245–250.

Moberg, P.J., Pearlson, G.D., Speedie, L.J., Lipsey, J.R. & Folstein, S.E. (1987) Olfactory recognition: differential impairments in early and late Huntington's and Alzheimer's disease. *J. Clin. Exp. Neuropsychol.*, **9**,650–664.

Mollicone, R., Trojan, J. & Oriol, R. (1985) Appearance of H and B antigens in primary sensory cells of the rat olfactory apparatus and inner ear. *Devel. Brain Res.*, **17**,275–279.

Monaghan, D.T. & Cotman, C.W. (1985) Distribution of N-methyl-D-aspartate-sensitive L-[^{3}H]glutamate-binding sites in rat brain. *J. Neuroscience*, **5**,2909–2919.

Monath, T.P., Cropp, C.B. & Harrison, A.K. (1983) Mode of entry of a neurotropic arbovirus into the central nervous system. *Lab. Invest.*, **48**,399–410.

Monti Graziadei, G.A. (1983) Experimental studies on the olfactory marker protein. III. The olfactory marker protein in the olfactory neuroepithelium lacking connections with the forebrain. *Brain Res.*, **262**,303–308.

Monti Graziadei, G.A. & Graziadei, P.P.C. (1979) Neurogenesis and neuron regeneration in the olfactory system of mammals. II. Degeneration and reconstitution of the olfactory sensory neurons after axotomy. *J. Neurocytol.*, **8**,197–213.

Monti Graziadei, G.A. & Graziadei, P.P.C. (1984) The olfactory organ: neural transplantation. In *Neural Transplants*. Eds., J.R. Sladek & D.M. Gash. Plenum Press, N.Y. pp. 167–186.

Monti Graziadei, G.A., Karlan, M.S., Bernstein, J.J. & Graziadei, P.P.C. (1980a) Reinnervation of the olfactory bulb after section of the olfactory nerve in monkey *(Saimiri sciureus)*. *Brain Res.*, **189**,343–354.

Monti Graziadei, G.A., Stanley, R.S. & Graziadei, P.P.C. (1980b) The olfactory marker protein in the olfactory system of the mouse during development. *Neuroscience*, **5**,1239–1252.

Morales, J.A., Herzog, S., Kompter, C., Frese, K. & Rott, R. (1988) Axonal transport of Borna disease virus along olfactory pathways in spontaneously and experimentally infected rats. *Med. Microbiol. Immunol.,* **177,**51–68.

Moran, D.T., Jafek, B.W. & Rowley, J.C. (1985a) Ultrastructure of the vomeronasal organ in man: a pilot study. *Chem. Senses,* **10,**420–421.

Moran, D.T., Jafek, B.W., Rowley, J.C. & Eller, P.M. (1985b) Electron microscopy of olfactory epithelia in two patients with anosmia. *Arch. Otolaryngol.,* **111,**122–126.

Moran, D.T., Rowley, J.C. & Jafek, B.W. (1982a) Electron microscopy of human olfactory epithelium reveals a new cell type: the microvillar cell. *Brain Res.,* **253,**39–46.

Moran, D.T., Rowley, J.C., Jafek, B.W. & Lovell, M.A. (1982b) The fine structure of the olfactory mucosa in man. *J. Neurocytol.,* **11,**721–746.

Mori, K. (1987a) Monoclonal antibodies (2C5 and 4C9) against lactoseries carbohydrates identify subsets of olfactory and vomeronasal receptor cells and their axons in the rabbit. *Brain Res.,* **408,**215–221.

Mori, K. (1987b) Membrane and synaptic properties of identified neurons in the olfactory bulb. *Progr. Neurobiol.,* **29,**275–320.

Mori, K., Fujita, S.C., Imamura, K. & Obata, K. (1985) Immunohistochemical study of subclasses of olfactory nerve fibers and their projections to the olfactory bulb in the rabbit. *J. Comp. Neurol.,* **242,**214–229.

Mori, K., Imamura, K., Fujita, S.C. & Obata, K. (1987) Projections of two subclasses of vomeronasal nerve fibers to the accessory olfactory bulb in the rabbit. *Neuroscience,* **20,**259–278.

Mori, K., Kishi, K. & Ojima, H. (1983) Distribution of dendrites of mitral, displaced mitral, tufted and granule cells in the rabbit olfactory bulb. *J. Comp. Neurol.,* **219,**339–355.

Morrison, E.E. & Costanzo, R.M. (1989) Scanning electron microscopic study of degeneration and regeneration in the olfactory epithelium after axotomy. *J. Neurocytol.,* **18,**393–405.

Morrison, E.E. & Costanzo, R.M. (1990) Morphology of the human olfactory epithelium. *J. Comp. Neurol.,* **297,**1–13.

Morrison, E.E. & Graziadei, P.P.C. (1983) Transplants of olfactory mucosa in the rat brain. I. A light microscopic study of transplant organization. *Brain Res.,* **279,**241–245.

Moulton, D. G. (1963) Electrical activity in the olfactory system of rabbits with indwelling electrodes. In *Olfaction and Taste. Proceedings of the First International Symposium.* Ed., Y. Zotterman. Macmillan Co., N.Y. pp. 71–84.

Moulton, D.G. (1965) Differential sensitivity to odors. *Cold Spring Harbor Symp. Quant. Biol.,* **30,**201–206.

Moulton, D.G. (1974) Dynamics of cell populations in the olfactory epithelium. *Ann. N.Y. Acad. Sci.,* **237,**52–61.

Moulton, D.G. (1975) Cell renewal in the olfactory epithelium of the mouse. In *Olfaction and Taste, V.* Eds., D.A. Denton & J.P. Coghlan. Academic Press, N.Y. pp. 111–114.

Moulton, D.G. (1976) Spatial patterning response to odors in the peripheral olfactory system. *Physiol. Rev.,* **56,**578–593.

Moulton, D.G. & Beidler, L.M. (1967) Structure and function in the peripheral olfactory system. *Physiol. Rev.,* **47,**1–52.

Moulton, D.G., Celebi, G. & Fink, R.P. (1970) Olfaction in mammals – two aspects; proliferation of cells in the olfactory epithelium and sensitivity to odours. In *Ciba Foundation Symposium on Taste and Smell in Vertebrates*. Eds., G.E.W. Wolstenholme & J. Knight. Churchill, London. pp. 227–250.

Mouradian, L.E. & Scott, J.W. (1988) Cytochrome oxidase staining marks dendritic zones of the rat olfactory bulb external plexiform layer. *J. Comp. Neurol.*, **271**,507–518.

Mozell, M.M. (1962) Olfactory mucosal and neural responses in the frog. *Amer. J. Physiol.*, **203**,353–358.

Mozell, M.M. (1964) Evidence for sorption as a mechanism of the olfactory analysis of odorants. *Nature*, **203**,1181–1182.

Mozell, M.M. (1966) The spatiotemporal analysis of odorants at the level of the olfactory sheet. *J. Gen. Physiol.*, **50**,25–41.

Mozell, M.M. (1970) Evidence for a chromatographic model of olfaction. *J. Gen. Physiol.*, **56**,46–63.

Mozell, M.M. & Jagodowicz, M. (1973) Chromatographic separation of odorants by the nose: retention times measured across *in vivo* olfactory mucosa. *Science*, **181**,1247–1249.

Mozell, M.M., Sheehe, P.R., Hornung, D.E., Kent, P.F., Youngentob, S.L. & Murphy, S.J. (1987) "Imposed" and "inherent" mucosal activity patterns. *J. Gen. Physiol.*, **90**,625–650.

Mugnaini, E., Oertel, W.H. & Wouterlood, F. (1984a) Immunocytochemical localization of GABA neurons and dopamine neurons in the rat main and accessory olfactory bulbs. *Neuroscience Lett.*, **47**,221–226.

Mugnaini, E., Wouterlood, F., Dahl, A.-L. & Oertel, W.H. (1984b) Immunocytochemical identification of GABAergic neurons in the main olfactory bulb of the rat. *Arch. Ital. Biol.*, **122**,83–113.

Muller, J.F. & Marc, R.E. (1984) Three distinct morphological classes of receptors in fish olfactory organs. *J. Comp. Neurol.*, **222**,482–495.

Müller-Schwarze, D. (1971) Pheromones in black-tailed deer, *Odocoileus hemionus columbianus*. *Anim. Behav.*, **19**,141–152.

Müller-Schwarze, D. (1979) Chemical signals in alarm behavior of deer. In *Chemical Signals in Vertebrates and Aquatic Invertebrates*. Plenum Press, N.Y. pp. 39–51.

Mulvaney, B.D. & Heist, H.E. (1970) Mapping of rabbit olfactory cells. *J. Anat.*, **107**,19–30.

Mulvaney, B.D. & Heist, H.E. (1971a) Centriole migration during regeneration and normal development of olfactory epithelium. *J. Ultrastruct. Res.*, **35**,274–281.

Mulvaney, B.D. & Heist, H.E. (1971b) Regeneration of rabbit olfactory epithelium. *Amer. J. Anat.*, **131**,241–252.

Mumby, S.M., Kahn, R.A., Manning, D.R. & Gilman, A.G. (1986) Antisera of designed specificity for subunits of guanine nucleotide-binding regulatory proteins. *Proc. Natl. Acad. Sci. USA*, **83**,265–269.

Münz, H., Claas, B., Stumpf, W.E. & Jennes, L. (1982) Centrifugal innervation of the retina by luteinizing hormone releasing hormone (LHRH)-immunoreactive telencephalic neurons in teleostean fishes. *Cell Tissue Res.*, **222**,313–323.

Murphy, C. (1987) Olfactory psychophysics. In *Neurobiology of Taste and Smell*. Eds., T.E. Finger & W.L. Silver. Wiley, N.Y. pp. 251–273.

Murphy, C., Gilmore, M.M., Seery, C.S., Salmon, D.P. & Lasker, B.R. (1990) Olfactory thresholds are associated with degree of dementia in Alzheimer's disease. *Neurobiol. Aging*, **11**,465–469.

Murphy, R. B. (1988) Membrane probes in the olfactory system. Biophysical aspects of initial events. In *Molecular Neurobiology of the Olfactory System.* Eds., F.L. Margolis & T.V. Getchell. Plenum Press, N.Y. pp. 121–142.

Mustaparta, H. (1971) Spatial distribution of receptor-responses to stimulation with different odours. *Acta Physiol. Scand.,* **82,**154–166.

Mykytowycz, R. (1970) The role of skin glands in mammalian communication. In *Communication by Chemical Signals.* Eds., J.W. Johnston, D.G. Moulton & A. Turk. Appleton-Century-Crofts, N.Y. pp. 327–360.

Nadi, N.S., Head, R., Grillo, M., Hempstead, J., Grannot-Reisfeld, N. & Margolis, F.L. (1981) Chemical deafferentation of the olfactory bulb: plasticity of the levels of tyrosine hydroxylase, dopamine and norepinephrine. *Brain Res.,* **213,**365–377.

Nadi, N.S., Hirsch, J.D. & Margolis, F.L. (1980) Laminar distribution of putative neurotransmitter amino acids and ligand binding sites in the dog olfactory bulb. *J. Neurochem.,* **34,**138–146.

Naessen, R. (1970) The identification and topographical localisation of the olfactory epithelium in man and other mammals. *Acta Otolaryngol.* **70,**51–57.

Naessen, R. (1971a) An enquiry on the morphological characteristics and possible changes with age in the olfactory region of man. *Acta Otolaryngol.* **71,**49–62.

Naessen, R. (1971b) The "receptor surface" of the olfactory organ (epithelium) of man and guinea pig. *Acta Otolaryngol.,* **71,**335–348.

Nagahara, Y. (1940) Experimentelle Studien über die histologischen Veränderungen des Geruchsorgans nach der Olfactoriusdurchschneidung. Beiträge zur Kenntnis des feineren Baus des Geruchsorgans. *Japan. J. Med. Sci. V, Pathol.,* **5,**165–199.

Naguro, T. & Breipohl, W. (1982) The vomeronasal epithelia of NMRI mouse. A scanning electron-microscopic study. *Cell Tissue Res.,* **227,**519–534.

Nakamura, T. & Gold, G.H. (1987) A cyclic nucleotide-gated conductance in olfactory receptor cilia. *Nature,* **325,**442–444.

Nakashima, T., Kimmelman, C.P. & Snow, J.B. (1984) Structure of human fetal and adult olfactory neuroepithelium. *Arch. Otolaryngol.,* **110,**641–646.

Nakashima, T., Kimmelman, C.P. & Snow, J.B. (1985) Vomeronasal organs and nerves of Jacobson in the human fetus. *Acta Otolaryngol. (Stockh.),* **99,**266–271.

Negus, V. (1958) *The Comparative Anatomy and Physiology of the Nose and Paranasal Sinuses.* E. & S. Livingstone Ltd., Edinburgh.

Neidle, A. & Kandera, J. (1974) Carnosine – an olfactory peptide. *Brain Res.,* **80,**359–364.

Nickell, W.T., Norman, A.B., Wyatt, L.M. & Shipley, M.T. (1991) Olfactory bulb DA receptors may be located on terminals of the olfactory nerve. *NeuroReport,* **2,**9–12.

Nickell, W.T. & Shipley, M.T. (1988) Two anatomically specific classes of candidate cholinoceptive neurons in the rat olfactory bulb. *J. Neuroscience,* **8,**4482–4491.

Nickell, W.T. & Shipley, M.T. (1992) Neurophysiology of the olfactory bulb. In *The Science of Olfaction.* Eds., M. Serby & K. Chorbor. Springer-Verlag, Berlin. pp.172–212.

Nicoll, R.S., Alger, B.E. & Jahr, C.E. (1980a) Enkephalin blocks inhibitory pathways in the vertebrate CNS. *Nature,* **287,**22–25.

Nicoll, R.S., Alger, B.E. & Jahr, C.E. (1980b) Peptides as putative excitatory neurotransmitters: carnosine, enkephalin, substance P and TRH. *Proc. Roy. Soc. London B,* **210,**133–149.

Noda, M. & Harada, Y. (1981) Development of olfactory epithelium in the mouse: scanning electron microscopy. *Biomed. Res. (Suppl.)* **2,**449–454.

Nomura, T. & Kurihara, K. (1987a) Liposomes as a model for olfactory cells: changes in membrane potential in response to various odorants. *Biochemistry,* **26,**6135–6140.

Nomura, T. & Kurihara, K. (1987b) Effects of changed lipid composition on responses of liposomes to various odorants: possible mechanism of odor discrimination. *Biochemistry,* **26,**6141–6145.

Novoselov, V.I., Bragin, A.G., Novikov, J.V., Nesterov, V.I. & Fesenko, E.E. (1984) Transplants of olfactory mucosa in the anterior chamber of the eye: morphology, electrophysiology and biochemistry. *Devel. Neuroscience,* **6,**317–324.

O'Connell, R.J., Costanzo, R.M. & Hildebrandt, J.D. (1990) Adenylyl cyclase activation and electrophysiological responses elicited in male hamster olfactory receptor neurons by components of female pheromones. *Chem. Senses,* **15,**725–739.

O'Connell, R.J. & Meredith, M. (1984) Effects of volatile and nonvolatile chemical signals on male behaviors mediated by the main and accessory olfactory systems. *Behav. Neuroscience,* **98,**1083–1093.

O'Connell, R.J. & Mozell, M. (1969) Quantitative stimulation of frog olfactory receptors. *J. Neurophysiol.,* **32,**51–63.

Oelschläger, H.A. (1989) Early development of the olfactory and terminalis systems in baleen whales. *Brain Behav. Evol.,* **34,**171–183.

Oelschläger, H.A. & Buhl, E.H. (1985) Development and rudimentation of the peripheral olfactory system in the harbor porpoise *Phocoena phocoena* (Mammalia: Cetacea). *J. Morphol.,* **184,**351–360.

Ogawa, K.S., Fujimoto, K. & Ogawa, K. (1987) Cytochemical localization of adenylate cyclase and guanylate cyclase in the parietal cell of guinea pig gastric gland. *Acta Histochem. Cytochem.,* **20,**185–195.

Okano, M. & Takagi, S.F. (1974) Secretion and electrogenesis of the supporting cell in the olfactory epithelium. *J. Physiol. (Lond.),* **242,**353–370.

Oland, L.A., Orr, G. & Tolbert, L.P. (1990) Construction of a protoglomerular template by olfactory axons initiates the formation of olfactory glomeruli in the insect brain. *J. Neuroscience,* **10,**2096–2112.

Oland, L.A. & Tolbert, L.P. (1987) Glial patterns during early development of antennal lobes of *Manduca sexta:* a comparison between normal lobes and lobes deprived of antennal axons. *J. Comp. Neurol.,* **255,**196–207.

Oland, L.A. & Tolbert, L.P. (1989) Patterns of glial proliferation during formation of olfactory glomeruli in an insect. *Glia,* **2,**10–24.

Oland, L.A., Tolbert, L.P. & Mossman, K.L. (1988) Radiation-induced reduction of the glial population during development disrupts the formation of olfactory glomeruli in an insect. *J. Neuroscience,* **8,**353–367.

Oley, N., DeHan, R.S., Tucker, D., Smith, J.C. & Graziadei, P.P.C. (1975) Recovery of structure and function following transection of the primary olfactory nerves in pigeons. *J. Comp. Physiol. Psychol.,* **88,**477–495.

Onoda, N. (1988a) Monoclonal antibody immunohistochemistry of rabbit olfactory receptor neurons during development.*Neuroscience,* **26,**1003–1012.

Onoda, N. (1988b) Monoclonal antibody immunohistochemistry of degenerative

and renewal patterns in rabbit olfactory receptor neurons following unilateral olfactory bulbectomy. *Neuroscience,* **26,**1013–1022.

Onoda, N. & Fujita, S.C. (1988) Monoclonal antibody immunohistochemistry
of adult rabbit olfactory structures. *Neuroscience,* **26,**993–1002.

Onoda, N. & Imamura, K. (1984) Striped pattern for cytochrome oxidase activity in the rabbit olfactory bulb. *Neuroscience Res.,* **1,**457–461.

Ophir, D. & Lancet, D. (1988) Expression of intermediate filaments and des-
moplakin in vertebrate olfactory mucosa. *Anat. Rec.,* **221,**754–760.

Oppenheim, R.W. (1991) Cell death during development of the nervous system.
Ann. Rev. Neuroscience, **14,**453–501.

Oppenheim, R.W., Chuwang, I.-W. & Maderut, J.L. (1978) Cell death of
motoneurons in chick embryo spinal cord. II. The differentiation of motoneurons prior to their induced degeneration following limb bud removal.
J. Comp. Neurol., **177,**87–112.

O'Rahilly, R. (1967) The early development of the nasal pit in staged human
embryos. *Anat. Rec.,* **157,**380.

Orona, E., Rainer, E.C. & Scott, J.W. (1984) Dendritic and axonal organization
of mitral and tufted cells in the rat olfactory bulb. *J. Comp. Neurol.,*
226,346–356.

Orona, E., Scott, J.W. & Rainer, E.C. (1983) Different granule cell populations
innervate superficial and deep regions of the external plexiform layer in rat
olfactory bulb. *J. Comp. Neurol.,* **217,**227–237.

Ottoson, D. (1956) Analysis of the electrical activity of the olfactory epithelium.
Acta Physiol. Scand. (Suppl. 122), **35,**1–83.

Pace, U., Hanski, E., Salomon, Y. & Lancet, D. (1985) Odorant-sensitive
adenylate cyclase may mediate olfactory reception. *Nature,* **316,**255–258.

Pace, U. & Lancet, D. (1986) Olfactory GTP-binding protein: signal-transducing
polypeptide of vertebrate chemosensory neurons. *Proc. Natl. Acad. Sci.
USA,* **83,**4947–4951.

Pager, J. (1983) Unit responses changing with behavioral outcome in the olfactory bulb of unrestrained rats. *Brain Res.,* **289,**87–98.

Panhuber, H. & Laing, D.G. (1987) The size of mitral cells is altered when rats
are exposed to an odor from their day of birth. *Devel. Brain Res.,* **34,**133–
140.

Panhuber, H., Laing, D.L., Willcox, M.E., Eagleson, G.K. & Pittman, E.A.
(1985) The distribution of the size and number of mitral cells in the olfactory
bulb of the rat. *J. Anat.,* **140,**297–308.

Panhuber, H., Mackay-Sim, A. & Laing, D.G. (1987) Prolonged odor exposure
causes severe cell shrinkage in the adult rat olfactory bulb. *Devel. Brain
Res.,* **31,**307–311.

Papi, F., Fiore, L., Fiaschi, V. & Benvenuti, S. (1973) An experiment for testing
the hypothesis of olfactory navigation of homing pigeons. *J. Comp.
Physiol.,* **83,**93–102.

Papka, R.E. & Matulionis, D.H. (1983) Association of substance P-
immunoreactive nerves with the murine olfactory mucosa. *Cell Tissue Res.,*
230,517–525.

Parker, G.H. (1922) *Smell, Taste and Allied Senses in the Vertebrates.* Lippincott,
Philadelphia.

Parkes, A.S. & Bruce, H.M. (1962) Pregnancy-block in female mice placed in
boxes soiled by males. *J. Reprod. Fertil.,* **4,**,303–308.

Paton, J.A. & Nottebohm, F.N. (1984) Neurons generated in the adult brain
are recruited into functional circuits. *Science,* **225,** 1046–1048.

260   *References*

Pearlman, S.J. (1934) Jacobson's organ (Organon vomero-nasale, Jacobsoni):its anatomy, gross, microscopic and comparative, with some observations as well on its function. *Ann. Otol. Rhinol. Laryngol.*, **43**,739–768.

Pearson, A.A. (1941) The development of the nervus terminalis in man. *J. Comp. Neurol.*, **75**,39–66.

Pearson, R.C.A., Esiri, M.M., Hiorns, R.W., Wilcock, G.K. & Powell, T.P.S. (1985) Anatomical correlates of the distribution of the pathological changes in the neocortex in Alzheimer disease. *Proc. Natl. Acad. Sci. USA*, **82**,4531–4534.

Pedersen, P.E. & Benson, T.E. (1986) Projection of septal organ receptor neurons to the main olfactory bulb in rats. *J. Comp. Neurol.*, **252**,555–562.

Pedersen, P.E. & Blass, E.M. (1981) Olfactory control over suckling in albino rats. In *Development of Perception. Vol. 1.* Eds., R.N. Aslin, J.R. Alberts & M.R. Petersen. Academic Press, N.Y. pp. 359–381.

Pedersen, P.E. & Blass, E.M. (1982) Prenatal and postnatal determinants of the 1st suckling episode in albino rats. *Devel. Psychobiol.*, **15**,349–355.

Pedersen, P.E., Williams, T. & Blass, E.M. (1982) Activation and odor conditioning of suckling behavior in 3-day old albino rats. *Anim. Behav. Proc.*, **8**,329–341.

Pelosi, P., Baldaccini, N.E. & Pisanelli, A.M. (1982) Identification of a specific olfactory receptor for 2-isobutyl–3-methoxypyrazine. *Biochem. J.*, **201**,245–248.

Persaud, K.C., DeSimone, J.A., Getchell, M.L., Heck, G.L. & Getchell, T.V. (1987) Ion transport across the frog olfactory mucosa: the basal and odorant-stimulated states. *Biochim. Biophys. Acta*, **902**,65–79.

Persaud, K.C., Heck, G.L., DeSimone, S.K., Getchell, T.V. & DeSimone, J.A. (1988) Ion transport across the frog olfactory mucosa: the action of cyclic nucleotides on the basal and odorant-stimulated states. *Biochim. Biophys. Acta*, **944**,49–62.

Pevsner, J., Hou, V., Snowman, A.M. & Snyder, S.H. (1990) Odorant-binding protein. *J. Biol. Chem.*, **265**,6118–6125.

Pevsner, J., Reed, R.R., Feinstein, P.G. & Snyder, S.H. (1988) Molecular cloning of odorant-binding protein: member of a ligand carrier family. *Science*, **241**,336–339.

Pevsner, J., Sklar, P.B. & Snyder, S.H. (1986) Odorant-binding protein: localization to nasal glands and secretions. *Proc. Natl. Acad. Sci. USA*, **83**,4942–4946.

Pevsner, J., Trifiletti, R.R., Strittmatter, S.M. & Snyder, S.H. (1985) Isolation and characterization of an olfactory receptor protein for odorant pyrazines. *Proc. Natl. Acad. Sci. USA*, **82**,3050–3054.

Pfeiffer, W. (1978) Heterocyclic compounds as releasers of the fright reaction in the giant danie, *Danio malabaricus (Jerdon)* (Cyprinidae, Ostariophysi, Pisces). *J. Chem. Ecol.*, **4**,665–673.

Pfeuffer, E., Mollner, S., Lancet, D. & Pfeuffer, T. (1989) Olfactory adenylyl cyclase. Identification and purification of a novel enzyme form. *J. Biol. Chem.*, **264**,18803–18807.

Piatt, J. (1951) An experimental approach to the problem of pallial differentiation. *J. Comp. Neurol.*, **94**,105–121.

Pilar, G. & Landmesser, L. (1976) Ultrastructural differences during embryonic cell death in normal and peripherally deprived ciliary ganglia. *J. Cell Biol.*, **68**,339–356.

Pinching, A.J. & Døving, K.B. (1974) Selective degeneration in the rat olfactory bulb following exposure to different odours. *Brain Res.*, **82**,195–204.

Pinching, A.J. & Powell, T.P.S. (1971a) The neuron types of the glomerular layer of the olfactory bulb. *J. Cell Sci.*, **9**,305–345.

Pinching, A.J. & Powell, T.P.S. (1971b) The neuropil of the glomeruli of the olfactory bulb. *J. Cell Sci.*, **9**,347–377.

Pinching, A.J. & Powell, T.P.S. (1971c) The neuropil of the periglomerular region of the olfactory bulb. *J. Cell Sci.*, **9**,379–409.

Piondron, P. (1976) Effet de la suppression de l'odorat, sans lésion des bulbes olfactifs, sur la sélectivité du comportement maternel de la Brebis. *Comp. Rend. Acad. Sci., Paris*, **282**,489–491.

Pissonnier, D., Thierry, J.C., Fabre-Nys, C., Poindron, P. & Keverne, E.B. (1985) The importance of olfactory bulb noradrenaline for maternal recognition in sheep. *Physiol. Behav.*, **35**,361–363.

Plendl, J. & Schmahl, W. (1988) Dolichos biflorus agglutinin: a marker of the developing olfactory system in the NMRI-mouse strain. *Anat. Embryol.*, **177**,459–464.

Polak, E.H., Fombon, A.M., Tilquin, C. & Punter, P.H. (1989a) Sensory evidence for olfactory receptors with opposite chiral selectivity. *Behav. Brain Res.*, **31**,199–206.

Polak, E.H., Shirley, S.G. & Dodd, G.H. (1989b) Concanavalin A reveals olfactory receptors which discriminate between alkane odorants on the basis of size. *Biochem. J.*, **262**,475–478.

Pomeroy, S.L., LaMantia, A.-S., & Purves, D. (1990) Postnatal construction of neural circuitry in the mouse olfactory bulb. *J. Neuroscience*, **10**,1952–1966.

Pommerville, J.C., Strickland, J.B. & Harding, K.E. (1990) Pheromone interactions and ionic communication in gametes of aquatic fungus *Allomyces macrogynus*. *J. Chem. Ecol.*, **16**,121–131.

Poston, M.R., Bailey, M.S., Schwarcz, R. & Shipley, M.T. (1991) Differential complementary localization of metabolic enzymes for quinolinic acid in olfactory bulb astrocytes. *J. Comp. Neurol.*, **310**,1–8.

Powers, J.B. & Winans, S.S. (1973) Sexual behavior in peripherally anosmic male hamsters. *Physiol. Behav.*, **10**,361–368.

Powers, J.B. & Winans, S.S. (1975) Vomeronasal organ: critical role in mediating sexual behavior of the male hamster. *Science*, **187**,961–963.

Price, J.L. (1973) An autoradiographic study of complementary laminar patterns of terminations of afferent fibers to the olfactory cortex. *J. Comp. Neurol.*, **150**,87–108.

Price, J.L. & Powell, T.P.S. (1970a) The morphology of the granule cells of the olfactory bulb. *J. Cell Sci.*, **7**,91–123.

Price, J.L. & Powell, T.P.S. (1970b) The synaptology of the granule cells of the olfactory bulb. *J. Cell Sci.*, **7**,125–155.

Price, J.L. & Powell, T.P.S. (1970c) An electron-microscopic study of the termination of the afferent fibres to the olfactory bulb from the cerebral hemisphere. *J. Cell Sci.*, **7**,157–187.

Price, J.L. & Powell, T.P.S. (1970d) The mitral and short axon cells of the olfactory bulb. *J. Cell Sci.*, **7**,631–651.

Price, S. (1977) Specific anosmia to geraniol in mice. *Neuroscience Lett.*, **4**,49–50.

Price, S. (1978) Anisole binding protein from dog olfactory epithelium. *Chem. Senses Flavour*, **3**,51–55.

Price, S. & Willey, A. (1987) Benzaldehyde binding protein from dog olfactory epithelium. *Ann. N.Y. Acad. Sci.*, **510**,561–564.

Price, S. & Willey, A. (1988) Effects of antibodies against odorant binding proteins on electrophysiological responses to odorants. *Biochim. Biophys. Acta*, **965**,127–129.

Punter, P.H., Menco, B.P.M. & Boelens, H. (1981) The efficacy of *n*-aliphatic alcohols and *n*-aliphatic fatty acids on various membrane systems with special reference to olfaction and taste. In *Odor Quality and Chemical Structure. ACS Symposium Series 148*. Eds., H.R. Moskowitz & C.B. Warren. American Chemical Soc., Washington. pp. 93–108.

Pyatkina, G.A. (1982) Development of the olfactory epithelium in man. *Z. Mikrosk. Anat. Forsch., Leipzig*, **96**,361–372.

Quinn, N.P., Rossor, M.N. & Marsden, C.D. (1987) Olfactory threshold in Parkinson's disease. *J. Neurol. Neurosurg. Psychiatry*, **50**,88–89.

Rafols, J.A. & Getchell, T.V. (1983) Morphological relations between the receptor neurons, sustentacular cells and Schwann cells in the olfactory mucosa of the salamander. *Anat. Rec.*, **206**,87–101.

Raisman, G. (1972) An experimental study of the projection of the amygdala to the accessory olfactory bulb and its relationship to the concept of a dual olfactory system. *Exp. Brain Res.*, **14**,395–408.

Raisman, G. (1985) Specialized neuroglial arrangement may explain the capacity of vomeronasal axons to reinnervate central neurons. *Neuroscience*, **14**,237–254.

Rakic, P. (1974) Neurons in rhesus monkey visual cortex: systematic relation between time of origin and eventual disposition. *Science*, **183**,425–427.

Rall, W., Shepherd, G.M., Reese, T.S. & Brightman, M.W. (1966) Dendrodendritic synaptic pathway for inhibition in the olfactory bulb. *Exper. Neurol.*, **14**,44–56.

Ramon y Cajal, S. (1911) *Histologie du Système Nerveux de l'Homme et des Vertébrés*. Maloine, Paris.

Rasmussen, L.E., Schmidt, M.J., Henneous, R., Groves, D. & Daves, G.D., Jr. (1982) Asian bull elephants: flehmen-like responses to extractable components in female elephant estrous urine. *Science*, **217**,159–162.

Reed, C.J., Lock, E.A. & De Matteis, F. (1986) NADPH:cytochrome P–450 reductase in olfactory epithelium. *Biochem. J.*, **240**,585–592.

Reese, T.S. (1965) Olfactory cilia in the frog. *J. Cell Biol.*, **25**,209–230.

Rehn, B., Panhuber, H., Laing, D.G. & Breipohl, W. (1988) Spine density on olfactory granule cell dendrites is reduced in rats reared in a restricted olfactory environment. *Devel. Brain Res.*, **40**,143–147.

Rehnberg, B.G. & Schreck, C.B. (1986) The olfactory *l*-serine receptor in coho salmon: biochemical specificity and behavioral response. *J. Comp. Physiol. A*, **159**,61–67.

Restrepo, D., Miyamoto, T., Bryant, B.P. & Teeter, J.H. (1990) Odor stimuli trigger influx of calcium into olfactory neurons of the channel catfish. *Science*, **249**,1166–1168.

Restrepo, D. & Teeter, J.H. (1990) Olfactory neurons exhibit heterogeneity in depolarization-induced calcium changes. *Amer. J. Physiol.*, **258**,C1051–C1061.

Revial, M.F., Duchamp, A. & Holley, A. (1978a) Odour discrimination by frog olfactory receptors: a 2nd study. *Chem. Senses*, **3**,7–21.

Revial, M.F., Duchamp, A., Holley, A. & MacLeod, P. (1978b) Frog olfaction:

odour groups, acceptor distribution, and receptor categories. *Chem. Senses,* **3**,23–33.

Revial, M.F., Sicard, G., Duchamp, A. & Holley, A. (1982) New studies on odour discrimination in the frog's olfactory receptor cells. I. Experimental results. *Chem. Senses,* **7**,175–190.

Revial, M.F., Sicard, G., Duchamp, A. & Holley, A. (1983) New studies on odour discrimination in the frog's olfactory receptor cells. II. Mathematical analysis of electrophysiological responses. *Chem. Senses,* **8**,179–190.

Reyes, P.R., Golden, G.T., Fagel, P.L., Zalewska, M., Fariello, R.G., Katz, L. & Carner, E. (1987) The prepiriform cortex in dementia of the Alzheimer type. *Arch. Neurol.,* **44**,644–645.

Reyher, C.K.H., Lübke, J., Larsen, W.J., Hendrix, G.M., Shipley, M.T. & Baumgarten, H.G. (1991) Olfactory bulb granule cell aggregates: morphological evidence for interperikaryal electrotonic coupling via gap junctions. *J. Neuroscience,* **11**,1485–1495.

Rezek, D.L. (1987) Olfactory deficits as a neurologic sign in dementia of the Alzheimer type. *Arch. Neurol.,* **44**,1030–1032.

Rhein, L.D. & Cagan, R.H. (1980) Biochemical studies of olfaction: isolation, characterization, and odorant binding activity of cilia from rainbow trout olfactory rosettes. *Proc. Natl. Acad. Sci. USA,* **77**,4412–4416.

Rhein, L.D. & Cagan, R.H. (1983) Biochemical studies of olfaction: binding specificity of odorants to a cilia preparation from rainbow trout olfactory rosettes. *J. Neurochem.,* **41**,569–577.

Ribak, C.E., Vaughn, J.E. & Barber, R.P. (1981) Immunocytochemical localization of GABA-ergic neurons at the electron microscopic level. *Histochem. J.,* **13**,555–582.

Ribak, C.E., Vaughn, J.E., Saito, K., Barber, R. & Roberts, E. (1977) Glutamate decarboxylase in neurons of the olfactory bulb. *Brain Res.,* **126**,1–18.

Roberts, E. (1986) Alzheimer's disease may begin in the nose and may be caused by aluminosilicates. *Neurobiol. Aging,* **7**,561–567.

Rochel, S. & Margolis, F.L. (1982) Carnosine release from olfactory bulb synaptosomes is calcium-dependent and depolarization-stimulated. *J. Neurochem.,* **38**,1505–1514.

Rodolfo-Masera, T. (1943) Sur l'esistenza di un particolare organo olfattivo nel setto nasale dela cavia e di altri roditori. *Arch. Ital. Anat. Embryol.,* **48**,157–212.

Rogers, K.E., Dasgupta, P., Gubler, U., Grillo, M., Khew-Goodall, Y.S. & Margolis, F.L. (1987) Molecular cloning and sequencing of a cDNA for olfactory marker protein. *Proc. Natl. Acad. Sci. USA,* **84**,1704–1708.

Romanoff, A.L. (1960) *The Avian Embryo. Structural and Functional Development.* Macmillan Co., N.Y. p. 317.

Roos, J., Roos, M., Schaeffer, C. & Aron, C. (1988) Sexual differences in the development of accessory olfactory bulb in the rat. *J. Comp. Neurol.,* **270**,121–131.

Rosselli-Austin, L. & Altman, J. (1979) The postnatal development of the main olfactory bulb of the rat. *J. Devel. Physiol.,* **1**,295–313.

Rosselli-Austin, L., Hamilton, K.H. & Williams, J. (1987) Early postnatal development of the rat accessory olfactory bulb. *Devel. Brain Res.,* **36**,304–308.

Rosser, A.E. & Keverne, E.B. (1985) The importance of central noradrenergic neurones in the formation of an olfactory memory in the prevention of pregnancy block. *Neuroscience*, **15**,1141–1147.

Rosser, A.E., Remfry, C.J. & Keverne, E.B. (1989) Restricted exposure of mice to primer pheromones coincident with prolactin surges blocks pregnancy by changing hypothalamic dopamine release. *J. Reprod. Fertil.*, **87**, 553–559.

Rottman, S.J. & Snowdon, C.T. (1972) Demonstration and analysis of an alarm pheromone in mice. *J. Comp. Physiol. Psychol.*, **81**,483–490.

Rowley, J.C., Moran, D.T. & Jafek, B.W. (1989) Peroxidase backfills suggest the mammalian olfactory epithelium contains a second morphologically distinct class of bipolar sensory neuron: the microvillar cell. *Brain Res.*, **502**,387–400.

Royet, J.P., Jourdan, F. & Ploye, H. (1989a) Morphometric modifications associated with early sensory experience in the rat olfactory bulb: I. Volumetric study of the bulbar layers. *J. Comp. Neurol.*, **289**,586–593.

Royet, J.P., Jourdan, F., Ploye, H. & Souchier, C. (1989b) Morphometric modifications associated with early sensory experience in the rat olfactory bulb: II. Stereological study of the population of olfactory glomeruli. *J. Comp. Neurol.*, **289**,594–609.

Royet, J.P., Sicard, G., Souchier, C. & Jourdan, F. (1987) Specificity of spatial patterns of glomerular activation in the mouse olfactory bulb: computer-assisted image analysis of 2-deoxyglucose autoradiograms. *Brain Res.*, **417**,1–11.

Rudy, J.W. & Cheatle, G.D. (1977) Odor aversion learning in neonatal rats. *Science*, **198**,845–846.

Russell, G.F. & Hills, J.I. (1971) Odor differences between enantiomeric isomers. *Science*, **172**,1043–1044.

Russell, M.J. (1976) Human olfactory communication. *Nature*, **260**,520–522.

Sakai, M., Yoshida, M., Karasawa, N., Teramura, M., Ueda, H. & Nagatsu, I. (1987) Carnosine-like immunoreactivity in the primary olfactory neuron of the rat. *Experientia*, **298**,300.

Samanen, D.W. & Forbes, W.B. (1984) Replication and differentiation of olfactory receptor neurons following axotomy in the adult hamster: a morphometric analysis of postnatal neurogenesis. *J. Comp. Neurol.*, **225**,201–211.

Sanides-Kohlrausch, C. & Wahle, P. (1990a) Morphology of neuropeptide Y-immunoreactive neurons in the cat olfactory bulb and olfactory peduncle: postnatal development and species comparison. *J. Comp. Neurol.*, **291**,468–489.

Sanides-Kohlrausch, C. & Wahle, P. (1990b) VIP- and PHI-immunoreactivity in olfactory centers of the adult cat. *J. Comp. Neurol.*, **294**,325–339.

Satou, M. (1990) Synaptic organization, local neuronal circuitry, and functional segregation of the teleost olfactory bulb. *Progr. Neurobiol.*, **34**,115–142.

Satou, M., Fujita, I., Ichikawa, M., Yamaguchi, K. & Ueda, K. (1983) Field potential and intracellular potential studies of the olfactory bulb in the carp: evidence for a functional separation of the olfactory bulb into lateral and medial subdivisions. *J. Comp. Physiol.*, **152**,319–333.

Saucier, D. & Astic, L. (1986) Analysis of the topographical organization of olfactory epithelium projections in the rat. *Brain Res. Bull.*, **16**,455–462.

Sauer, F.C. (1935) Mitosis in the neural tube. *J. Comp. Neurol.*, **62**,377–405.

Scalia, F. & Winans, S.S. (1975) The differential projections of the olfactory bulb and accessory olfactory bulb in mammals. *J. Comp. Neurol.,* **161,**31–56.

Schell, F.M., Burghardt, G.M., Johnston, A. & Coholich, C. (1990) Analysis of chemicals from earthworms and fish that elicit prey attach by ingestively naive garter snakes *(Thamnophis). J. Chem. Ecol.,* **16,**67–77.

Schiffman, S.S. (1983) Taste and smell in disease. *New Engl. J. Med.,* **308,**1275–1279.

Schild, D. (1989) Whole-cell currents in olfactory receptor cells of *Xenopus laevis. Exp. Brain Res.,* **78,**223–232.

Schmidt, A., Naujoks-Manteuffel, C. & Roth, G. (1988) Olfactory and vomeronasal projections and the pathway of the nervus terminalis in ten species of salamanders. *Cell Tissue Res.,* **251,**45–50.

Schmiedel-Jakob, I., Michel, W.C., Anderson, P.A.V. & Ache, B.W. (1990) Whole cell recording from lobster olfactory receptor cells: multiple ionic bases for the receptor potential. *Chem. Senses,* **15,**397–405.

Schneider, D., Kasang, G. & Kaissling, K.-E. (1968) Bestimmung der Riechschwelle von *Bombyx mori* mit Tritium-markiertem Bombykol. *Naturwissensch.,* **55,**395.

Schneider, S.P. & Macrides, F. (1978) Laminar distributions of interneurons in the main olfactory bulb of the adult hamster. *Brain Res. Bull.,* **3,**73–82.

Schneider, S.P. & Scott, J.W. (1983) Orthodromic response properties of rat olfactory bulb mitral and tufted cells correlate with their projection patterns. *J. Neurophysiol.,* **50,**358–378.

Schoenfeld, T.A. & Macrides, F. (1984) Topographic organization of connections between the main olfactory bulb and pars externa of the anterior olfactory nucleus in the hamster. *J. Comp. Neurol.,* **227,**121–135.

Schoenfeld, T.A., Marchand, J.E. & Marcrides, F. (1985) Topographic organization of tufted cell axonal projections in the hamster main olfactory bulb: an intrabulbar association system. *J. Comp. Neurol.,* **235,**503–518.

Schultz, E.W. (1941) Regeneration of olfactory cells. *Proc. Soc. Exper. Biol. Med.,* **46,**41–43.

Schultz, E.W. (1960) Repair of the olfactory mucosa with special reference to regeneration of olfactory cells (sensory neurons). *Amer. J. Path.,* **37,**1–19.

Schwanzel-Fukuda, M., Bick, D. & Pfaff, D.W. (1989) Luteinizing hormone-releasing hormone (LHRH)-expressing cells do not migrate normally in an inherited hypogonadal (Kallmann) syndrome. *Mol. Brain Res.,* **6,**311–326.

Schwanzel-Fukuda, M. & Pfaff, D.W. (1989) Origin of luteinizing hormone-releasing neurons. *Nature,* **338,**161–164.

Schwanzel-Fukuda, M. & Pfaff, D.W. (1990) The migration of luteinizing hormone-releasing hormone (LHRH) neurons from the medial olfactory placode into the medial basal forebrain. *Experientia,* **46,**956–962.

Schwanzel-Fukuda, M. & Silverman, A.J. (1980) The nervus terminalis of the guinea pig: a new luteinizing hormone-releasing hormone (LHRH) neuronal system. *J. Comp. Neurol.,* **191,**213–225.

Schwartz Levey, M., Chikaraishi, D.M. & Kauer, J.S. (1991) Characterization of potential precursor populations in the mouse olfactory epithelium using immunocytochemistry and autoradiography. *J. Neuroscience,* **11,**3556–3564.

Schwob, J.E., Farber, N.B. & Gottlieb, D.I. (1986) Neurons of the olfactory epithelium in adult rats contain vimentin. *J. Neuroscience,* **6,**208–217.

266   *References*

Schwob, J.E. & Gottlieb, D.I. (1986) The primary olfactory projection has two chemically distinct zones. *J. Neuroscience*, **6**,3393–3404.

Schwob, J.E. & Price, J.L. (1984) The development of axonal connections in the central olfactory system of rats. *J. Comp. Neurol.*, **223**,177–202.

Schwob, J.E. & Szumowski, K.E. (1989) Olfactory sensory neurons are tropically dependent on the olfactory bulb for their prolonged survival. *Soc. Neuroscience Abstr.*, **15**,749.

Scott, J.W. (1981) Electrophysiological identifications of mitral and tufted cells and distributions of their axons in olfactory system of the rat. *J. Neurophysiol.*, **46**,918–931.

Scott, J.W. & Harrison, T.A. (1987) The olfactory bulb: anatomy and physiology. In *Neurobiology of Taste and Smell*. Eds., T.E. Finger & W.L. Silver. Wiley, N.Y. pp. 151–178.

Scott, J.W., McBride, R.L. & Schneider, S.P. (1980) The organization of projections from the olfactory bulb to the piriform cortex and olfactory tubercle in the rat. *J. Comp. Neurol.*, **194**,519–534.

Scott, J.W., McDonald, J.K. & Pemberton, J.L. (1987) Short axon cells of the rat olfactory bulb display NADPH-diaphorase activity, neuropeptide Y-like immunoreactivity and somatostatin-like immunoreactivity. *J. Comp. Neurol.*, **260**,378–391.

Segovia, S. & Guillamón, A. (1982) Effects of sex steroids on the development of the vomeronasal organ in the rat. *Devel. Brain Res.*, **5**,209–212.

Segovia, S., Orenzanz, L.M., Valencia, A. & Guillamón, A. (1984) Effects of sex steroids on the development of the accessory olfactory bulb in the rat: a volumetric study. *Devel. Brain Res.*, **16**,312–314.

Selway, R. & Keverne, E.B. (1990) Hippocampal lesions are without effect on olfactory memory formation in the context of pregnancy block. *Physiol. Behav.*, **47**,249–252.

Senba, E., Daddona, P.E. & Nagy, J.I. (1987) Adenosine deaminase-containing neurons in the olfactory system of the rat during development. *Brain Res. Bull.*, **18**,635–648.

Senf, W., Menco, B.P.M., Punter, P.H. & Duyvesteyn, P. (1980) Determination of odor affinities based on the dose-response relationships of the frog's electro-olfactogram. *Experientia*, **36**,213–215.

Senut, M.C., Menetrey, D. & Lamour, Y. (1989) Cholinergic and peptidergic projections from the medial septum and the nucleus of the diagonal band of Broca to dorsal hippocampus, cingulate cortex and olfactory bulb: a combined wheat germ agglutinin-apohorseradish peroxidase-gold immunohistochemical study. *Neuroscience*, **30**,385–403.

Serby, M. (1987) Olfactory deficits in Alzheimer's disease. *J. Neural Transm. (Suppl.)*, **24**,69–77.

Serby, M., Corwin, J., Conrad, P. & Rotrosen, J. (1985) Olfactory dysfunction in Alzheimer's disease and Parkinson's disease. *Amer. J. Psychiatry*, **142**,781–782.

Serby, M., Larson, P. & Kalkstein, D. (1991) The nature and course of olfactory deficits in Alzheimer's disease. *Amer. J. Psychiatry*, **148**,357–360.

Seroogy, K.B., Brecha, N. & Gall, C. (1985) Distribution of cholecystokinin-like immunoreactivity in the rat main olfactory bulb. *J. Comp. Neurol.*, **239**,373–383.

Sharp, F.R., Kauer, J.S. & Shepherd, G.M. (1975) Local sites of activity-related glucose metabolism in rat olfactory bulb during olfactory stimulation. *Brain Res.*, **98**,596–600.

Sharp, F.R., Kauer, J.S. & Shepherd, G.M. (1977) Laminar analysis of 2-deoxyglucose uptake in olfactory bulb and olfactory cortex of rabbit and rat. *J. Neurophysiol.*, **40**,800–813.

Shepherd, G.M. (1972) Synaptic organization of the mammalian olfactory bulb. *Physiol. Rev.*, **52**,864–917.

Shibuya, T., Aihara, Y. & Tonosaki, K. (1977) Single cell responses to odors in the reptilian olfactory bulb. In *Food Intake and Chemical Senses*. Eds., Y. Katsuki, M. Sato, S. Takagi & Y. Oomura. Univ. of Tokyo Press, Tokyo. pp. 23–32.

Shinoda, K., Shiotani, Y. & Osawa, Y. (1989) "Necklace olfactory glomeruli" form unique components of the rat primary olfactory system. *J. Comp. Neurol.*, **284**,362–373.

Shipley, M.T. (1985) Transport of molecules from nose to brain: transneuronal anterograde and retrograde labeling in the rat olfactory system by wheat germ agglutinin-horseradish peroxidase applied to the nasal epithelium. *Brain Res. Bull.*, **15**,129–142.

Shipley, M.T. & Adamek, G.D. (1984) The connections of the mouse olfactory bulb: a study using orthograde and retrograde transport of wheat germ agglutinin conjugated to horseradish peroxidase. *Brain Res. Bull.*, **12**,669–688.

Shipley, M.T. & Costanzo, R. (1984) Olfactory bulb cytochrome oxidase (CO) staining patterns suggest that glomeruli are functional units. *Soc. Neuroscience Abstr.*, **10**,118.

Shipley, M.T., Halloran, F.J. & DeLaTorre, J. (1985) Surprisingly rich projection from locus coeruleus to the olfactory bulb in the rat. *Brain Res.*, **329**,294–299.

Shipley, M.T. & Reyes, P. (1991) Anatomy of the human olfactory bulb and central olfactory pathways. In *The Human Sense of Smell*. Eds., D.G. Laing, R.L. Doty & W. Breipohl. Springer-Verlag, Berlin. pp. 29–60.

Shirley, S.G., Polak, E. & Dodd, G.H. (1983) Chemical-modification studies on rat olfactory mucosa using a thiol-specific reagent and enzymatic iodination. *Eur. J. Biochem.*, **132**,485–494.

Shirley, S.G., Polak, E., Mather, R.A. & Dodd, G.H. (1987) The effect of concanavalin A on the rat electro-olfactogram. *Biochem. J.*, **245**,175–184.

Shirley, S.G., Robinson, C.J., Dickinson, K., Aujla, R. & Dodd, G. (1986) Olfactory adenylate cyclase of the rat. Stimulation by odorants and inhibition by Ca^{2+}. *Biochem. J.*, **240**,605–607.

Sicard, G. (1985) Olfactory discrimination of structurally related molecules: receptor cell responses to camphoraceous odorants. *Brain Res.*, **326**,203–212.

Sicard, G. & Holley, A. (1984) Receptor cell responses to odorants: similarities and differences among odorants. *Brain Res.*, **292**,283–296.

Silver, W.L. (1987) The common chemical sense. In *Neurobiology of Taste and Smell*. Eds., T.E. Finger & W.L. Silver. Wiley, N.Y. pp. 65–87.

Silver, W.L., Arzt, A.H. & Mason, J.R. (1988) A comparison of the discriminatory ability and sensitivity of the trigeminal and olfactory systems to chemical stimuli in the tiger salamander. *J. Comp. Physiol. A*, **164**,55–66.

Silver, W.L., Farley, L.G., Womble, M. & Finger, T.E. (1989) The effect of neonatal capsaicin administration on trigeminal nerve fibers in the nasal cavity. *Chem. Senses*, **14**,748.

Silver, W.L., Mason, J.R., Adams, M.A. & Smeraski, C.A. (1986) Nasal trigeminal chemoreception: responses to *n*-aliphatic alcohols. *Brain Res.*, **376**,221–229.

Silver, W.L., Mason, J.R., Marshall, D.A. & Maruniak, J.A. (1985) Rat trigeminal, olfactory and taste responses after capsaicin desentization. *Brain Res.*, **333**,45–54.

Silverman, J.D. & Kruger, L. (1989) Calcitonin-gene-related peptide (CGRP) immunoreactive innervation of the rat head with emphasis on the specialized sensory structures. *J. Comp. Neurol.*, **280**,303–330.

Simmons, P.A. & Getchell, T.V. (1981a) Neurogenesis in olfactory epithelium: loss and recovery of transepithelial voltage transients following olfactory nerve section. *J. Neurophysiol.*, **45**,516–528.

Simmons, P.A. & Getchell, T.V. (1981b) Physiological activity of newly differentiated olfactory receptor neurons correlated with morphological recovery from olfactory nerve section in the salamander. *J. Neurophysiol.* **45**,529–549.

Simmons, P.A., Rafols, J.A. & Getchell, T.V. (1981) Ultrastructural changes in olfactory receptor neurons following olfactory nerve section. *J. Comp. Neurol.*, **197**,237–257.

Singer, A.G., Agosta, W.C., O'Connell, R.J., Pfaffman, C., Bowen, D.V. & Field F.H. (1976) Dimethyl disulfide: an attractant pheromone in hamster vaginal secretion. *Science*, **191**,948–949.

Singer, A.G., Clancy, A.N., Macrides, F. & Agosta, W. (1984) Chemical studies of hamster vaginal discharge: effects of endocrine ablation and protein digestion on behaviorally active macromolecular fractions. *Physiol. Behav.*, **33**,645–651.

Singer, A.G., Macrides, F., Clancy, A.N. & Agosta, W.C. (1986) Purification and analysis of a proteinaceous aphrodisiac pheromone from hamster vaginal discharge. *J. Biol. Chem.*, **261**,13323–13326.

Singh, D.N.P. & Nathaniel, E.J.H. (1977) Postnatal development of mitral cell perikaryon in the olfactory bulb of the rat. A light and ultrastructural study. *Anat. Rec.*, **189**,413–432.

Singh, P.J. & Tobach, E. (1975) Olfactory bulbectomy and nursing behavior in rat pups. *Devel. Psychobiol.*, **8**,151–164.

Skeen, L.C. (1977) Odor-induced patterns of deoxyglucose consumption in the olfactory bulb of the tree shrew, *Tupaia glis*. *Brain Res.*, **124**,147–153.

Skeen, L.C., Due, B.R. & Douglas, F.E. (1985) Effects of early anosmia on two classes of granule cells in developing mouse olfactory bulbs. *Neuroscience Lett.*, **54**,301–306.

Skeen, L.C., Due, B.R., & Douglas, F.E. (1986) Neonatal sensory deprivation reduces tufted cell number in mouse olfactory bulbs. *Neuroscience Lett.*, **63**,5–10.

Sklar, P.B., Anholt, R.R.H. & Snyder, S.H. (1986) The odorant-sensitive adenylate cyclase of olfactory receptor cells. *J. Biol. Chem.*, **261**,15538–15543.

Smart, I. (1971) Location and orientation of mitotic figures in the developing mouse olfactory epithelium. *J. Anat.*, **109**,243–251.

Smith, C.G. (1951) Regeneration of sensory olfactory epithelium and nerves in adult frogs. *Anat. Rec.*, **109**,661–671.

Smith, R., Preston, R.R., Schulz, S., Gagnon, M.L. & Van Houten, J. (1987) Correlations between cyclic AMP binding and chemoreception in *Paramecium*. *Biochim. Biophys. Acta*, **928**,171–178.

Smuts, M.S. (1977) Concanavalin A binding to the epithelial surface of the developing mouse olfactory placode. *Anat. Rec.*, **188**,29–37.

Snyder, S.H., Sklar, P.B., Hwang, P.M. & Pevsner, J. (1989) Molecular mechanisms of olfaction. *Trends Neuroscience*, **12**,35–38.

Snyder, S.H., Sklar, P.B. & Pevsner, J. (1988a) Olfactory receptor mechanisms: odorant-binding protein and adenylate cyclase. In *Molecular Neurobiology of the Olfactory System,* Eds., F.L. Margolis & T.V. Getchell. Plenum Press, N.Y. pp. 3–24.

Snyder, S.H., Sklar, P.B. & Pevsner, J. (1988b) Molecular mechanisms of olfaction. *J. Biol. Chem.,* **261,**13971–13974.

Soll, D.R. (1990) Behavioral studies into the mechanism of eukaryotic chemotaxis. *J. Chem. Ecol.,* **16,**133–150.

Sorensen, P.W., Hara, T.J. & Stacey, N.E. (1987) Extreme sensitivity of mature and gonadally-regressed goldfish to a potent steroidal pheromone, 17α,20β-dihydroxy–4-pregnen–3-one. *J. Comp. Physiol. A,* **160,**305–313.

Sorensen, P.W., Hara, T.J., Stacey, N.E. & Dulka, J.G. (1990) Extreme olfactory specificity of male goldfish to the preovulatory steroidal pheromone 17α,20β-dihydroxy–4-pregnen–3-one. *J. Comp. Physiol. A,* **166,**373–383.

Sorensen, P.W., Hara, T.J., Stacey, N.E. & Goetz, F.W. (1988) F prostaglandins function as potent olfactory stimulants comprising the postovulatory female sex pheromone in goldfish. *Biol. Reprod.,* **39,**1039–1050.

Sprague, G.F., Blair, L.C. & Thorner, J. (1983) Cell interactions and regulation of cell type in the yeast *Saccharomyces cerevisiae. Ann. Rev. Microbiol.,* **37,**623–660.

Springer, A.D. (1983) Centrifugal innervation of goldfish retina from ganglion cells of the nervus terminalis. *J. Comp. Neurol.,* **214,**404–415.

Stabell, O.B. (1984) Homing and olfaction in salmonids: a critical review with special reference to the Atlantic salmon. *Biol. Rev.,* **59,**333–388.

Stacey, N.E. & Kyle, A.L. (1983) Effects of olfactory tract lesions on sexual and feeding behavior in goldfish. *Physiol. Behav.,* **30,**621–628.

Stacey, N.E. & Sorensen, P.W. (1986) 17α,20β-dihydroxy–4-pregnen–3-one: a steroidal primer pheromone which increases milt volume in the goldfish, *Carassius auratus. Can. J. Zool.,* **64,**2412–2417.

Stahl, B., Distel, H. & Hudson, R. (1990) Effects of reversible nare occlusion on the development of the olfactory epithelium in the rabbit nasal septum. *Cell Tissue Res.,* **259,**275–281.

Steinbrecht, R.A. (1969) Comparative morphology of olfactory receptors. In *Olfaction and Taste III.* Ed., C. Pfaffmann. Rockefeller Univ. Press, N.Y. pp. 3–33.

Steinlen, S., Klumpp, S. & Schultz, J.E. (1990) Guanylate cyclase in olfactory cilia from rat and pig. *Biochim. Biophys. Acta,* **1054,**69–72.

Stell, W.K., Walker, S.E., Chohan, K.S. & Ball, A.K. (1984) The goldfish nervus terminalis: a luteinizing hormone-releasing hormone and molluscan cardioexcitatory peptide immunoreactive olfactoretinal pathway. *Proc. Natl. Acad. Sci. USA,* **81,**940–944.

Stewart, W.B., Kauer, J.S. & Shepherd, G.M. (1979) Functional organization of rat olfactory bulb analysed by the 2-deoxyglucose method. *J. Comp. Neurol.,* **185,**715–734.

Stickrod, G., Kimble, D.P. & Smotherman, W.P. (1982) In utero taste/odor aversion conditioning in the rat. *Physiol. Behav.,* **28,**5–7.

Stoddart, D.M. (1979) Some responses of a free living community of rodents to the odors of predators. In *Chemical Signals in Vertebrates and Aquatic Invertebrates.* Eds., D. Müller-Schwarze & R. M. Silverstein. Plenum, N.Y. pp. 1–10.

Stone, H. (1969) Effect of ethmoidal nerve stimulation on olfactory bulbar electrical activity. In *Olfaction and Taste: Proceedings of the Third Inter-*

national Symposium. Ed., C. Pfaffmann. Rockefeller Univ. Press, N.Y. pp. 216–220.

Stone, H., Carregal, E.J.A. & Williams, B. (1966) The olfactory-trigeminal response to odorants. *Life Sci.,* **5,**2195–2201.

Stone, H., Williams, B. & Carregal, E.J.A. (1968) The role of the trigeminal nerve in olfaction. *Exper. Neurol.,* **21,**11–19.

Stout, R.P. & Graziadei, P.P.C. (1980) Influence of the olfactory placode on the development of the brain in *Xenopus laevis* (Daudin). I. Axonal growth and connections of the transplanted placode. *Neuroscience,* **5,**2175–2186.

Street, S.F. (1937) The differentiation of the nasal area of the chick embryo in grafts. *J. Exp. Zool.,* **77,**49–85.

Stroop, W.G., Rock, D.L. & Fraser, N.W. (1984) Localization of herpes simplex virus in the trigeminal and olfactory systems of the mouse central nervous system during acute and latent infections by in situ hybridization. *Lab. Invest.,* **51,**27–38.

Struble, R.G. & Walters, C.P. (1982) Light microscopic differentiation of two populations of rat olfactory bulb granule cells. *Brain Res.,* **236,**237–251.

Stryer, L. & Bourne, H.R. (1986) G proteins: a family of signal transducers. *Ann. Rev. Cell Biol.,* **2,**391–419.

Sullivan, R.M. & Hall, W.G. (1988) Reinforcers in infancy. *Devel. Psychobiol.,* **21,**215–223.

Sullivan, R.M. & Hall, W.G. (1988) Reinforcers in infancy: classical conditioning using stroking or intra-oral infusions of milk as a UCS. *Devel. Psychobiol.,* **21,**215–223.

Sullivan, R.M. & Leon, M. (1986) Early olfactory learning induces an enhanced olfactory bulb response in young rats. *Devel. Brain Res.,* **27,**278–282.

Sullivan, R.M., Wilson, D.A. & Leon, M. (1989) Norepinephrine and learning-induced plasticity in infant rat olfactory system. *J. Neuroscience,* **9,**3996–4006.

Sumner, D. (1964) Post-traumatic anosmia. *Brain,* **87,**107–120.

Sutterlin, A.M. & Sutterlin, N. (1971) Electrical responses of the olfactory epithelium of Atlantic salmon *(Salmo salar). J. Fish. Res. Board Can.,* **29,**565–572.

Suzuki, N. (1977) Intracellular responses of lamprey olfactory receptors to current and chemical stimulation. In *Food Intake and Chemical Senses.* Eds., T. Katsuki, M. Sato, S. Takagi & Y. Omura. Univ. of Tokyo Press, Tokyo. pp. 13–22.

Suzuki, N. (1978) Effects of different ionic environments on the responses of single olfactory receptors in the lamprey. *Comp. Biochem. Physiol.,* **61A,**461–467.

Suzuki, N. (1984) Anterograde fluorescent labeling of olfactory receptor neurons by Procion and Lucifer dyes. *Brain Res.,* **311,**181–185.

Suzuki, N. & Tucker, D. (1971) Amino acids as olfactory stimuli in freshwater catfish, *Ictalurus catus (Linn.). Comp. Biochem. Physiol.,* **40A,**399–404.

Suzuki, Y. & Takeda, M. (1991) Basal cells in the mouse olfactory epithelium after axotomy: immunohistochemical and electron microscopic studies. *Cell Tissue Res.,* **266,**239–245.

Sydor, W., Teitelbaum, Z., Blacher, R., Sun, S., Benz, W. & Margolis, F.L. (1986) Amino acid sequence of a unique neuronal protein: rat olfactory marker protein. *Arch. Biochem. Biophys.,* **249,**351–362.

Szabo, K. & Mendoza, A.S. (1988) Developmental studies on the rat vomeronasal organ: vascular pattern and neuroepithelial differentiation. I. Light microscopy. *Devel. Brain Res.,* **39,**253–258.

Takagi, S.F. (1989) *Human Olfaction.* Univ. of Tokyo Press.

Takagi, S.F., Aoki, K., Lino, M. & Yajima, T. (1969a) The electropositive potential in the normal and degenerating olfactory epithelium. In *Olfaction and Taste, Vol. III.* Ed., C. Pfaffmann. Rockefeller Univ. Press, N.Y. pp. 92–108.

Takagi, S.F., Kitamura, H., Imai, K. & Takeuchi, H. (1969b) Further studies on the roles of sodium and potassium in the generation of the electro-olfactogram. *J. Gen. Physiol.,* **53,**115–130.

Takagi, S.F., Wyse, G.A., Kitamura, H. & Ito, K. (1968) The roles of sodium and potassium ions in the generation of the electro-olfactogram. *J. Gen. Physiol.,* **51,**552–578.

Takagi, S.F. & Yajima, T. (1964) Electrical responses to odours of degenerating olfactory epithelium. *Nature,* **202,**1220.

Takagi, S.F. & Yajima, T. (1965) Electrical activity and histological change in the degenerating olfactory epithelium. *J. Gen. Physiol.,* **48,**559–569.

Takahashi, S., Iwanaga, T., Takahashi, Y., Nakano, Y. & Fujita, T. (1984) Neuron-specific enolase and S–100 protein in the olfactory mucosa of human fetuses. *Cell Tissue Res.,* **238,**231–234.

Talamo, B.R., Rudel, R., Kosik, K.S., Lee, V.M.-Y., Neff, S., Adelman, L. & Kauer, J. (1989) Pathological changes in olfactory neurons in patients with Alzheimer's disease. *Nature,* 337,736–739.

Taniguchi, K. & Mikami, S. (1985) Fine structure of the epithelia of the vomeronasal organ of horse and cattle. A comparative study. *Cell Tissue Res.,* **240,**41–48.

Taniguchi, K. & Mochizuki, K. (1982) Morphological studies on the vomeronasal organ in the golden hamster. *Japan. J. Vet. Sci.,* **44,**419–426.

Taniguchi, K. & Mochizuki, K. (1983) Comparative morphological studies on the vomeronasal organ in rats, mice, and rabbits. *Japan. J. Vet. Sci.,* **45,**67–76.

Taniguchi, K., Taniguchi, K. & Mochizuki, K. (1982) Developmental studies on the vomeronasal organ in the golden hamster. *Japan. J. Vet. Sci.,* **44,**709–716.

Teicher, M.H. & Blass, E.M. (1976) Suckling in newborn rats: eliminated by nipple lavage, reinstated by pup saliva. *Science,* **193,**422–425.

Teicher, M.H. & Blass, E.M. (1977) First suckling response of the newborn albino rat: the roles of olfaction and amniotic fluid. *Science,* **198,**635–636.

Teicher, M.H., Stewart, W.B., Kauer, J.S. & Shepherd, G.M. (1980) Suckling pheromone stimulation of a modified glomerular region in the developing rat olfactory bulb revealed by the 2-deoxyglucose method. *Brain Res.,* **194,**530–535.

Thomas, L. (1974) A fear of pheromones. In *The Lives of a Cell.* Viking Press, N.Y. pp. 16–19.

Thommesen, G. (1978) The spatial distribution of odor induced potentials in the olfactory bulb of char and trout (Salmonidae). *Acta Physiol. Scand.,* **102,**205–217.

Thommesen, G. (1982) Specificity and distribution of receptor cells in the olfactory mucosa of char *(Salmo alpinus L.). Acta Physiol. Scand.,* **115,**47–56.

Thommesen, G. (1983) Morphology, distribution, and specificity of olfactory receptor cells in salmonid fishes. *Acta Physiol. Scand.,* **117,**241–249.

Thommesen, G. & Døving, K.B. (1977) Spatial distribution of the EOG in the rat; a variation with odour quality. *Acta Physiol. Scand.*, **99**,270–280.

Thornhill, R.A. (1970) Cell division in the olfactory epithelium of the lamprey, *Lampetra fluviatilis. Z. Zellforsch.*, **109**,147–157.

Tobach, E., Rouger, Y. & Schneirla, T.C. (1967) Development of olfactory function in the rat pup. *Amer. Zool.*, **7**,792–793.

Tolbert, L.P. & Oland, L.A. (1989) A role for glia in the development of organized neuropilar structures. *Trends Neuroscience*, **12**,70–75.

Tolbert, L.P., Oland, L.A. & Or, G. (1989) Intercellular interactions among insect olfactory cells in culture. *Soc. Neuroscience Abstr.*, **15**,445.

Tomlinson, A.H. & Esiri, M.M. (1983) Herpes simplex encephalitis. Immunohistological demonstration of spread of virus via olfactory pathways in mice. *J. Neurol. Sci.*, **60**,473–484.

Tonosaki, K. & Shibuya, F.T. (1979) Action of some drugs on gecko olfactory bulb mitral cell responses to odor stimulation. *Brain Res.*, **167**,180–184.

Trombley, P.Q. & Westbrook, G.L. (1991) Voltage-gated currents in identified rat olfactory receptor neurons. *J. Neuroscience*, **11**,435–444.

Trotier, D. (1986) A patch-clamp analysis of membrane currents in salamander olfactory receptor cells. *Pflügers Arch.*, **407**,589–595.

Trotier, D. (1990) Channel activities in amphibian olfactory receptor cells. In *Proceedings of the Tenth International Symposium on Olfaction and Taste.* Ed., K.B. Døving. GCS A/S, Oslo. pp. 73–84.

Trotier, D. & MacLeod, P. (1983) Intracellular recordings from salamander olfactory receptor cells. *Brain Res.*, **268**,225–237.

Trotier, D. & MacLeod, P. (1986) Intracellular recordings from salamander olfactory supporting cells. *Brain Res.*, **374**,205–211.

Trotier, D. & MacLeod, P. (1987) The amplification process in olfactory receptor cells. *Ann. N.Y. Acad. Sci.*, **510**,677–679.

Tucker, D. (1963) Olfactory, vomeronasal and trigeminal receptor responses to odorants. In *Olfaction and Taste. Proceedings of the First International Symposium.* Ed., Y. Zotterman. Macmillan Co., N.Y. pp. 45–69.

Tucker, D. (1965) Electrophysiological evidence for olfactory function in birds. *Nature*, **207**,34–36.

Tucker, D. (1971) Nonolfactory responses from the nasal cavity: Jacobson's organ and the trigeminal system. In *Handbook of Sensory Physiology, Vol. IV, Chemical Senses, 1, Olfaction.* Ed., L.M. Beidler. Springer-Verlag, Berlin. pp. 151–181.

Twomey, J.A., Barker, C.M., Robinson, G. & Howell, D.A. (1979) Olfactory mucosa in herpes simplex encephalitis. *J. Neurol. Neurosurg. Psychiatry*, **42**,983–987.

Vaccarezza, O.L., Sepich, L.N. & Tramezzani, J.H. (1981) The vomeronasal organ of the rat. *J. Anat.*, **132**,167–185.

Vale, R.D. (1987) Intracellular transport using microtubule based motors. *Ann. Rev. Cell Biol.*, **3**,347–378.

Vallee, R.B., Shpetner, H.S. & Paschal, B.M. (1989) The role of dynein in retrograde axonal transport. *Trends Neuroscience*, **12**,66–70.

Van Campenhout, E. (1937) Le développement du systeme nerveux cranien chez le poulet. *Arch. Biol. (Liège)*, **48**,611–666.

Van Haastert, P.J.M., Jannsens, P.M.W. & Erneux, C. (1991) Sensory transduction in eukaryonts. A comparison between *Dictyostelium* and vertebrate cells. *Eur. J. Biochem.*, **195**,289–303.

Van Houten, J. & Preston, R.R. (1987) Chemoreception in single-celled organisms. In *Neurobiology of Taste and Smell.* Eds., T.E. Finger & W.L. Silver. Wiley, N.Y. pp. 11–38.

Van Ooteghem, S., Schumacher, S. & Shipley, M.T. (1984) Development of a cholinergic pathway to the rat olfactory bulb: AChE appears two weeks after cholinergic fibers reach the bulb. *Soc. Neuroscience Abstr.*, **10**,143.

Venneman, W., Van Nie, C.J. & Tibboel, D. (1982) Developmental abnormalities of the olfactory bulb: a comparative study of the pig and chick embryo. *Teratology*, **26**,65–70.

Verhaagen, J., Greer, C.A. & Margolis, F.L. (1990a) B–50/GAP43 gene expression in the rat olfactory system during postnatal development and aging. *Eur. J. Neuroscience*, **2**,397–407.

Verhaagen, J., Oestreicher, A.B., Gispen, W.H. & Margolis, F.L. (1989) The expression of the growth associated protein B50/GAP43 in the olfactory system of neonatal and adult rats. *J. Neuroscience*, **9**,683–691.

Verhaagen, J., Oestreicher, A.B., Grillo, M., Khew-Goodall, Y.-S., Gispen, W.H. & Margolis, F.L. (1990b) Neuroplasticity in the olfactory system: differential effects of central and peripheral lesions of the primary olfactory pathway on the expression of B–50/GAP43 and the olfactory marker protein. *J. Neuroscience Res.*, **26**,31–44.

Verwoerd, C.D.A. & Van Oostrum, C.G. (1979) Cephalic neural crest and placodes. *Adv. Anat. Embryol. Cell Biol.*, **58**,1–75.

Viereck, C., Tucker, R.P. & Matus, A. (1989) The adult olfactory system expresses microtubule-associated proteins found in the developing brain. *J. Neuroscience*, **9**,3547–3557.

Vodyanoy, V. & Murphy, R.B. (1983) Single-channel fluctuations in bimolecular lipid membranes induced by rat olfactory epithelial homogenates. *Science*, **220**,717–719.

Vodyanoy, V. & Vodyanoy, I. (1987) ATP and GTP are essential for olfactory response. *Neuroscience Lett.*, **73**,253–258.

Vogt, R.G. (1987) The molecular basis of pheromone reception: its influence on behavior. In *Pheromone Biochemistry.* Eds., G.D. Prestwich & G.L. Blomquist. Academic Press, N.Y. pp. 385–431.

Vogt, R.G., Köhne, A.C., Dubnau, J.T. & Prestwich, G.D. (1989) Expression of pheromone binding proteins during antennal development in the Gypsy moth *Lymantria dispar. J. Neuroscience*, **9**,3332–3346.

Vogt, R.G., Prestwich, G.D. & Lerner, M.R. (1991a) Odorant-binding-protein subfamilies associate with distinct classes of olfactory receptor neurons in insects. *J. Neurobiol.*, **22**,74–84.

Vogt, R.G., Prestwich, G.D. & Riddiford, L.M. (1988) Sex pheromone receptor proteins. *J. Biol. Chem.*, **263**,3952–3959.

Vogt, R.G., & Riddiford, L.M. (1981) Pheromone binding and inactivation by moth antennae. *Nature*, **293**,161–163.

Vogt, R.G., Rybczynski, R. & Lerner, M.R. (1991b) Molecular cloning and sequencing of general odorant-binding proteins GOBP1 and GOBP2 from the tobacco hawk moth *Manduca sexta:* comparisons with other insect OBPs and their signal peptides. *J. Neuroscience*, **11**,2972–2984.

Voigt, J.M., Guengerich, F.P. & Baron, J. (1985) Localization of a cytochrome P–450 isozyme (cytochrome P–450 PB-B) and NADPH-cytochrome P–450 reductase in rat nasal mucosa. *Cancer Lett.*, **27**,241–247.

Vollrath, M., Altmannsberger, M., Weber, K. & Osborn, M. (1985) An ultrastructural and immunohistological study of the rat olfactory epithelium: unique properties of olfactory sensory cells. *Differentiation*, **29**,243–253.

274   *References*

Walker, D.G., Breipohl, W., Simon-Taha, A., Lincoln, D., Lobie, P.E. & Aragon, J.G. (1990) Cell dynamics and maturation within the olfactory epithelium proper of the mouse – a morphometric analysis. *Chem. Senses,* **15,**741–753.

Walker, J.C., Walker, D.B., Tambiah, C.R. & Gilmore, K.S. (1986) Olfactory and nonolfactory odor detection in pigeons: elucidation by a cardiac acceleration paradigm. *Physiol. Behav.,* **38,**575–580.

Wallraff, H.G. (1988) Olfactory deprivation in pigeons: examination of methods applied in homing experiments. *Comp. Biochem. Physiol.,* **89A,**621–629.

Wang, D., Chen, P., Jiang, X.C. & Halpern, M. (1988) Isolation from earthworms of a proteinaceous chemoattractant to garter snakes. *Arch. Biochem. Biophys.,* **267,**459–466.

Wang, R.T. & Halpern, M. (1980a) Light and electron microscopic observations on the normal structure of the vomeronasal organ of garter snakes. *J. Morphol.,* **164,**47–67.

Wang, R.T. & Halpern, M. (1980b) Scanning electron microscopic studies of the surface morphology of the vomeronasal epithelium and olfactory epithelium of garter snakes. *Amer. J. Anat.,* **157,**399–428.

Wang, R.T. & Halpern, M. (1988) Neurogenesis in the vomeronasal epithelium of adult garter snakes: 3. Use of H^3-thymidine autoradiography to trace the genesis and migration of bipolar neurons. *Amer. J. Anat.,* **183,**178–185.

Ward, C.D., Hess, W.A. & Calne, D.B. (1983) Olfactory impairment in Parkinson's disease. *Neurology,* **33,**943–946.

Warner, M.D., Peabody, C.A. & Berger, P.A. (1988) Olfactory deficits and Down's syndrome. *Biol. Psychiatry,* **23,**833–836.

Warner, M.D., Peabody, C.A., Flattery, J.J. & Tinklenberg, J.R. (1986) Olfactory deficits and Alzheimer's disease. *Biol. Psychiatry,* **21,**116–118.

Waterman, R.E. & Meller, S.M. (1973) Nasal pit formation in the hamster; a transmission and scanning electron microscopic study. *Devel. Biol.,* **34,**255–266.

Wenzel, B.M. (1968) Olfactory prowess of the kiwi. *Nature,* **220,**1133–1134.

Wenzel, B.M. (1985) Olfactory behavior in procellariform birds. *Natl. Geo. Res. Rep.,* **18,**779–788.

Wenzel, B.M. & Sieck, M.H. (1972) Olfactory perception and bulbar electrical activity in several avian species. *Physiol. Behav.,* **9,**287–294.

Williams, R. & Rush, R.A. (1988) Electron microscopic immunocytochemical localization of nerve growth factor in developing mouse olfactory neurons. *Brain Res.,* **463,**21–27.

Wilson, D.A., Guthrie, K.M. & Leon, M. (1990) Modification of olfactory bulb synaptic inhibition by early unilateral olfactory deprivation. *Neuroscience Lett.,* **116,**250–256.

Wilson, D.A. & Leon, M. (1987) Evidence of lateral synaptic interactions in olfactory bulb output cell responses to odors. *Brain Res.,* **417,**175–180.

Wilson, D.A. & Leon, M. (1988a) Spatial patterns of olfactory bulb single-unit responses to learned olfactory cues in young rats. *J. Neurophysiol.,* **59,**1770–1782.

Wilson, D.A. & Leon, M. (1988b) Noradrenergic modulation of olfactory bulb excitability in the postnatal rat. *Devel. Brain Res.,* **42,**69–75.

Wilson, D.A., Sullivan, R.M. & Leon, M. (1985) Odor familiarity alters mitral cell response in the olfactory bulb of neonatal rats. *Devel. Brain Res.,* **22,**314–317.

Wilson, D.A., Sullivan, R.M. & Leon, M. (1987) Single-unit analysis of postnatal olfactory learning: modified olfactory bulb output response patterns to learned attractive odors. *J. Neuroscience*, **7**,3154–3162.

Wilson, K.C.P. & Raisman, G. (1980) Age-related changes in the neurosensory epithelium of the mouse vomeronasal organ: extended period of postnatal growth in size and evidence for rapid cell turnover in the adult. *Brain Res.*, **185**,103–113.

Winans, S.S. & Powers, J.B. (1977) Olfactory and vomeronasal deafferentation of male hamsters: histological and behavioral analyses. *Brain Res.*, **126**,325–344.

Winans, S.S. & Scalia, F. (1970) Amygdaloid nucleus: new afferent input from the vomeronasal organ. *Science*, **170**,330–332.

Wirsig, C.R. & Getchell, T.V. (1986) Amphibian terminal nerve: distribution revealed by LHRH and AChE markers. *Brain Res.*, **385**,10–21.

Wirsig, C.R. & Leonard, C.M. (1986a) The terminal nerve projects centrally in the hamster. *Neuroscience*, **19**,709–717.

Wirsig, C.R. & Leonard, C.M. (1986b) Acetylcholinesterase and luteinizing hormone-releasing hormone distinguish separate populations of terminal nerve neurons. *Neuroscience*, **19**,719–740.

Wirsig, C.R. & Leonard, C.M. (1987) Terminal nerve damage impairs the mating behavior of the male hamster. *Brain Res.*, **417**,293–303.

Witkin, J.W. (1987) Immunocytochemical demonstration of luteinizing hormone-releasing hormone in optic nerve and nasal region of fetal rhesus macaque. *Neuroscience Lett.*, **79**,73–77.

Witkin, J.W. & Silverman, A.J. (1983) Luteinizing hormone-releasing hormone (LHRH) in rat olfactory systems. *J. Comp. Neurol.*, **218**:426–432.

Woo, C.C., Coopersmith, R. & Leon, M. (1987) Localized changes in olfactory bulb morphology associated with early olfactory learning. *J. Comp. Neurol.*, **263**,113–125.

Woo, C.C. & Leon, M. (1987) Sensitive period for neural and behavioral response development to learned odors. *Devel. Brain Res.*, **36**,309–313.

Worley, P.F., Baraban, J.M., DeSouza, E.B. & Snyder, S.H. (1986a) Mapping second messenger systems in the brain: differential localizations of adenylate cyclase and protein kinase C. *Proc. Natl. Acad. Sci. USA*, **83**,4053–4057.

Worley, P.F., Baraban, J.M., Van Dop, C., Neer, E.J. & Snyder, S.H. (1986b) G_o, a guanine nucleotide-binding protein: immunohistochemical localization in rat brain resembles distribution of second messenger systems. *Proc. Natl. Acad. Sci. USA*, **83**,4561–4565.

Wray, S., Nieburgs, A. & Elkabes, S. (1989) Spatiotemporal cell expression of luteinizing hormone-releasing hormone in the prenatal mouse: evidence for an embryonic origin in the olfactory placode. *Devel. Brain Res.*, **46**,309–318.

Wright, R.H. (1982) *The Sense of Smell*. CRC Press, Boca Raton, FL.

Wysocki, C.J. (1979) Neurobehavioral evidence for the involvement of the vomeronasal system in mammalian reproduction. *Neuroscience Biobehav. Rev.*, **3**,301–341.

Wysocki, C.J. (1989) Vomeronasal chemoreception: its role in reproductive fitness and physiology. In *Neural Control of Reproductive Function*. Eds., J.M. Lakoski, R.R. Perez-Polo & D.K. Rassin. Alan R. Liss, N.Y. pp. 545–566.

Wysocki, C.J. & Beauchamp, G. (1984) Ability to smell androstenone is genetically determined. *Proc. Natl. Acad. Sci. USA*, **81**,4899–4902.

Wysocki, C.J. & Meredith, M. (1987) The vomeronasal system. In *Neurobiology of Taste and Smell*. Eds., T.E. Finger & W.M. Silver. Wiley, N.Y. pp. 125–150.

Wysocki, C.T., Wellington, T.L. & Beauchamp, G.K. (1980) Access of urinary nonvolatiles to the mammalian vomeronasal organ. *Science*, **207**,781–783.

Wysocki, C.J., Whitney, G. & Tucker, D. (1977) Specific anosmia in the laboratory mouse. *Behav. Genet.*, **7**,171–188.

Yamagishi, M., Hasegawa, S., Nakano, Y., Takahashi, S. & Iwanaga, T. (1989) Immunohistochemical analysis of the olfactory mucosa by use of antibodies to brain proteins and cytokeratin. *Ann. Otol. Rhinol. Laryngol.*, **98**,384–388.

Yamamoto, C. & Matsui, S. (1976) Effect of stimulation of excitatory nerve tract on release of glutamic acid from olfactory cortex slices in vitro. *J. Neurochem.*, **26**,487–491.

Yamamoto, M. (1982) Comparative morphology of the peripheral olfactory organ in teleosts. In *Chemoreception in Fishes*, Ed., T.J. Hara. Elsevier, Amsterdam. pp. 39–59.

Yamamoto, M. & Ueda, K. (1977) Comparative morphology of fish olfactory epithelium. I. Salmoniformes. *Bull. Jpn. Soc. Sci. Fish.*, **43**,1163–1176.

Yamazaki, K., Beauchamp, G.K., Egorov, I.K., Bard, J., Thomas, L. & Boyse, E.A. (1983a) Sensory distinction between H–2^b and H–2^{bml} mutant mice. *Proc. Natl. Acad. Sci. USA*, **80**,5685–5688.

Yamazaki, K., Beauchamp, G.K., Imai, Y., Bard, J., Phelan, S.P., Thomas, L. & Boyse, E.A. (1990) Odortypes determined by the major histocompatibility complex in germfree mice. *Proc. Natl. Acad. Sci. USA*, **87**,8413–8416.

Yamazaki, K., Beauchamp, G.K., Wysocki, C.J., Bard, J., Thomas, L. & Boyse, E.A. (1983b) Recognition of H–2 types in relation to the blocking of pregnancy in mice. *Science*, **221**,186–188.

Yamazaki, K., Yamaguchi, M., Baranoski, L., Bard, J., Boyse, E.A. & Thomas, L. (1979) Recognition among mice: evidence from the use of a Y-maze differentially scented by congenic mice of different major histocompatibility types. *J. Exp. Med.*, **150**,755–760.

Yeo, J.A.G. & Keverne, E.B. (1986) The importance of vaginal-cervical stimulation for maternal behavior in the rat. *Physiol. Behav.*, **37**,23–26.

Yntema, C.L. (1955) Ear and nose. In *Analysis of Development*. Eds., B.H. Willier, P.A. Weiss & V. Hamburger. W.B. Saunders, Philadelphia. pp. 415–428.

Záborsky, L., Carlsen, J., Brashear, H.R. & Heimer, L. (1986) Cholinergic and GABAergic afferents to the olfactory bulb in the rat with special emphasis on the projection neurons in the nucleus of the horizontal limb of the diagonal band. *J. Comp. Neurol.*, **243**,488–509.

Zeiske, E., Theisen, B. & Gruber, S.H. (1987) Functional morphology of the olfactory organ of two carcharhinid shark species. *Can. J. Zool.*, **65**,2406–2412.

Zheng, L.-M. & Jourdan, F. (1988) Atypical olfactory glomeruli contain original olfactory axon terminals: an ultrastructural horseradish peroxidase study in the rat. *Neuroscience*, **26**,367–378.

Zheng, L.-M., Pfaff, D.W. & Schwanzel-Fukuda, M. (1990) Synaptology of luteinizing hormone-releasing hormone (LHRH)-immunoreactive cells in the nervus terminalis of the gray short-tailed opossum. *J. Comp. Neurol.*, **295**,327–337.

Zheng, L.-M., Ravel, N. & Jourdan, F. (1987) Topography of centrifugal ace-
tylcholinesterase-positive fibres in the olfactory bulb of the rat: evidence
for original projections in atypical glomeruli. *Neuroscience*, **23,**1083–1093.
Ziegelberger, G. (1989) Cyclic nucleotides in the olfactory system of the moths
Antheraea polyphemus and *Bombyx mori*. In *ISOT X. Proceedings of the
Tenth International Symposium on Olfaction and Taste*. Ed., K. Døving.
GCS A/S, Oslo. pp. 85–91.
Ziegelberger, G., Van den Berg, M.J., Kaissling, K.-E., Klumpp, S. & Schultz,
J.E. (1990) Cyclic GMP levels and guanylate cyclase activity in pheromone
sensitive antennae of the silkmoths *Antheraea polyphemus* and *Bombyx
mori*. *J. Neuroscience*, **10,**1217–1225.
Zielinski, B.S., Getchell, M.L. & Getchell, T.V. (1989a) Ultrastructural evidence for peptidergic innervation of the apical region of frog olfactory
epithelium. *Brain Res.*, **492,**361–365.
Zielinski, B.S., Getchell, M.L., Wenokur, R.L. & Getchell, T.V. (1989b) Ultrastructural localization and identification of adrenergic and cholinergic
nerve terminals in the olfactory mucosa. *Anat. Rec.*, **225,**232–245.
Zippel, H.P., Meyer, D.L. & Knaust, M. (1988) Peripheral and central post-
lesion plasticity in the olfactory system of the goldfish: behavior and morphology. In *Post-Lesion Neural Plasticity*. Ed., H. Flohr. Springer-Verlag,
Berlin. pp. 577–591.
Zusho, H. (1982) Posttraumatic anosmia. *Arch. Otolaryngol.*, **108,**90–92.
Zwilling, E. (1934) Induction of the olfactory placode by the forebrain in *Rana
pipiens*. *Proc. Soc. Exper. Biol. Med.*, **31,**933–935.
Zwilling, E. (1940) An experimental analysis of the development of the *Anuran*
olfactory organ. *J. Exp. Zool.*, **84,**291–318.

Index

accessory olfactory bulb, 40, 54, 70, 74, 122, 127, 129, 132, 134, 161–3, 181–4, 199, 203, 205
acetylcholine receptor, 41, 185, 211
acetylcholinesterase, 72, 73, 74, 141, 185, 198
adenosine deaminase, 178, 180
adenylate cyclase, 35, 42, 78, 91–5, 98, 178, 200, 209
alarm substance, 23
alcoholism, 116
Allomyces macrogynus, 6
Alzheimer's disease, 38, 67, 116–17
amino acid, 6, 53, 79–80, 87, 91, 95, 164
amygdala (amygdaloid), 122–3, 128, 144–6, 158–61, 163, 203, 204
annelid, 9
anosmia, 114, 115, 116, 117, 203
antenna, 2, 9, 10, 18, 76, 77–8, 86, 189–90
anterior commissure, 74, 134, 145–6, 153, 158, 205
anterior olfactory nucleus, 66, 134, 144–6, 153, 158–60, 163, 206
  pars externa, 146, 158
antibody, 44–9, 54–5, 59, 71, 73, 83, 150, 177, 179
aortic body, 5
aphrodisin, 128–9
arthropod, 9–10, 76
aspartate
  as stimulus, 6
  as transmitter, 147, 155, 161, 202
astrocyte, 68, 135, 184, 193
ATP, 65, 92
ATPase, 64, 65, 198
axon
  of mitral cell, 140, 142–6
  of periglomerular cell, 135
  of sensory neuron, 18, 19–20, 37–8, 54, 57–8, 60–1, 63–8, 70, 108, 125, 134, 141, 175, 191–2
  of short axon cell, 137, 139, 140
  of tufted cell, 137, 140, 144–6
axonal transport, 64–7, 212

B–50, 38, 58
bacteria, 5–6
basal cell, 24, 25, 26, 27, 43, 48–9, 55, 56–9, 62, 124, 174, 211
  globose basal cell, 25, 26, 43, 48–9, 211
basement membrane, 24, 42, 49, 124
benzodiazepine, 41
Blanes cell, 140
Bombyx mori, 78
Bowman's gland, 15, 19, 25, 26, 42, 43, 44, 46, 49–50, 62, 71, 124, 177, 210

Caenorhabditis elegans, 8
Cajal cell, 140
calcium, 7, 64, 92–3, 96, 101, 110, 112
calmodulin, 92–4
capsaicin, 71
carbonic anhydrase, 54
carnosine, 39–40, 55, 154–5, 179, 210
carotid body, 5
carvone, 76, 82
cerebellum, 135, 157
cerebral vesicle, 167, 168, 175, 181, 183
cerebrum (cerebral cortex), 20, 132, 133–4, 189
CGRP, 71, 160
channel, 35, 91–6, 101–12
cholecystokinin, 148, 150, 153
cilia
  biochemistry, 39
  ciliary necklace, 35–6
  function, 35, 50, 51, 53–4, 65, 78–9, 90, 91–7, 101–2, 108–9, 111
  genesis of, 175–8, 192–3
  structure, 25, 29, 30, 31–6, 43, 58, 60, 69, 86, 125–7
cockroach, 2
coding, 193, 208–9
concanavalin A, 60, 85
conjugation, 6
corticotropin-releasing factor (CRF), 147, 151, 155
cribriform plate, 38, 72, 116, 125, 192
cyclic AMP (cAMP), 6, 7, 88, 91–5, 97, 98, 99, 109, 178, 201, 205, 209

279

cyclic GMP (cGMP), 7, 92, 95, 109
cytochrome oxidase, 138–41, 198
cytochrome P-450, 46

dauer larva, 8
degeneration, 67
dendrite
  of granule cell, 139, 151, 185, 196–7
  of mitral cell, 135, 139, 142–4, 146,
    147, 163, 165, 185, 187, 198
  of periglomerular cell, 135, 183
  of sensory cell, 18, 27–8, 31, 54, 58,
    76–8, 91–2, 109, 110, 175–8
  of short axon cell, 137, 139, 140
  spines on dendrite, 135, 151
diacylglycerol, 7, 95–7
diagonal band, horizontal limb of, 66,
  158–60, 163
Dictyostelium discoideum, 7, 88, 94
diencephalon, 74, 132, 175, 188
dimethyl disulfide, 128–9
disparlure, 77
dopamine, 147, 150–1, 153, 156, 190, 198,
  204
dopamine D2 receptor, 42, 151, 156, 211
Down's syndrome, 42
dynein, 65
dynorphin B, 160

elasmobranch, 3, 11, 13, 31
electro-olfactogram (EOG), 78, 85, 91,
  100–3, 179, 192, 194
encephalitis, 65–6
enkephalin, 150, 153, 157
ensheathing cell, 19–20, 60–1, 63, 68, 135,
  171, 175, 177
entorhinal cortex, 144–6, 158–60
epidermal growth factor, 210
Escherichia coli, 6
estrous cycle, 128
ethmoid
  bone, 17, 38, 72
  nerve, 71, 113, 114
external plexiform layer (EPL), 136–9,
  143, 147, 151–2, 158, 161, 163, 165,
  182, 184, 193

fila olfactoria, 38
fish, 3, 11, 13, 14, 21–2, 53, 63, 67, 72,
  163–6
flehmen, 120–2, 129
folic acid, 6, 7, 94
freeze-fracture, 35–6, 42–4, 86
frog, 42, 50, 51, 55, 64, 71–2

G-protein, 35, 42, 78, 87, 92–7, 99, 178,
  200, 207

GABA, 150–1, 153, 155–7, 159, 190, 201–
  2, 203–4
galanin, 160
GAP-43, 38, 58
glia, 63, 68, 135, 183–4, 189–90, 212
glial fibrillary acidic protein (GFAP), 68
globose basal cell, *see* basal cell
glomerulus (glomeruli), 52, 54, 55, 68,
  113, 135–7, 141, 143, 148, 151, 153–
  6, 158, 161, 165, 184, 188–90, 193,
  195, 199, 208
  modified glomerular complex, 141, 195
glutamate, 147, 155, 160, 202
glutamic acid decarboxylase, 150
goldfish, 21–2, 55, 73
Golgi cell, 140
granule cell, 55, 136–40, 141, 144, 147,
  150–2, 160, 161, 163, 165, 181–6,
  195, 211
growth associated protein, 38
guanylate cyclase, 95

heat shock protein, 48
hippocampus, 158, 203, 206
homing, 21
horizontal cell, 140
Huntington's chorea, 116
hyposmia, 115
hypothalamus, 72, 74, 117, 122, 128, 175,
  177, 204

immunoglobulin, 83–4
inositol trisphosphate, 7, 95–8, 111
insect, 9–10, 18, 77–8, 95, 184, 209
internal (inner) plexiform layer (IPL),
  136, 139, 145, 146, 152, 161
intramembranous particle, 35–6, 43, 44,
  69, 86, 90, 99, 178
ion channel, 35, 91–6, 101–12
IP$_3$, *see* inositol trisphosphate
isovaleric acid, 76

Jacobson's organ, 118
jellyfish, 8

Kallmann's syndrome, 117
kinesin, 65
Korsakoff's psychosis, 116

lamina propria, 25, 28, 49, 71–2, 120, 124
lateral inhibition, 149–50
lateral olfactory tract, 134, 158, 162, 195,
  198
learning, 22, 23, 199–206
lectin, 54, 85, 86
LHRH, 72–4, 117, 129, 160, 175, 177
liposome, 89
lobster, 2, 97

locus coeruleus, 66, 156, 158–61, 163, 199, 201–2, 203, 204

macrosmatic, 17, 172
major histocompatibility complex (MHC), 84
Manduca sexta, 78, 189–90
Masera, organ of, 18, 69–70
maternal–infant relationship, 20–1, 195, 199–202, 213
mating, 21–2, 127–30, 202–3
medial septum, 74, 158–61, 194
membrane, 35, 64, 65, 77–92, 104–12
memory, 203–6
microsmatic, 17
microvilli, 31, 42–7, 51, 53–4, 78, 101, 125–7, 178, 179
mitral cell, 52, 68, 113, 135, 138–9, 142–58, 161, 164–5, 180–3, 195–6, 214–15
mollusk, 9
moth, 2, 10, 77–78, 86, 95, 189–90
mucopolysaccharide, 49–50
mucus, 24, 35, 42, 46, 49, 51–2, 75–8, 91, 103, 105, 112, 114, 209–10

naris occlusion, 197–9, 205–6
nasopalatine canal, 118–20, 172
nasopalatine nerve, 71, 113, 120
nematode, 8
nest-finding, 21
neural plate, 168–71
neuroblastoma, 89
neurofilament, 38–9, 61, 65
neuron cell adhesion molecule (NCAM), 38, 179
neuron-specific enolase, 38
neuropeptide Y, 153
neurotransmitter, 65, 107, 146–7, 150, 154–7, 200, 210
nipple search behavior, 20, 157, 180, 195, 213
NMDA receptor, 147, 204
noradrenergic fibers, 66, 201–4
norepinephrine, 157, 199–201, 203–5

odorant
  binding protein (OBP), 76–8, 209
  stimulus, 42, 50, 51–3, 55, 75–8, 81, 89, 91, 95, 98–9, 102–3, 114, 148–9, 178, 194
olfactory
  bulb, 38, 52, 54, 55, 59, 60, 66, 68–9, 70, 72, 74, 108, 114, 117, 127–8, 132–66, 180, 181–96, 199, 208
  cortex, 20, 134, 144–7, 152, 157–61, 195, 206
  marker protein (OMP), 39–40, 54, 59, 178–80, 192

nerve, 20, 38, 63–8, 135, 171, 175, 181, 190–2, 211–12
  peduncle, 153
  placode, 74, 117, 167–71, 176–7
  tubercle, 66, 144–6, 158–61
oxytocin, 202

palate, 171–3
Paramecium, 6
parisin, 7
Parkinson's disease, 67, 116
pars externa, *see* anterior olfactory nucleus
patch clamp, 92, 95, 104–8
periglomerular cell, 135–7, 141, 147, 150–1, 153, 155–7, 160, 181–6, 190
pheromone
  in ducks, 16
  in fishes, 22, 73–4, 129, 213
  in hamsters, 128–9
  in invertebrates, 6–7, 8, 10, 21, 77–8, 88, 90
  in mice, 23, 203, 204–5, 208
  in snakes, 130
phosphodiesterase, 92
phosphoinositide, 91, 95–8, 209
phospholipase C, 7, 95–8
pituitary gland, 72, 74, 117, 129, 202, 204
placode, *see* olfactory placode
potassium, 64, 101, 103, 109, 110–12
predator–prey relationship, 22–3
pregnancy block, 130, 163, 202–5, 208, 213
prolactin, 204–5
prostaglandin, 22
Purkinje cell, 157
pyriform cortex, 20, 66, 144–6, 147, 158–61

quinolinic acid, 193

raphé nucleus, 158–61, 163
receptor, 4, 39, 76–93, 98–110, 193, 207, 208, 213
regeneration, 57, 64, 67–8
reptile, 13, 16, 22, 50, 124–5, 130–1
respiratory epithelium, 35, 62, 69, 115, 127
respiratory system, 11, 13, 15–16
retina, 55, 69, 72, 73, 81, 154
ruffed cell, 166

Saccharomyces cerevisiae, 3
salmon, 21, 53, 79, 164
Schwann cell, 19–20, 60, 63–8, 175, 177
sensillum, 2, 10, 18, 76–7
sensory deprivation, 197–9
septal organ, 18, 69–70

septum (nasal), 18, 28, 46, 69, 74, 118, 121, 125, 167, 171–2, 180
serotonin, 185
serotonergic fiber, 66
sexual behavior, 21–2, 73–4, 127–30, 164
shark, 3, 11
short axon cell, 137, 139–40, 146, 152–3
single odor environment, 196–7
sirenin, 6–7
slime mold, 7, 94
snail, 2, 9, 19
snake, 22, 50, 124–5, 130–1
sodium, 64, 101, 107, 109, 110
somatostatin, 153
squalene, 130
stereospecificity, 80–83
stimulus, 75–81, 148–9, 178
stria terminalis, 161
subependymal layer, 136, 140, 183, 190
substance P, 50, 71–2, 147, 190
suckling behavior, 20, 184, 195, 202
supporting cell, 24, 25, 26, 27, 28–9, 42–8, 54, 101, 103, 124, 169, 176, 177, 178, 179, 198, 209–10
SUS-1, 44–5
sustentacular cell, *see* supporting cell
synapse
    corticobulbar, 158–61
    dendrodendritic, 135, 137, 146, 150, 151, 152, 155–7, 186, 202, 204, 205
    granule/mitral, *see* dendrodendritic
    mitral/granule, *see* dendrodendritic
    in olfactory cortex, 146, 158
    reciprocal, 135, 150, 151, 155–7, 186, 205
    sensory/mitral, 68, 143–4, 153, 179, 185–6
    sensory/periglomerular, 153, 155
    short axon/granule, 152
    short axon/periglomerular, 152

taenia tecta, 158, 160
taste, 4–5, 9

telencephalon, 74, 153, 175, 181, 188
terminal nerve, 18, 19, 72–4, 117, 128, 129, 175, 177, 214
thalamus, 20, 69
thymidine, 56–7, 59
tight junction, 28–31, 44, 72, 177
tongue
    in development, 172–3
    flicking, 22, 120, 130
transduction, 5–8, 35, 42, 86, 87, 91–100, 108, 178, 207, 208, 209
transplantation, 188–9
trigeminal nerve, 18, 19, 70–2, 113–15, 214
trout, 79–80
tufted cell, 135–9, 142–58, 161, 165, 183, 190
turbinate, 13, 16, 17, 46, 172, 180
turtle, 21
2-deoxyglucose (2DG), 140–1, 148, 149, 158, 195, 197, 198, 199
tyrosine hydroxylase, 150–1, 190–1, 198

ubiquitin, 48
urine, 84, 120, 129, 130, 202, 204, 206

vasoactive intestinal peptide, 147, 153
ventricle, 181–2, 183
vimentin, 61
virus, 65–7, 69, 70, 72, 117, 212
vomeronasal
    organ, 18, 21, 22, 31, 39, 54, 68, 74, 75, 118–31, 134, 161–3, 171, 177, 178, 183, 203, 204, 208
    pump, 120

whale, 11, 132

yeast, 6, 88

zinc sulfate, 56, 62, 66, 102, 128, 154
zona incerta, 158, 160